Foundations of Engineering Mechanics

Series Editors: V.I. Babitsky, J. Wittenburg

Foundations of Engineering Mechanics

Series Editors: Vladimir I. Babitsky, Loughborough University, UK
Jens Wittenburg, Karlsruhe University, Germany

Further volumes of this series can be found on our homepage: springer.com

Slivker, V.I.
Mechanics of Structural Elements, 2007
ISBN 978-3-540-44718-4

Elsoufiev, S.A.
Strength Analysis in Geomechanics, 2007
ISBN 978-3-540-37052-9

Awrejcewicz, J., Krysko, V.A.,
Krysko, A.V.
Thermo-Dynamics of Plates and Shells, 2007
ISBN 978-3-540-37261-8

Wittbrodt, E., Adamiec-Wojcik, I.,
Wojciech, S.
Dynamics of Flexible Multibody Systems, 2006
ISBN 3-540-32351-1

Aleynikov, S.M.
Spatial Contact Problems in Geotechnics, 2006
ISBN 3-540-25138-3

Skubov, D.Y., Khodzhaev, K.S.
Non-Linear Electromechanics, 2006
ISBN 3-540-25139-1

Feodosiev, V.I.,
Advanced Stress and Stability Analysis Worked Examples, 2005
ISBN 3-540-23935-9

Lurie, A.I.
Theory of Elasticity, 2005
ISBN 3-540-24556-1

Sosnovskiy, L.A.,
TRIBO-FATIGUE · Wear-Fatigue Damage and its Prediction, 2005
ISBN 3-540-23153-6

Andrianov, I.V., Awrejcewicz, J.,
Manevitch, L.I. (Eds.)
Asymptotical Mechanics of Thin-Walled Structures, 2004
ISBN 3-540-40876-2

Ginevsky, A.S., Vlasov, Y.V.,
Karavosov, R.K.
Acoustic Control of Turbulent Jets, 2004
ISBN 3-540-20143-2,

Kolpakov, A.G.
Stressed Composite Structures, 2004
ISBN 3-540-40790-1

Shorr, B.F.
The Wave Finite Element Method, 2004
ISBN 3-540-41638-2

Svetlitsky, V.A.
Engineering Vibration Analysis - Worked Problems 1, 2004
ISBN 3-540-20658-2

Babitsky, V.I., Shipilov, A.
Resonant Robotic Systems, 2003
ISBN 3-540-00334-7

Le xuan Anh,
Dynamics of Mechanical Systems with Coulomb Friction, 2003
ISBN 3-540-00654-0

Nagaev, R.F.
Dynamics of Synchronising Systems, 2003
ISBN 3-540-44195-6

Neimark, J.I.
Mathematical Models in Natural Science and Engineering, 2003
ISBN 3-540-43680-4

Perelmuter, A.V., Slivker, V.I.
Numerical Structural Analysis, 2003
ISBN 3-540-00628-1

Lurie, A.I.,
Analytical Mechanics, 2002
ISBN 3-540-42982-4

Manevitch, L.I., Andrianov, I.V.,
Oshmyan, V.G.
Mechanics of Periodically Heterogeneous Structures, 2002
ISBN 3-540-41630-7

Babitsky, V.I., Krupenin, V.L.
Vibration of Strongly Nonlinear Discontinuous Systems, 2001
ISBN 3-540-41447-9

Landa, P.S.
Regular and Chaotic Oscillations, 2001
ISBN 3-540-41001-5

(Continued after index)

Konstantin Naumenko Holm Altenbach

Modeling of Creep for Structural Analysis

With 77 Figures and 5 Tables

Springer

Series Editors:

V.I. Babitsky
University Loughborough
Dept. Mechanical Engineering
Loughborough
Leics. LE11 3TU
UK

J. Wittenburg
Universität Karlsruhe
Fakultät Maschinenbau
Institut für Technische Mechanik
Kaiserstr. 12
76128 Karlsruhe
Germany

Authors:

Konstantin Naumenko
Martin-Luther-Universität
Halle-Wittenberg
Zentrum für Ingenieurwissenschaften
Lehrstuhl für Technische Mechanik
06099 Halle
Germany

Holm Altenbach
Martin-Luther-Universität
Halle-Wittenberg
Zentrum für Ingenieurwissenschaften
Lehrstuhl für Technische Mechanik
06099 Halle
Germany

ISSN print edition: 1612-1384

ISBN-10: 3-540-70834-0
ISBN-13: 978-3-540-70834-6 **Springer Berlin Heidelberg New York**

Library of Congress Control Number: 2007920935

Springer is a part of Springer Science+Business Media

springer.com

Typesetting by the authors and SPi
Cover-Design: deblik, Berlin
Printed on acid-free paper SPIN: 11764410 62/3100/SPi - 5 4 3 2 1 0

Preface

Many engineering structures are subjected to high temperature and mechanical loadings over a long time of operation. Examples include structural components of power plants, chemical refineries and heat engines. Design procedures and residual life assessments for pipework systems, rotors, turbine blades, etc. require to take into account creep and damage processes. The aim of "creep modeling for structural analysis" is the development of methods to simulate and analyze the time-dependent changes of stress and strain states in engineering structures up to the critical stage of creep rupture.

The scope of this book is related to the fields "creep mechanics" [62, 242] and "continuum creep and damage mechanics" [138]. The principal subjects of creep mechanics are the formulation of constitutive equations for creep in structural materials under multi-axial stress states; the application of structural mechanics models of beams, plates, shells and three-dimensional solids and the utilization of procedures for the solution of non-linear initial-boundary value problems. They have become traditional since the pioneering texts written in 1960s by Odqvist and Hult [243], Hult [142] and Rabotnov [272]. These classical books provide a first collection of solutions to the creep problems for elementary structures such as rods, beams and circular plates based on the simple constitutive models like the Norton-Bailey equation. The results illustrate the basic features of creep in structures: time dependent deformations, relaxation and redistribution of stresses and creep buckling. Furthermore, the introduction of internal or hidden state variables to characterize processes accompanying creep deformation has been established. The monograph of Penny and Mariott [257] (first edition in 1971) concentrates on simplified methods and empirical relationships which are useful for the design procedures. The monographs of Kraus [178], Malinin [208] and Boyle and Spence [80], published in 1980s, introduce new constitutive models with hardening/recovery and damage variables and initiate the use of advanced numerical methods for the structural analysis. The monographs published by Lemaitre and Chaboche [191] and Skrzypek and Ganczarski [298] in 1990s, although only partly connected with the creep mechanics, designate the framework of continuum thermodynamics to derive constitutive models, present the advanced techniques for testing materials under multi-axial non-proportional loading conditions and overview the developments of continuum damage mechanics. The recently published book by Betten [62] addresses the application of the theory of tensor valued functions to the constitutive modeling of anisotropic creep and damage processes.

Creep problems in materials and structures are widely discussed in various conferences and scientific papers. The International Union of Theoretical and Applied Mechanics (IUTAM) organizes every ten years the symposium "Creep in Structures" (1960 - Stanford, 1970 - Gothenburg, 1980 - Leicester, 1990 - Cracow, 2000 - Nagoya) to establish new and fundamental topics on creep and bring together scientists and engineers from the fundamental research and applications. The proceedings show an essential progress in modeling and understanding creep phenomena starting from the physical and microstructural aspects of creep and creep-damage up to the structural design procedures.

During the last decade many advances and new results in the field of creep mechanics were presented in conference proceedings and scientific papers. Examples include: the interlinks with materials science in formulation of constitutive equations to consider different deformation and damage mechanisms over a wide range of stresses and temperature; the application of tensor-valued state variables to account stress state effects and deformation/damage induced anisotropy; the assessment of models for beams, plates and shells in creep and damage related structural analysis; the development and verification of creep-damage material subroutines for the use in general purpose finite element codes; the application of the finite element method to the creep analysis of real engineering structures; the consideration of processing conditions, such as welding or induction bending of pipes, as they influence the subsequent creep behavior in structures.

The objective of this book is to review some of the classical and recently proposed approaches to the modeling of creep for structural analysis applications as well as to extend the collection of available solutions of creep problems by new, more sophisticated examples.

In Chapter 1 we discuss basic features of the creep behavior in materials and structures and present an overview of various approaches to the modeling of creep.

Chapter 2 collects constitutive models that describe creep processes under multiaxial stress states. The starting point of the engineering creep theory is the introduction of the inelastic strain, the creep potential, the flow rule, the equivalent stress and internal state variables. Constitutive models of isotropic secondary creep based on the von Mises-Odqvist creep potential are introduced. To account for stress state effects creep potentials that include three invariants of the stress tensor are discussed. Consideration of material symmetries provide restrictions for the creep potential. A novel direct approach to find scalar valued arguments of the creep potential for the given group of material symmetries is proposed. Transverse isotropy and orthotropic symmetry are two important types of symmetries in the creep mechanics [62]. For these two cases appropriate invariants of the stress tensor, equivalent stress and strain expressions as well as constitutive equations are derived.

Further extensions of the classical creep theory are related to processes accompanying creep deformation. Primary creep and transient creep effects can be described by the introduction of hardening state variables. The time and strain hardening models as well as the back stress concept are examined as they predict multi-axial creep behavior. Tertiary creep and long term strength can be characterized by the introduction of damage state variables. A systematic review of different types of constitutive equations with damage variables and corresponding evolution equations is presented. Stress state effects and damage induced anisotropy are discussed in detail.

Chapter 3 deals with the application of constitutive models to the description of creep for several structural materials. Constitutive and evolution equations, response functions and material constants are presented according to recently published experimental data. Furthermore a new model for anisotropic creep in a multi-pass weld metal is presented.

In Chapter 4 we discuss structural mechanics problems. We start with a summary of governing equations describing creep in three-dimensional solids. Several simplifying assumptions are made in order to illustrate the basic ideas of initial-boundary value problems, direct variational methods and time step algorithms. Then various structural mechanics models of beams, plates and shells are reviewed and evaluated in the context of their applicability to creep problems. An emphasis is placed on effects of transverse shear deformation, boundary layers and geometrical nonlinearities.

Constitutive models with scalar damage variables are incorporated into the ABAQUS and ANSYS finite element codes by means of user-defined material subroutines. To verify the developed subroutine several benchmark problems are presented. For these problems special numerical solutions based on the Ritz method are obtained. Finite element solutions for the same problems are performed to illustrate that the subroutines are correctly coded and implemented. Furthermore these benchmarks are used to study the applicability of the developed subroutines over a wide range of element types including shell and solid elements. Based on several examples, the influence of the mesh size on the accuracy of solutions is demonstrated. An example for a spatial steam pipeline is presented. Results are compared with the data from engineering practice discussed in the literature.

Appendices A - C is a summary of the direct tensor notation and basic tensor operations used throughout the text. This notation has an advantage of a clear, compact and coordinate free representation of constitutive models and initial-boundary value problems. The theory of anisotropic tensor functions and invariants is discussed in detail. A novel approach to derive the basic set of functionally independent invariants for vectors and second rank tensors for the given symmetry group is presented. The invariants are found as integrals of a generic partial differential equation (basic equation for invariants).

Several chapters of this book have grown out of our lectures and lecture notes on fundamentals of continuum mechanics, mechanics of materials and finite element modeling for graduate level students. Many results presented originate from scientific and academic exchange projects. We wish to acknowledge financial support from the German Research Foundation (DFG), German Academic Exchange Service (DAAD) and the State Saxony-Anhalt. This book is partly based on the

Habilitation thesis of the first author [232]. We thank Professors R. Kienzler and J. Skrzypek for a lot of critical comments and suggestions to the manuscript. We would like to acknowledge Professors J. Betten, O. T. Bruhns, T. Hyde, Z. L. Kowalewski, E. Krempl, O. K. Morachkowski, N. Ohno and P. A. Zhilin for many fruitful discussions which stimulated our research in creep mechanics. For the careful reading of the manuscript we thank Mr. A. Al-Yafawi and Mrs. B. Renner.

Halle (Saale),
Summer/Autumn 2006
K. Naumenko and H. Altenbach

Contents

1 Introduction

Creep is the progressive time-dependent inelastic deformation under constant load and temperature. *Relaxation* is the time-dependent decrease of stress under the condition of constant deformation and temperature. For many structural materials, for example steel, both the creep and the relaxation can be observed above a certain critical temperature. The creep process is accompanied by many different slow microstructural rearrangements including dislocation movement, ageing of microstructure and grain-boundary cavitation.

The above definitions of creep and relaxation are related to the case of uni-axial homogeneous stress states realized in standard material testing. Under *creep in structures* one understands time-dependent changes of strain and stress states taking place in structural components as a consequence of external loading and temperature. Examples of these changes include progressive deformations, relaxation and redistribution of stresses, local reduction of material strength, etc. Furthermore, the strain and stress states are inhomogeneous and multi-axial in most cases. The scope of "creep modeling for structural analysis" is to develop a tool which allows to simulate the time-dependent behavior in engineering structures up to the critical state of creep rupture.

In Chapter 1 we discuss basic features of creep behavior of materials and structures and present the state of the art within the framework of creep modeling.

1.1 Creep Phenomena in Structural Materials

The analysis of the material behavior can be based on different experimental observations, for example, macroscopic and microscopic. The engineering approach is related to the stress-strain analysis of structures and mostly based on the standard mechanical tests. In this section we discuss basic features of the creep behavior according to recently published results of creep testing under uni-axial and multi-axial stress states.

1.1.1 Uni-Axial Creep

Uni-axial creep tests belong to the basic experiments of the material behavior evaluation. A standard cylindrical tension specimen is heated up to the temperature

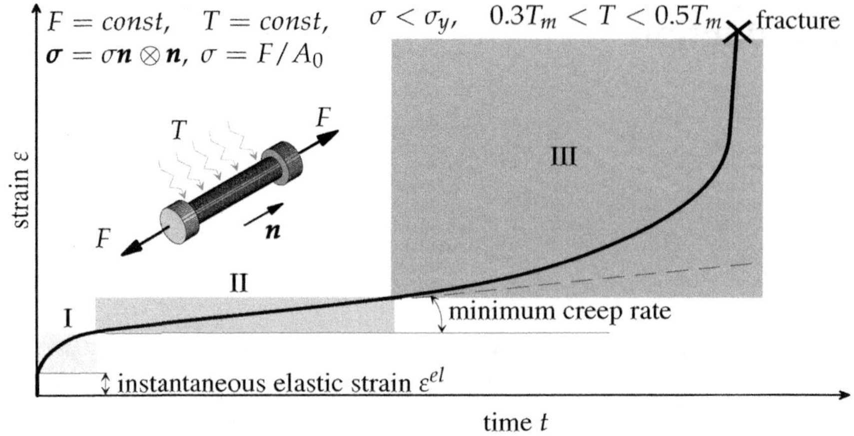

Fig. 1.1 Strain vs. time curve under constant load F and temperature T. I - primary creep, II - secondary creep, III - tertiary creep

$T = (0.3 - 0.5)\,T_m$ (T_m is the melting temperature of the material) and loaded by a tensile force F. The value of the normal stress in the specimen σ is usually much less than the yield limit of the material σ_y. The instantaneous material response is therefore elastic. The load and the temperature are kept constant during the test and the axial engineering strain ε is plotted versus time t. A typical creep curve for a metal is schematically shown in Fig. 1.1. The instantaneous response can be characterized by the strain value ε^{el}. The time-dependent response is the slow increase of the strain ε with a variable rate. Following Andrade [98], three stages can be considered in a typical creep curve: the first stage (primary or reduced creep), the second stage (secondary or stationary creep) and the third stage (tertiary or accelerated creep). During the primary creep stage the creep rate decreases to a certain value (minimum creep rate). The secondary stage is characterized by the approximately constant creep rate. During the tertiary stage the strain rate increases. At the end of the tertiary stage creep rupture of the specimen occurs.

A number of properties can be deduced from the uni-axial creep curve. These are the duration of the stages, the value of minimum creep rate, the time to fracture and the strain value before fracture[1].

The shape of the creep curve and the duration of the creep stages depend strongly on the stress and temperature values, Fig. 1.2. The dependencies on stress and temperature are of primary interest to an engineer designing some structure or machine. In order to obtain mechanical properties of the material, series of creep tests are usually performed for different stress and temperature values. From the resulting families of creep curves one can obtain the minimum creep rate vs. stress curve, the minimum creep rate vs. temperature curve, the creep rate vs. time curve and the stress vs. time to fracture curve (long term strength curve). The ranges of stress and temperature should be selected according to the ranges expected in the structure

[1] The fracture strain is sometimes related to the ductility of the material.

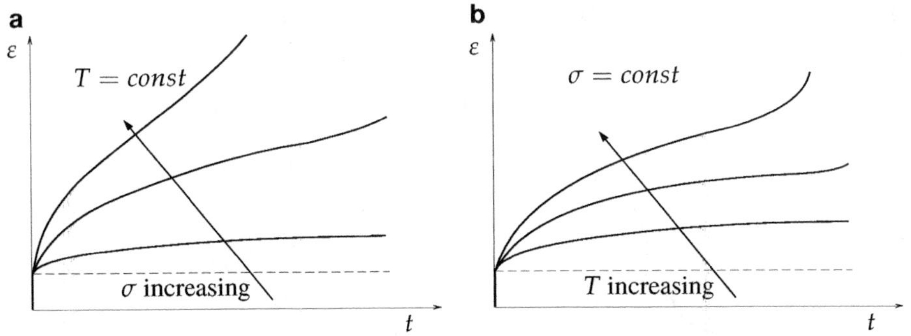

Fig. 1.2 Influence of stress and temperature on the creep behavior. **a** Stress dependence, **b** temperature dependence

during the service. Examples of the above mentioned curves for various materials can be found in monographs [80, 178, 208, 243, 241, 257] and many papers related to the experimental analysis of creep, e.g. [108, 146, 147, 165].

Two additional forms of the time-dependent stress-strain behavior are creep recovery and stress relaxation, Fig 1.3. Creep recovery is usually observed, when after a certain period of time the load is spontaneously removed, Fig. 1.3b. After unloading the strain drops about the value ε^{el} (recovery of the elastic strain). Then the strain slowly decreases down to the permanent (irrecoverable) value ε^{pm}, whereas ε^{rec} is the recovered inelastic strain. A typical stress relaxation curve is shown in Fig. 1.3c. Stress relaxation is observed when the strain is held constant in time ($\varepsilon = const$). A uni-axial specimen is instantaneously deformed to the strain value $\varepsilon^{el} = \sigma/E$, where E is the Young's modulus. During the test the load is continuously decreased in such a way that the initial strain remains constant. A threshold of the initial stress (strain) exists below which the relaxation is not observable.

In many cases it is convenient to introduce the inelastic (creep) strain ε^{cr} as the difference between the measured strain ε and the calculated elastic strain ε^{el}. The creep curves can be presented as creep strain vs. time curves, Fig. 1.3a and b. In the case of relaxation it is usually assumed, e.g. [208, 309], that the total zero strain rate is the sum of the elastic and the creep strain rates

$$\dot{\varepsilon} = \frac{\dot{\sigma}}{E} + \dot{\varepsilon}^{cr} = 0$$

According to this assumption the creep strain with a decaying rate develops during the relaxation test, Fig 1.3c.

In addition to creep and relaxation, many different tests under variable loading and/or strain conditions are discussed in the literature. Examples for the creep curves under stepwise loading are presented in [116, 208] among others. In this case the creep test starts under a certain value of the load. After reaching steady-state creep rate the load is rapidly increased (decreased) and kept constant over a period of time (holding time). Such tests allow to analyze transient creep effects, e.g. the duration of primary creep after the rapid change of loading. Furthermore, they indicate that

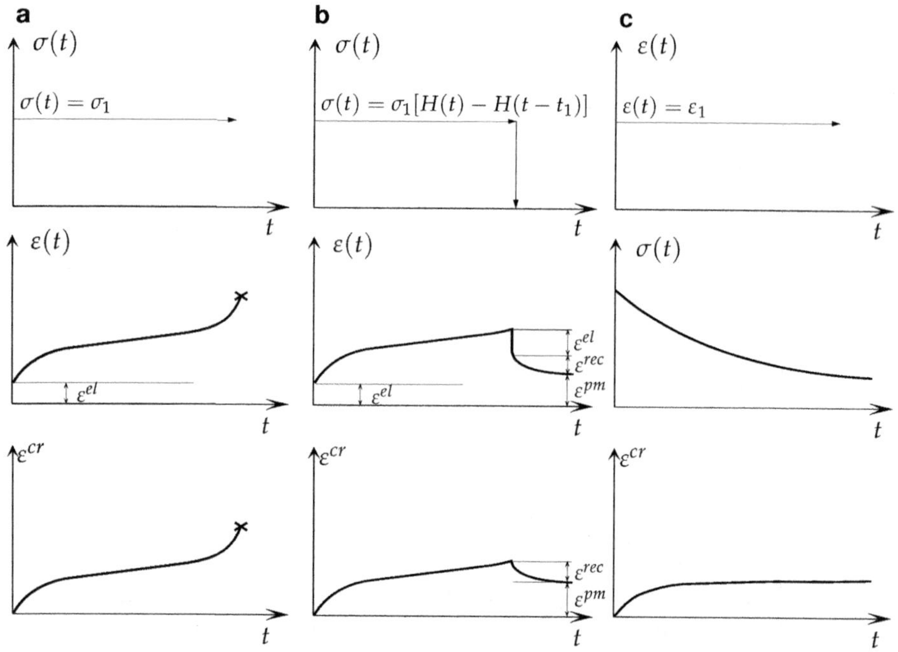

Fig. 1.3 Different forms of time-dependent stress-strain behavior under constant temperature. **a** Creep at constant stress, **b** creep recovery ($H(t)$ denotes the Heaviside function), **c** stress relaxation

the steady-state creep rate in the current loading step depends not only on the value of the applied stress but also on the loading history (e.g. the number of previous stress cycles, the holding time, etc.).

A periodically varied load causes cyclic creep response. The periodic stress can be characterized by the amplitude σ^a, the period τ_c and the mean stress σ^m. Two typical cases of the periodic loading are presented in Figs 1.4a and b. Let us assume that the maximum stress $\sigma^{max} = \sigma^m + \sigma^a$ is much less than the yield limit of the material. Creep behavior for the case of periodic loading with a holding time is schematically illustrated in Fig. 1.4c. Here the mean stress σ^m, the amplitude σ^a, the rate of loading/unloading and the holding time influence the creep response. The case of harmonic loading is shown in Fig. 1.4b. Such loading is important in those engineering applications, where technological or operational conditions (non-stationary flow, combustion, acoustic action, etc.) cause the development of forced vibrations. The harmonic stress variation can be described as follows

$$\sigma = \sigma^m(1 + \hat{A}\sin\Omega t), \quad \hat{A} = \frac{\sigma^a}{\sigma^m}, \quad \Omega = \frac{2\pi}{\tau_c} = 2\pi f, \qquad (1.1.1)$$

where $\hat{A}$ is the stress cycle assymmetry parameter. Creep behavior under harmonic loading (1.1.1) with frequencies $f > 1 \ldots 2$ Hz has been studied in [47, 185, 310, 311]. For this cyclic loading condition primary, secondary and tertiary stages can

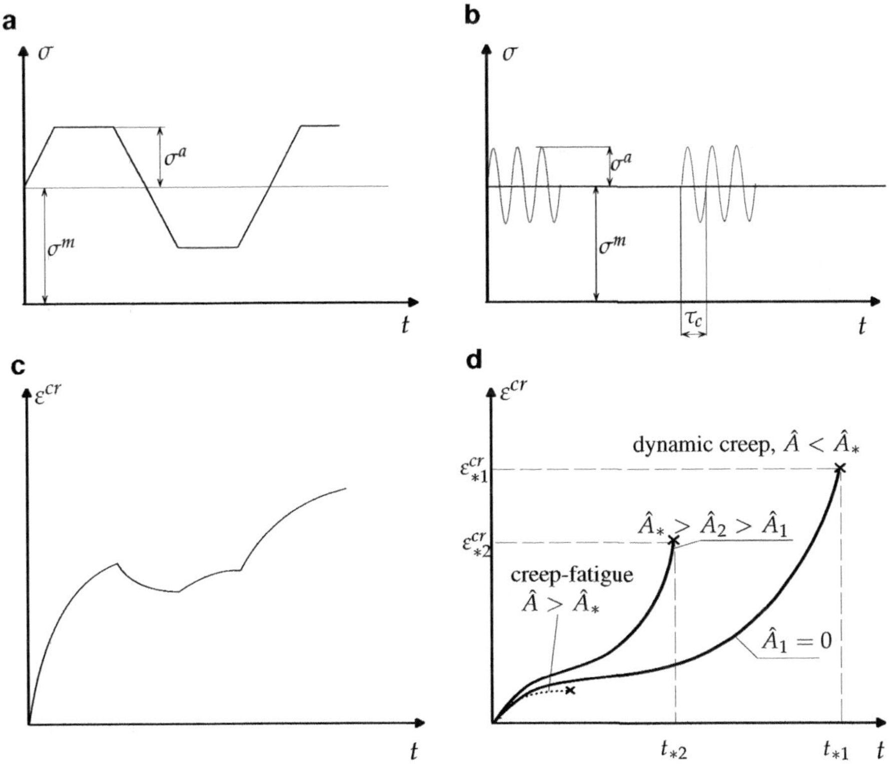

Fig. 1.4 Types of loading and corresponding cyclic creep curves. **a** Cyclic loading with holding, **b** harmonic loading with high frequency, **c** creep response for cyclic loading with holding, **d** different responses for high frequent loading

be observed similarly to the static case, Fig. 1.4d. Furthermore, the shape of the cyclic creep curve is geometrically similar to the static one caused by the stress $\sigma = \sigma^m$, but the creep rate is rather higher and the time to fracture is essentially smaller. It was found that creep under fast cyclic loading is not sensitive to the frequency of stress variation, e.g. [312]. In contrast, the stress cycle asymmetry parameter $\hat{A}$ has significant influence on the creep rate. For a number of investigated materials a material constant $\hat{A}_*$ has been found which is termed as the critical value of the stress cycle asymmetry parameter. If $\hat{A} < \hat{A}_*$ the high cyclic creep process is similar to the static one with increased creep rate and decreased time to fracture. If $\hat{A} > \hat{A}_*$ such a behavior is not observable, and fracture takes place as a consequence of creep-fatigue interaction. Following [185, 272, 312], the processes of high-frequency cyclic creep are classified as: dynamic creep for $\hat{A} < \hat{A}_*$ and high cyclic creep for $\hat{A} > \hat{A}_*$. Creep curves for both cases are schematically presented in Fig. 1.4d.

Typical stress-strain diagrams, obtained from a strain controlled test under constant strain rate and temperature are illustrated in Fig. 1.5. It is obvious that the

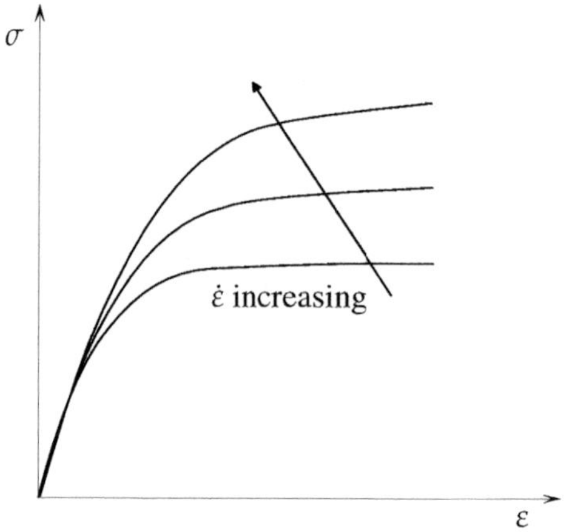

Fig. 1.5 Strain rate dependence of the stress-strain behavior

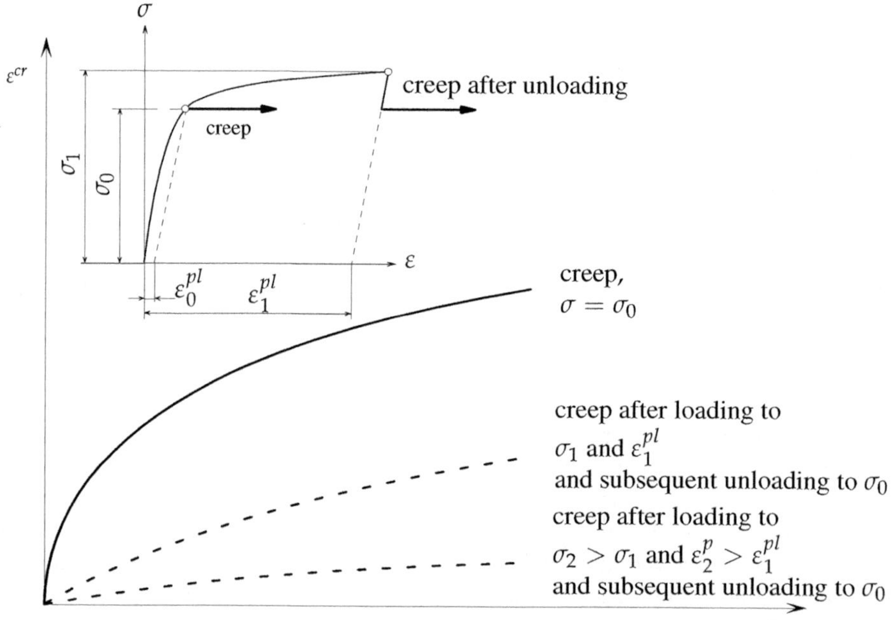

Fig. 1.6 Effect of initial plastic strain on creep behavior, for details see [160]

stress-strain behavior depends significantly on the value of the strain rate. Various examples of experimental data for steels obtained from the strain controlled tests are presented in [181, 309].

Creep behavior is highly sensitive to the type of material processing (e.g. plastic forming, heat treatment). As an example, let us illustrate the effect of spontaneous plastic pre-strain on the subsequent creep behavior, Fig. 1.6. The first creep curve (solid line) is a typical creep curve under the constant stress σ_0. The dotted lines present the second and the third creep curves after spontaneous loading to the stresses σ_1 and $\sigma_2 > \sigma_1$ leading to small plastic strains ε_1^{pl} and $\varepsilon_2^{pl} > \varepsilon_1^{pl}$, respectively, and subsequent unloading to the stress σ_0. The creep rate after the loading to the plastic strain is significantly lower compared to the creep rate of the "virgin" material. The effect of reduction in creep rate becomes stronger with the increase of the prior plastic strain. Effects of this type are sometimes termed as "plasticity-creep" or "creep-plasticity" interactions, e.g. [151, 179, 181, 214].

Various materials show anisotropic creep behavior. Examples are: directionally solidified nickel-based superalloys, e.g. [332], fiber reinforced materials, e.g. [280, 281], deep drawing sheets, e.g. [54, 61], and multi-pass weld metals [144]. In these cases series of uni-axial creep tests for specific loading directions are performed in order to establish the material behavior. The number of the required tests and the corresponding loading directions are dictated according to the assumed symmetries of the material microstructure.

1.1.2 Multi-Axial Creep and Stress State Effects

Experimental data obtained from uni-axial tests allow to establish basic features of the creep behavior and to find relations between strain rate, stress, temperature, time, etc. However, most structural members are subjected to multi-axial stress and strain conditions. In order to analyze the influence of the stress state on the time dependent material behavior, multi-axial creep tests are required.

Various techniques have been developed to test materials under multi-axial loading conditions. Examples of multi-axial specimens for creep testing are: thin-walled tubes subjected to axial force and torque, e.g. [173], two- and three-dimensional cruciform specimens subjected to axial forces, e.g. [289, 290], circumferentially notched specimens subjected to axial force, e.g. [149, 258].

Figure 1.7 shows a thin-walled tube under the axial force and torque with the magnitudes F and M, respectively. Let r_m be the mean radius of the cross section, h the wall thickness and L the gauge length. With the local cylindrical basis $\boldsymbol{e}_r$, $\boldsymbol{e}_\varphi$ and $\boldsymbol{k}$, as shown in Fig. 1.7, the stress state can be characterized by the following tensor

$$\boldsymbol{\sigma} = \sigma \boldsymbol{k} \otimes \boldsymbol{k} + \tau(\boldsymbol{e}_\varphi \otimes \boldsymbol{k} + \boldsymbol{k} \otimes \boldsymbol{e}_\varphi), \quad \sigma = \frac{F}{2\pi r_m h}, \quad \tau = \frac{M}{2\pi r_m^2 h} \tag{1.1.2}$$

The deviatoric part of the stress tensor is

$$\boldsymbol{s} = \sigma\left(\boldsymbol{k} \otimes \boldsymbol{k} - \frac{1}{3}\boldsymbol{I}\right) + \tau(\boldsymbol{e}_\varphi \otimes \boldsymbol{k} + \boldsymbol{k} \otimes \boldsymbol{e}_\varphi), \tag{1.1.3}$$

where $\boldsymbol{I}$ is the second rank unit tensor, see Sect. A.4. As a stress measure which allows to compare different multi-axial creep tests let us use the von Mises equivalent stress σ_{vM} which is defined as follows

$$\sigma_{vM} = \sqrt{\frac{3}{2}\boldsymbol{s} \cdot\cdot \boldsymbol{s}} = \sqrt{\sigma^2 + 3\tau^2}$$

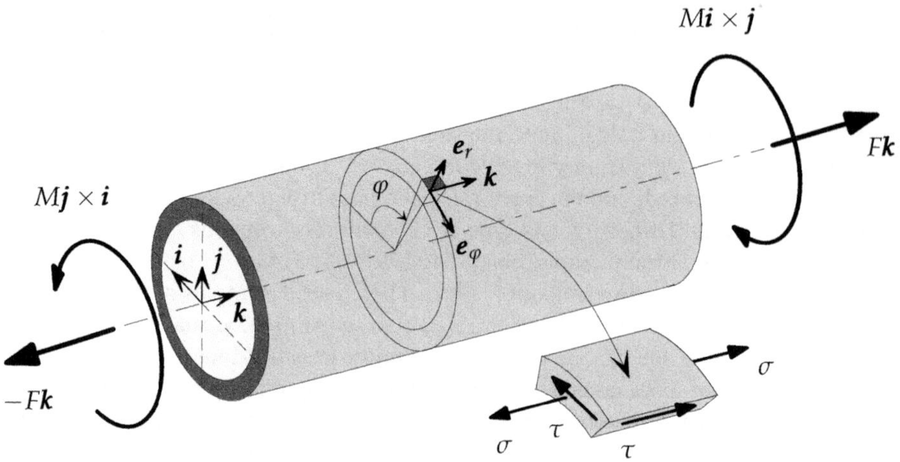

Fig. 1.7 Thin-walled tube for multi-axial creep tests

From the measured elongation ΔL and the angle of twist ϕ_T the axial strain ε_L and the shear strain γ can be computed

$$\varepsilon_L = \frac{\Delta L}{L}, \quad \gamma = \frac{r_m \phi_T}{L}$$

Assuming that the material behavior is isotropic, the strain state in a tube can be characterized by the following tensor

$$\boldsymbol{\varepsilon} = \varepsilon_L \boldsymbol{k} \otimes \boldsymbol{k} + \varepsilon_Q (\boldsymbol{I} - \boldsymbol{k} \otimes \boldsymbol{k}) + \frac{1}{2}\gamma(\boldsymbol{e}_\varphi \otimes \boldsymbol{k} + \boldsymbol{k} \otimes \boldsymbol{e}_\varphi),$$

where $\varepsilon_Q = \Delta r_m / r_m$ is the transverse normal strain. The creep strain tensor is defined as the difference between the strain tensor $\boldsymbol{\varepsilon}$ which includes the measurable quantities and the tensor of initial elastic strains which can be calculated from Hooke's law. As a result we obtain

$$\begin{aligned}\boldsymbol{\varepsilon}^{cr} &= \left(\varepsilon_L + 2\varepsilon_Q - \frac{1-2\nu}{E}\sigma\right)\frac{1}{3}\boldsymbol{I} + \left(\varepsilon_L - \varepsilon_Q - \frac{(1+\nu)}{E}\sigma\right)(\boldsymbol{k} \otimes \boldsymbol{k} - \frac{1}{3}\boldsymbol{I}) \\ &+ \frac{1}{2}\left(\gamma - \frac{2(1+\nu)}{E}\tau\right)(\boldsymbol{k} \otimes \boldsymbol{e}_\varphi + \boldsymbol{e}_\varphi \otimes \boldsymbol{k}),\end{aligned} \tag{1.1.4}$$

where ν is the Poisson's ratio. The basic assumption related to the multi-axial creep behavior is the volume constancy during the creep deformation, e.g. [241, 243]. In this case the following relations should be satisfied

$$\operatorname{tr} \boldsymbol{\varepsilon} = \operatorname{tr} \boldsymbol{\varepsilon}^{el} \quad \Rightarrow \quad \varepsilon_L + 2\varepsilon_Q = \frac{1-2\nu}{E}\sigma$$

From (1.1.4) follows

$$\boldsymbol{\varepsilon}^{cr} = \frac{3}{2}\left(\varepsilon_L - \frac{1}{E}\sigma\right)(\boldsymbol{k}\otimes\boldsymbol{k} - \frac{1}{3}\boldsymbol{I}) + \frac{1}{2}\left(\gamma - \frac{2(1+\nu)}{E}\tau\right)(\boldsymbol{k}\otimes\boldsymbol{e}_\varphi + \boldsymbol{e}_\varphi\otimes\boldsymbol{k})$$

Under the condition of stationary loading the creep rate tensor is

$$\dot{\boldsymbol{\varepsilon}} = \dot{\boldsymbol{\varepsilon}}^{cr} = \frac{3}{2}\dot{\varepsilon}_L(\boldsymbol{k}\otimes\boldsymbol{k} - \frac{1}{3}\boldsymbol{I}) + \frac{1}{2}\dot{\gamma}(\boldsymbol{k}\otimes\boldsymbol{e}_\varphi + \boldsymbol{e}_\varphi\otimes\boldsymbol{k}) \qquad (1.1.5)$$

The von Mises equivalent creep rate is defined by

$$\dot{\varepsilon}_{vM} = \sqrt{\frac{2}{3}\dot{\boldsymbol{\varepsilon}}\cdot\cdot\,\dot{\boldsymbol{\varepsilon}}} = \sqrt{\dot{\varepsilon}_L^2 + \frac{1}{3}\dot{\gamma}^2}$$

The results of creep tests on tubes are usually presented as: strains ε_L and γ vs. time curves, e.g. [139, 151, 160], creep strains

$$\varepsilon_L^{cr} = \varepsilon_L - \frac{\sigma}{E}, \quad \gamma^{cr} = \gamma - \frac{2(1+\nu)}{E}\tau$$

vs. time curves, e.g [224, 255, 245], von Mises equivalent creep strain

$$\varepsilon_{vM}^{cr} = \sqrt{\frac{2}{3}\boldsymbol{\varepsilon}^{cr}\cdot\cdot\,\boldsymbol{\varepsilon}^{cr}} = \sqrt{(\varepsilon_L^{cr})^2 + \frac{1}{3}(\gamma^{cr})^2}$$

vs. time curves, e.g. [173, 175], and the so-called specific dissipation work

$$q(\bar{t}) = \int_0^{\bar{t}} \dot{\boldsymbol{\varepsilon}}\cdot\cdot\,\boldsymbol{s}dt = \int_0^{\bar{t}} (\dot{\varepsilon}_L\sigma + \dot{\gamma}\tau)dt$$

vs. time curves [303, 304].

Figure 1.8 illustrates typical results of creep testing under constant von Mises stress σ_{vM}. Sketches of creep curves are presented for the case of tension under the normal stress $\sigma = \sigma_{vM}$ and torsion under the shear stress $\tau = \sigma_{vM}/\sqrt{3}$. For many structural materials the kind of the stress state (e.g. tension or torsion) has negligible influence on the primary and secondary creep behavior. However, this is not the case for the tertiary creep and the long term strength. Tubular specimen subjected to tension usually exhibit much shorter lifetime and lower ductility if compared to the case of pure torsion. This stress state effect has been observed for copper in [173] and for austenitic steels in [236, 318], for example.

Many results of creep tests under combined tension-torsion loading are published. Figure 1.9a shows the plot of the equation $\sigma^2 + 3\tau^2 = \sigma_{vM}^2 = const$ with respect to coordinates σ and $\sqrt{3}\tau$. Different stress states leading to the same fixed value of the von Mises stress can be conveniently characterized by the angle α (stress state angle). The corresponding values for the normal and the shear stress can be then calculated as follows

$$\sigma = \sigma_{vM}\cos\alpha, \quad \tau = \sigma_{vM}\frac{\sin\alpha}{\sqrt{3}}$$

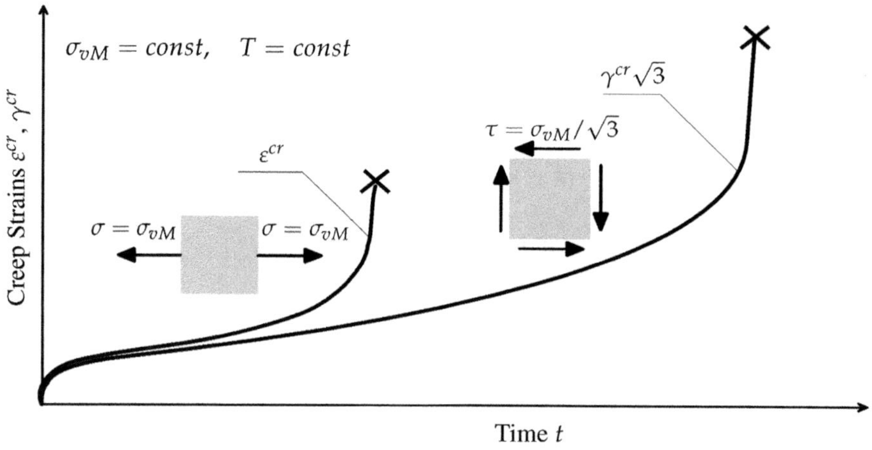

Fig. 1.8 Stress state effect of tertiary creep

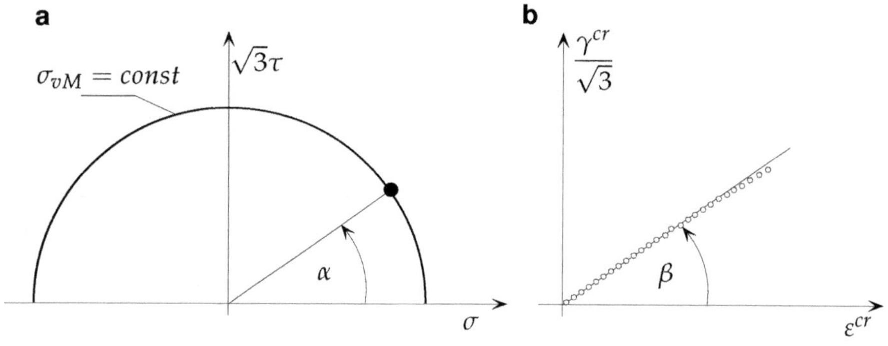

Fig. 1.9 Creep response under combined tension-torsion loading. **a** Plane stress state, **b** strain trajectory

For example, $\alpha = 0$ corresponds to the case of tension and $\alpha = \pi/2$ to the case of torsion. $0 < \alpha < \pi/2$ characterizes the combined loading case. The loading conditions realized in creep tests can be classified as follows

a) stationary σ_{vM} and different but stationary α,
b) time-varying (e.g. stepwise or cyclic) σ_{vM} under fixed α,
c) time-varying α under fixed σ_{vM} and
d) both σ_{vM} and α are time-varying.

The loading cases a) and b) are called simple or proportional loadings, while the cases c) and d) are classified as non-proportional loadings. The results of creep tests under the combined loading can be conveniently presented as $\gamma^{cr}/\sqrt{3}$ vs. ε^{cr} curves (so-called strain trajectories), e.g. [224, 235]. A sketch of such a curve for the loading case a) is presented in Fig. 1.9b. For many metals and alloys, e.g. [224, 235, 252], the direction of the strain trajectory characterized by the angel β, Fig. 1.9b, coincides with the direction of the applied stress state characterized by the angle α. According to this result one can assume that the creep rate tensor is coaxial and collinear with the stress deviator, i.e. $\dot{\boldsymbol{\varepsilon}} = \lambda \boldsymbol{s}$. Taking into account (1.1.3) and (1.1.5) the following relations can be obtained

$$\frac{3}{2}\dot{\varepsilon}_L = \lambda\sigma, \quad \frac{1}{2}\dot{\gamma} = \lambda\tau \quad \Rightarrow \quad \frac{\dot{\varepsilon}_L}{\dot{\gamma}/\sqrt{3}} = \frac{\sigma}{\sqrt{3}\tau}$$

In many cases, experimental results show that the above relations are well satisfied, e.g. [139, 224, 235, 252].

Non-coincidence of the strain-trajectory and the stress state angles indicates the anisotropy of the creep behavior. Anisotropic creep may be caused either by the initial anisotropy of the material microstructure as a result of material processing or by the anisotropy induced during the creep process. Examples for anisotropic tension-torsion creep are presented for a directionally solidified nickel-based superalloy in [246] and for a fiber-reinforced material in [280, 281]. The trajectories of creep strains presented in [160] for austenitic steel tubes demonstrate that initial small plastic pre-strain causes the anisotropy of subsequent creep behavior. The deformation induced anisotropy may be observed in creep tests under non-proportional loading conditions. The effects of the induced anisotropy are usually related to anisotropic hardening, [252, 160, 235], and damage processes, e.g. [224].

Another stress state effect is the different creep behavior under tensile and compressive loadings. Examples are presented for several alloys in [109, 201, 309, 349], for polymers in [166, 193], and for ceramics in [261]. Experimental results show that for the same value of stress in tension and compression, the value of the creep rate under tension is significantly greater than the corresponding absolute value under compression. This effect indicates that besides the von Mises equivalent stress, additional characteristics of the stress state (e.g. the mean stress) may influence the creep process.

1.2 Creep in Engineering Structures

Creep in structures is a variety of time-dependent changes of strain and stress states including progressive deformations, relaxation and redistribution of stresses, local reduction of the material strength. To illustrate these processes let us consider a beam with a rectangular cross section. We assume that the beam is heated up to a certain temperature, clamped at the ends and uniformly loaded as shown in Fig. 1.10a. The loading is moderate leading to spontaneous elastic deformation of the beam. Let the maximum deflection of the beam in the reference "elastic" state be w_0 and the maximum bending stress be σ_0. Furthermore, let us assume that creep curves of the material under uni-axial tension and compression are as sketched in Fig. 1.10b. Here the time to fracture of a uni-axial specimen loaded by the tensile stress with the magnitude σ_0 (the magnitude of maximum reference bending stress in the beam) is specified by t_f. The tertiary creep stage is stress state dependent,

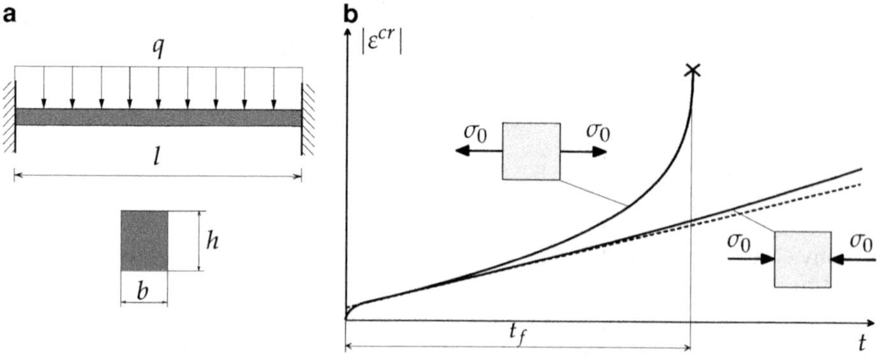

Fig. 1.10 Uniformly loaded clamped beam. **a** Geometry and loading, **b** sketch of the assumed creep curves under tension and compression

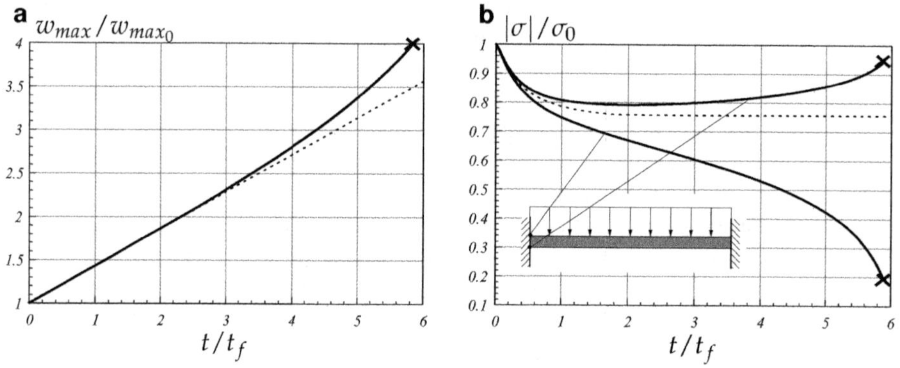

Fig. 1.11 Creep of a uniformly loaded clamped beam. **a** Normalized maximum deflection vs. normalized time, **b** normalized maximum bending stress vs. time

i.e. for the same stress magnitudes in tension and compression the creep rate under tension is much greater then the corresponding absolute value under compression. The dotted line in 1.10b shows the idealized creep curve having only the stress state independent secondary stage.

Creep processes in a beam under the constant load q and the assumed material behavior are the progressive deformation which may be characterized by the maximum deflection vs. time curve, Fig. 1.11a, the relaxation of the bending stresses, Fig. 1.11b, and the stress redistributions, Fig. 1.12. The results illustrated in Figs 1.11 and 1.12 are obtained from the finite element calculation [231]. Here let us discuss some basic features of creep in the case of the non-homogeneous stress and strain states. First let us explain origins of the simultaneous increase of deformations and the relaxation of stresses. For this purpose we assume that the beam deforms in such a manner that every cross section behaves like a rigid plane, i.e. it may only translate and rotate about the axis which is orthogonal to the plane of bending. Furthermore, we assume that mechanical interactions between the cross sections are only due to forces and moments. The above assumptions are the basis

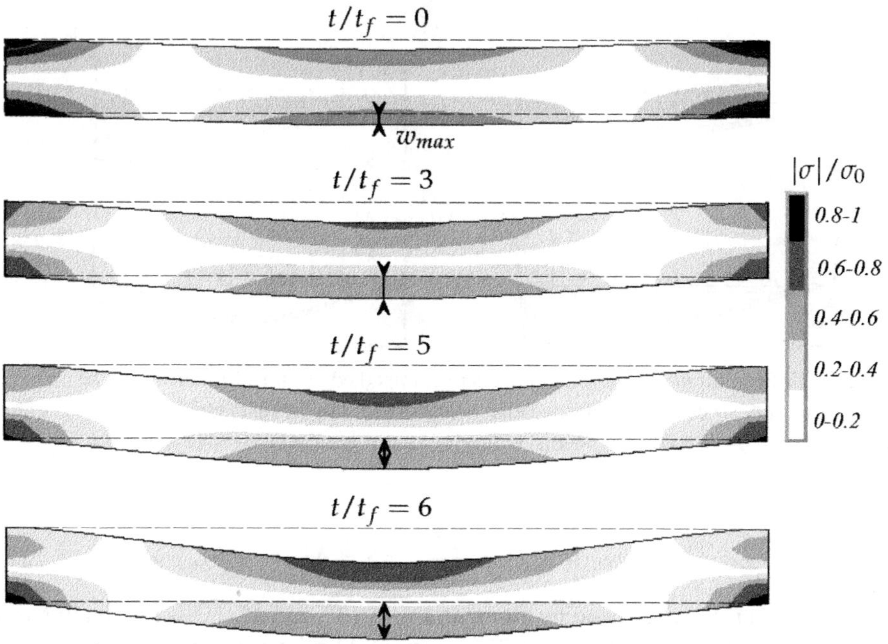

Fig. 1.12 Distributions of the normalized bending stress at different time steps

of various theories of beams. Let us note that the results presented in Figs 1.11 and 1.12 are obtained without these assumptions. However, one may show that they are well satisfied [231].

Figure 1.13a is a sketch of the elastic deformation of the beam in the reference state. In Figure 1.13b the clamped edges are replaced by the pin supports and the moments M_0. These moments must be applied in order to fix the zero cross section rotations at the ends. As a result of creep, the deformations of the beam increase in time. If the moments M_0 are kept constant then, after a period of time the beam would have a deformed shape as sketched in Fig. 1.13c. In this case the angles of cross section rotations at the ends increase in time. In order to keep the zero cross section rotations the moments must be relaxed, Fig. 1.13d. If the material behaves as shown in Fig. 1.10b by the dotted line, a steady state exists, for which the moments do not depend on time and the deflection increases with a constant rate. The steady-state solutions for the maximum deflection and bending stress are presented in Fig. 1.11 by dotted lines. The rate of maximum deflection, the maximum bending moment and the maximum bending stress in the steady state can be estimated according to the elementary beam theory [80, 208, 241].

The next feature is the redistribution of bending moments during the creep process. The origin can be explained based on equilibrium conditions. As an example let us write down the equilibrium condition for the moments considering a half of

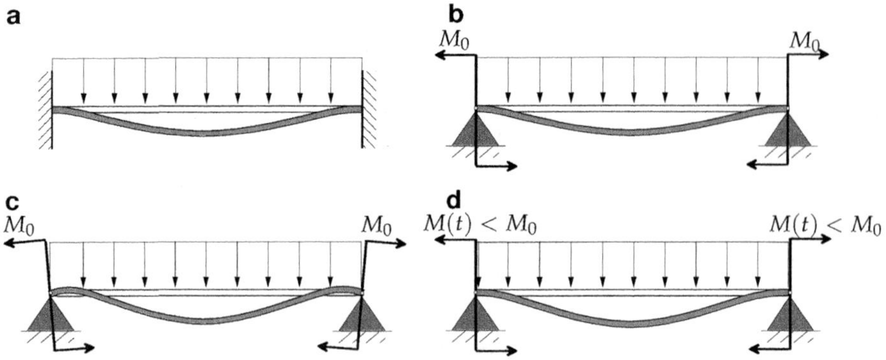

Fig. 1.13 Relaxation of bending moments in clamped edges. **a** Deformed elastic beam in the reference state, **b** equivalent elastic beam with simple supports and edge moments, **c** "crept" beam under constant edge moments, **d** "crept" beam under relaxed edge moments

the beam

$$M(t) + M_m(t) = \frac{ql^2}{8} \quad \Rightarrow \quad \dot{M}_m = -\dot{M},$$

where $M(t)$ is the edge bending moment, Fig. 1.13c, and $M_m(t)$ is the bending moment in the middle cross section. The moment M relaxes down as a consequence of creep process. The above equilibrium condition states that the moment M_m increases. The rate of increase is equal to the rate of relaxation.

Similar considerations explain the redistribution of bending stresses. For the sake of brevity assume that the beam is simply supported, i.e. $M(t) = M_0 = 0$. In this statically determined case the bending moments in all cross sections remain constant during the creep process. However, the stresses in the points of cross sections redistribute essentially. The outer tensile and compressive layers exhibit the highest creep rates due to the maximum stress values in the reference state. Therefore they will show the highest relaxation rates at the beginning of the creep process. The redistribution of stresses over the cross section is enhanced by the essential non-linearity of the creep rate with respect to the stress magnitude. Steady state creep solutions for bending stresses are discussed in [80, 208, 241].

Results presented in Fig. 1.12 show that the distributions of absolute values of the bending stresses are non-symmetrical with respect to the beam centerline. This is the consequence of the assumed stress state dependent tertiary creep behavior, Fig. 1.10. Tensile layers of the beam cross section "creep" with higher rates compared to compressive layers.

Creep fracture originates in outer tensile layers of the clamped cross sections [231]. These layers exhibit, however, the lowest values of stresses at the final stage of creep process, Figs 1.11b and 1.12. This result can be explained by material damage processes (e.g. grain boundary cavitation and ageing of microstructure) accompanying creep deformation. These processes develop over time with the rates determined not only by the reference stress values but also by the complete loading history. A damaged material has lower ultimate stress compared to the virgin one. Outer tensile layers of the clamped cross sections are places with the highest "damage grade".

The above discussed features of creep are common for many structures operating under high-temperature conditions. Examples are structural components of power plants, chemical refineries or heat engines, e.g. [127]. Design of pipework systems, rotors, turbine blades, etc. requires the consideration of creep. Creep processes may cause excessive deformations, damage, buckling, crack initiation and growth.

Different types of creep failure in the recent years are discussed in the literature. Examples of critical structural members include pipe bends [186], welds [297], turbine blade root fixings [127], etc. The possibilities to analyze a structural prototype in the laboratory are limited by the long duration of tests and related costs. Furthermore, examinations of creep and damage states in a structure during the service (e.g. replicas) can be only made at specific outer surface positions and after certain periods of time. The modeling of creep processes in structures is therefore an essential contribution to optimal design and residual life assessment. Furthermore it contributes to understanding and analysis of time-dependent deformations, stress redistributions and damage growth under given temperature and loading conditions.

1.3 Basic Approaches to Creep Modeling

The basic approaches to the description of creep behavior can be classified as follows. The *empirical modeling* is the study of correlations between the creep rate, stress, temperature and time. In addition, extrapolation methods are developed to predict time-dependent deformations and life time of a structure based on experimental data from short-term uni-axial creep tests. The aim of this approach is to derive simple formulae for an estimation of the structural behavior under creep conditions. An example is the Monkman-Grant relation which states that the product of the minimum creep rate and the time to fracture is a constant. Many different empirical relations of this type are reviewed in [257]. They are useful in early stages of design for a robust prediction of the components operation life. It should be noted that the empirical approach provides one-dimensional relations. The dependencies of creep behavior on the type of stress state are not discussed. Furthermore, the possibility of stress redistributions cannot be considered.

Within the *materials science modeling*, creep is characterized by a variety of microstructural rearrangements. According to assumed scenarios of transport processes in the microscale (diffusion of vacancies, climb and glide of dislocations, etc.) equations for the creep rate are derived. The form of the specific rate equation depends on the assumed deformation and damage mechanisms for specific stress and temperature ranges, e.g. [120]. Many diverse equations of this type are reviewed in [119, 159, 228]. In addition, kinetic equations for internal state variables are discussed. Examples for these variables include dislocation density, [113], internal (back) stress, e.g. [119], and various damage parameters associated with ageing and cavitation processes, [104]. The aim of this approach is to provide correlations between quantities characterizing the type of microstructure and processing (grain size, types of alloying and hardening, etc.) and quantities characterizing the material behavior, e.g. the creep rate. Furthermore, the mechanisms based classification of different forms of creep equations including different stress and temperature functions is helpful in the structural analysis. However, the models proposed within the materials science are usually one-dimensional and operate with scalar-valued quantities like magnitudes of stress and strain rates.

The *micromechanical models* deal with discrete simulations of material behavior for a representative volume element with geometrically idealized microstructure. Simplifying assumptions are made for the behavior of constituents and their interactions, for the type of the representative volume element and for the exerted boundary conditions. Examples include numerical simulations of void growth in a power law creeping matrix material, e.g. [323, 326], crack propagation through a power law creeping multi-grain model, e.g. [248, 325], stress redistributions between constituents in a binary medium, e.g. [233]. Micromechanical models contribute to understanding creep and damage processes in heterogeneous systems. With respect to engineering applications the micromechanical approach suffers, however, from significant limitations. One of them is that a typical high-temperature structural material, for example steel, has a complex composition including dislocation structures, grain boundaries, dispersion particles, precipitates, etc. A reliable micromechanical description of creep would therefore require a rather complex model of a multiphase medium with many evolving and interacting constituents.

The objective of *continuum mechanics modeling* is to investigate creep in idealized three-dimensional solids. The idealization is related to the hypothesis of a continuum, e.g. [134]. The approach is based on balance equations and assumptions regarding the kinematics of deformation and motion. Creep behavior is described by means of constitutive equations which relate deformation processes and stresses. Details of topological changes of microstructure like subgrain size or mean radius of carbide precipitates are not considered. The processes associated with these changes like hardening, recovery, ageing and damage can be taken into account by means of hidden or internal state variables and corresponding evolution equations, [62, 191, 272, 298]. Creep constitutive equations with internal state variables can be applied to structural analysis. Various models and methods recently developed within the mechanics of structures can be extended to the solution of creep problems. Examples are theories of rods, plates and shells as well as direct variational methods, e.g. [9, 62, 80, 208, 262, 299]. Numerical solutions by the finite element method combined with various time step integration techniques allow to simulate time dependent structural behavior up to critical state of failure. Examples of recent studies include circumferentially notched bars [136], pipe weldments [140] and thin-walled tubes [182]. In these investigations qualitative agreements between the theory and experiments carried out on model structures have been established. Constitutive equations with internal state variables have been found to be mostly suited for the creep analysis of structures [140]. However, it should be noted that this approach requires numerous experimental data of creep for structural materials over a wide range of stress and temperature as well as different stress states.

2 Constitutive Models of Creep

Analysis of creep in engineering structures requires the formulation and the solution of an initial-boundary value problem including the balance equations and the constitutive assumptions. Equations describing the kinematics of three-dimensional solids as well as balance equations of continuum mechanics are presented in various monographs and textbooks, e.g. [32, 39, 48, 61, 111, 134, 184, 205]. In what follows we discuss constitutive equations for the description of creep behavior in three-dimensional solids.

The starting point of the engineering creep theory is the introduction of the inelastic strain, the creep potential, the flow rule, the equivalent stress and internal state variables, Sect. 2.1. In Sect. 2.2 we discuss constitutive models of secondary creep. We start with the von Mises-Odqvist creep potential and the flow rule widely used in the creep mechanics. To account for stress state effects creep potentials that include three invariants of the stress tensor are introduced. Consideration of material symmetries provide restrictions for the creep potential. A novel direct approach to find scalar valued arguments of the creep potential for the given group of material symmetries is proposed. For several cases of material symmetry appropriate invariants of the stress tensor, equivalent stress and strain expressions as well as constitutive equations for anisotropic creep are derived. In Sect. 2.3 we review experimental foundations and models of transient creep behavior under different multi-axial loading conditions. Section 2.4 is devoted to the description of tertiary creep under multi-axial stress states. Various models within the framework of continuum damage mechanics are discussed.

All equations are presented in the direct tensor notation. This notation guarantees the invariance with respect to the choice of the coordinate system and has the advantage of clear and compact representation of constitutive assumptions, particularly in the case of anisotropic creep. The basic rules of the direct tensor calculus as well as some new results for basic sets of invariants with respect to different symmetry classes are presented in Appendices A - C.

2.1 General Remarks

The modeling of creep under multi-axial stress states is the key step in the adequate prediction of the long term structural behavior. Such a modeling requires the introduction of tensors of stress, strain, strain rate and corresponding inelastic parts. Usually, they are discussed within the framework of continuum mechanics starting from fundamental balance equations. One of the most important and fundamental questions is that of the definition (or even the existence) of a measure of the inelastic strain and the decomposition of the total strain into elastic and irreversible parts within the material description. From the theoretical point of view this is still a subject of many discussions within the non-linear continuum mechanics, e.g. [49, 50, 229, 253].

In engineering mechanics, these concepts are often introduced based on intuitive assumptions, available experimental data and applications. Therefore, a lot of formulations of multi-axial creep equations can be found in the literature. In what follows some of them will be discussed. First let us recall several assumptions usually made in the creep mechanics [62, 242].

The assumption of infinitesimal strains allows to neglect the difference between the true stresses and strains and the engineering stresses and strains. According to the continuum mechanics there are no differences between the Eulerian and the Lagrangian approaches within the material description. Creep equations in the geometrical non-linear case (finite strains) are discussed in the monograph [70], for example. Finite strain equations based on rheological models are presented in the monographs [180, 253]. The linearized equations of continuum creep mechanics can be used in the majority of engineering applications because structures are usually designed such that the displacements and strains arising as a consequence of the applied loading do not exceed the prescribed small values. The exception is the case of thin-walled shells, where geometrical non-linearities must be considered even if strains are infinitesimal, see Sect. 4.4.

The assumption of the classical non-polar continuum restricts the class of materials. The equations of motion within the continuum mechanics include the balance of momentum and the balance of angular momentum, e.g. [111]. These equations introduce the stress and the moment stress tensors. Polar materials are those which are characterized by constitutive equations with respect to both tensors (in general, they are non-symmetric). In addition, the rotation degrees of freedom, i.e. the rotation tensor and the angular velocity, are introduced as independent quantities. Models of polar continua found application to granular or porous materials [100, 107, 220], fiber suspensions [25, 112], or other media with changing microstructure. At present, the moment stress tensor and the anti-symmetric part of the stress tensor are not considered in the engineering creep theories. The reason for this is the higher order complexity of the models and as a consequence increased effort for the identification of material characteristics and the numerical solutions.

The assumption of isothermal conditions makes it possible to decouple the thermal and the mechanical problem. Furthermore, heat transfer problems are not considered. The influence of the constant temperature on the creep rate is described by an Arrhenius function, see Sect. 2.2.3. Coupled thermo-mechanical problems of creep and damage are discussed in [298], where the influence of creep cavitation on thermal conductivity is considered.

In this chapter we shall use the following notation. Let $\boldsymbol{\sigma}$ be the Cauchy stress tensor and $\boldsymbol{\varepsilon}$ be the tensor of infinitesimal strains as they are defined in [32, 61, 205], among others. Let the symmetric second rank tensor $\dot{\boldsymbol{\varepsilon}}^{cr}$ be the tensor of the rate of infinitesimal inelastic strains induced by the creep process. For the infinitesimal strains one can assume the additive split of the total strain rate into elastic and creep parts, i.e. $\dot{\boldsymbol{\varepsilon}} = \dot{\boldsymbol{\varepsilon}}^{el} + \dot{\boldsymbol{\varepsilon}}^{cr}$. The constitutive equation relating the stress tensor and the elastic part of the strain tensor can be formulated according to the generalized Hooke's law [32, 59, 129, 205] and will be introduced later. Creep deformation is accompanied by various microstructural changes having different influences on the strain rate. The current state of the material microstructure is determined by the entire previous history of the creep process. It can be characterized by a set of additional field variables termed as internal or hidden state variables. In this chapter we shall discuss internal state variables characterizing the states of hardening/recovery and damage. In order to distinguish between the hardening and damage mechanisms we shall specify the "internal hardening variables" by H_i and the "internal damage variables" by ω_j. The number of such variables and the corresponding evolution equations (ordinary differential equations with respect to the time variable) is dictated by the knowledge of creep-damage mechanisms for a specified metal or alloy, the availability of experimental data on creep and long term strength as well as the type of the structural analysis application. In some cases the internal state variables must be introduced as tensors of different rank in order to include effects of the deformation or damage induced anisotropy.

Constitutive equations of multi-axial creep are usually based on the concept of the creep potential and the flow rule. The associated flow rule has the origin in the engineering theory of plasticity. The basic assumptions of this theory are:

- The existence of a yield condition (creep condition, see [59], for example) expressed by the equation $F(\boldsymbol{\sigma}) = 0$, where F is a scalar valued function. In the general case one can presume that F depends not only on the stress tensor but also on the internal state variables and the temperature [208, 272], i.e. the yield condition has a form

$$F(\boldsymbol{\sigma}, H_i, \omega_j, T) = 0, \quad i = 1, \ldots, n, \quad j = 1, \ldots, m \tag{2.1.1}$$

- The existence of a flow potential as a function of the stress tensor $\Phi(\boldsymbol{\sigma})$.

The flow rule (sometimes called the normality rule) is the following assumption for the inelastic strain rate tensor

$$\dot{\boldsymbol{\varepsilon}}^{in} = \dot{\eta}\frac{\partial \Phi}{\partial \boldsymbol{\sigma}}, \tag{2.1.2}$$

where $\dot{\eta}$ is a scalar factor. In the special case that the flow potential coincides with the yield function i.e. $\Phi = F$ (2.1.2) represents the associated flow rule. With respect to the variation of the stress tensor $\delta\boldsymbol{\sigma}$ one distinguishes between the cases of elastic state, unloading from an elastic-plastic state, neutral loading and loading, i.e.

$$\begin{cases} F(\boldsymbol{\sigma}) < 0, & \text{elastic state} \\ F(\boldsymbol{\sigma}) = 0, \quad \text{and} \quad \delta F = \delta\boldsymbol{\sigma} \cdot\cdot \dfrac{\partial F}{\partial \boldsymbol{\sigma}} < 0 & \text{unloading} \\ F(\boldsymbol{\sigma}) = 0, \quad \text{and} \quad \delta F = \delta\boldsymbol{\sigma} \cdot\cdot \dfrac{\partial F}{\partial \boldsymbol{\sigma}} = 0 & \text{neutral loading} \\ F(\boldsymbol{\sigma}) = 0, \quad \text{and} \quad \delta F = \delta\boldsymbol{\sigma} \cdot\cdot \dfrac{\partial F}{\partial \boldsymbol{\sigma}} > 0 & \text{loading} \end{cases}$$

For work hardening materials $\dot{\eta} > 0$ is set in the case of loading/neutral loading, otherwise $\dot{\eta} = 0$, see e.g. [207]. Further details of the flow theory as well as different arguments leading to (2.1.2) can be found in textbooks on theory of plasticity, e.g. [141, 154, 156, 164, 207, 212, 299].

Within the creep mechanics the flow theory is usually applied without the concept of the yield stress or yield condition. This is motivated by the fact that creep is a thermally activated process and the material starts to creep even under low and moderate stresses lying below the yield limit. Furthermore, at high temperatures $0.5T_m < T < 0.7T_m$ the main creep mechanism for metals and alloys is the diffusion of vacancies, e.g. [120]. Under this condition the existence of a yield or a creep limit cannot be verified experimentally. In [191], p. 278 it is stated that "the concept of a loading surface and the loading-unloading criterion which was used in plasticity is no longer necessary". In monographs [59, 62, 207, 208, 257] the flow rule is applied as follows

$$\dot{\boldsymbol{\varepsilon}}^{cr} = \dot{\eta} \frac{\partial \Phi}{\partial \boldsymbol{\sigma}}, \quad \dot{\eta} > 0 \tag{2.1.3}$$

Equation (2.1.3) states the "normality" of the creep rate tensor to the surfaces $\Phi(\boldsymbol{\sigma}) = const$. The scalar factor $\dot{\eta}$ is determined according to the hypothesis of the equivalence of the dissipation power [4, 62]. The dissipation power is defined by $P = \dot{\boldsymbol{\varepsilon}}^{cr} \cdot\cdot \boldsymbol{\sigma}$. It is assumed that $P = \dot{\varepsilon}^{cr}_{eq} \sigma_{eq}$, where $\dot{\varepsilon}^{cr}_{eq}$ is an equivalent creep rate and σ_{eq} is an equivalent stress. The equivalent measures of stress and creep rate are convenient to compare experimental data under different stress states (see Sect. 1.1.2). From the above hypothesis follows

$$\dot{\eta} = \frac{P}{\dfrac{\partial \Phi}{\partial \boldsymbol{\sigma}} \cdot\cdot \boldsymbol{\sigma}} = \frac{\dot{\varepsilon}^{cr}_{eq} \sigma_{eq}}{\dfrac{\partial \Phi}{\partial \boldsymbol{\sigma}} \cdot\cdot \boldsymbol{\sigma}} \tag{2.1.4}$$

The equivalent creep rate is defined as a function of the equivalent stress according to the experimental data for uni-axial creep as well as creep mechanisms operating for the given stress range. An example is the power law stress function

$$\dot{\varepsilon}^{cr}_{eq}(\sigma_{eq}) = a\sigma^n_{eq} \tag{2.1.5}$$

Another form of the flow rule without the yield condition has been proposed by Odqvist, [241, 243]. The steady-state creep theory by Odqvist, see [241], p. 21 is based on the variational equation $\delta W = \delta\boldsymbol{\sigma} \cdot\cdot \dot{\boldsymbol{\varepsilon}}^{cr}$ leading to the flow rule

$$\dot{\boldsymbol{\varepsilon}}^{cr} = \frac{\partial W}{\partial \boldsymbol{\sigma}}, \tag{2.1.6}$$

where the scalar valued function $W(\boldsymbol{\sigma})$ plays the role of the creep potential[1]. In order to specify the creep potential, the equivalent stress $\sigma_{eq}(\boldsymbol{\sigma})$ is introduced. Taking into account that $W(\boldsymbol{\sigma}) = W(\sigma_{eq}(\boldsymbol{\sigma}))$ the flow rule (2.1.6) yields

$$\dot{\boldsymbol{\varepsilon}}^{cr} = \frac{\partial W}{\partial \sigma_{eq}} \frac{\partial \sigma_{eq}}{\partial \boldsymbol{\sigma}} = \dot{\varepsilon}^{cr}_{eq} \frac{\partial \sigma_{eq}}{\partial \boldsymbol{\sigma}}, \quad \dot{\varepsilon}^{cr}_{eq} \equiv \frac{\partial W}{\partial \sigma_{eq}} \tag{2.1.7}$$

The creep potential $W(\sigma_{eq})$ is defined according to experimental data of creep under uni-axial stress state for the given stress range. An example is the Norton-Bailey-Odqvist creep potential

$$W = \frac{\sigma_0}{n+1} \left(\frac{\sigma_{vM}}{\sigma_0} \right)^{n+1}, \tag{2.1.8}$$

widely used for the description of steady-state creep of metals and alloys. In (2.1.8) σ_0 and n are material constants and σ_{vM} is the von Mises equivalent stress. Below we discuss various restrictions on the potentials, e.g. the symmetries of the creep behavior and the inelastic incompressibility.

In order to compare the flow rules (2.1.3) and (2.1.6) let us compute the dissipation power. From (2.1.7) it follows

$$P = \dot{\boldsymbol{\varepsilon}}^{cr} \cdot\cdot\, \boldsymbol{\sigma} = \frac{\partial W}{\partial \sigma_{eq}} \frac{\partial \sigma_{eq}}{\partial \boldsymbol{\sigma}} \cdot\cdot\, \boldsymbol{\sigma} = \dot{\varepsilon}^{cr}_{eq} \frac{\partial \sigma_{eq}}{\partial \boldsymbol{\sigma}} \cdot\cdot\, \boldsymbol{\sigma},$$

We observe that the equivalence of the dissipation power follows from (2.1.7) if the equivalent stress satisfies the following partial differential equation

$$\frac{\partial \sigma_{eq}}{\partial \boldsymbol{\sigma}} \cdot\cdot\, \boldsymbol{\sigma} = \sigma_{eq} \tag{2.1.9}$$

Furthermore, in this case the flow rules (2.1.3) and (2.1.6) lead to the same creep constitutive equation. Many proposed equivalent stress expressions satisfy (2.1.9).

The above potential formulations originate from the works of Richard von Mises, where the existence of variational principles is assumed in analogy to those known from the theory of elasticity (the principle of the minimum of the complementary elastic energy, for example). Richard von Mises wrote [328]: "Die Formänderung regelt sich derart, daß die pro Zeiteinheit von ihr verzehrte Arbeit unverändert bleibt gegenüber kleinen Variationen der Spannungen innerhalb der Fließgrenze. Da die Elastizitätstheorie einen ähnlichen Zusammenhang zwischen den Deformationsgrößen und dem elastischen Potential lehrt, so nenne ich die Spannungsfunktion F auch das "plastische Potential" oder "Fließpotential"." It can be shown that the variational principles of linear elasticity are special cases of the energy balance equation (for isothermal or adiabatic processes), see e.g. [204], p. 148,

[1] The dependence on the temperature is dropped for the sake of brevity.

for example. Many attempts have been made to prove or to motivate the potential formulations within the framework of irreversible thermodynamics. For quasi-static irreversible processes various extremum principles (e.g. the principle of least irreversible force) are stipulated in [347]. Based on these principles and additional arguments like material stability, the potential formulations and the flow rules (2.1.1) and (2.1.6) can be verified. In [191], p. 63 a complementary dissipation potential as a function of the stress tensor as well as the number of additional forces conjugate to internal state variables is postulated, whose properties, e.g. the convexity, are sufficient conditions to satisfy the dissipation inequality. In [212] theories of plasticity and visco-plasticity are based on the notion of the dissipation pseudo-potentials. However, as far as we know, the flow rules (2.1.1) and (2.1.6) still represent the assumptions confirmed by various experimental observations of steady-state creep in metals rather than consequences of the fundamental laws. The advantage of variational statements is that they are convenient for the formulation of initial-boundary value problems and for the numerical analysis of creep in engineering structures. The direct variational methods (for example, the Ritz method, the Galerkin method, the finite element method) can be applied for the numerical solution.

Finally, several creep theories without creep potentials may be found in the literature. In the monograph [253] various constitutive equations of elastic-plastic and elastic-visco-plastic behavior in the sense of rheological models are discussed without introducing the plasticity, creep or dissipation potentials. For example, the models of viscous flow of isotropic media known from rheology, e.g. [126, 276], can be formulated as the relations between two coaxial tensors

$$\boldsymbol{\sigma} = G_0 \boldsymbol{I} + G_1 \dot{\boldsymbol{\varepsilon}} + G_2 \dot{\boldsymbol{\varepsilon}} \cdot \dot{\boldsymbol{\varepsilon}} \tag{2.1.10}$$

or

$$\dot{\boldsymbol{\varepsilon}} = H_0 \boldsymbol{I} + H_1 \boldsymbol{\sigma} + H_2 \boldsymbol{\sigma} \cdot \boldsymbol{\sigma}, \tag{2.1.11}$$

where G_i are functions of invariants of $\dot{\boldsymbol{\varepsilon}}$ while H_i depend on invariants of $\boldsymbol{\sigma}$. The application of the dissipative inequality provides restrictions imposed on G_i or H_i. The existence of the potential requires that G_i or H_i must satisfy certain integrability conditions [62, 205].

2.2 Secondary Creep

Secondary or stationary creep is for many applications the most important creep model. After a relatively short transient period the material creeps in such a manner that an approximate equilibrium between hardening and softening processes can be assumed. This equilibrium exists for a long time and the long term behavior of a structure can be analyzed assuming stationary creep processes. In this section several models of secondary creep are introduced. The secondary or stationary creep assumes constant or slowly varying loading and temperature conditions. Furthermore, the stress tensor is assumed to satisfy the condition of proportional loading, i.e. $\boldsymbol{\sigma}(t) = \varphi(t)\boldsymbol{\sigma}_0$, where $\varphi(t)$ is a slowly varying function of time and $\boldsymbol{\sigma}_0$ is a constant tensor.

2.2.1 Isotropic Creep

In many cases creep behavior can be assumed to be isotropic. In what follows the classical potential and the potential formulated in terms of three invariants of the stress tensor are introduced.

2.2.1.1 Classical Creep Equations. The starting point is the Odqvist flow rule (2.1.6). Under the assumption of the isotropic creep, the potential must satisfy the following restriction

$$W(\boldsymbol{Q}\cdot\boldsymbol{\sigma}\cdot\boldsymbol{Q}^T) = W(\boldsymbol{\sigma}) \tag{2.2.1}$$

for any symmetry transformation $\boldsymbol{Q}$, $\boldsymbol{Q}\cdot\boldsymbol{Q}^T = \boldsymbol{I}$, $\det\boldsymbol{Q} = \pm 1$. From (2.2.1) it follows that the potential depends only on the three invariants of the stress tensor (see Sect. C.2). Applying the principal invariants

$$\begin{aligned} J_1(\boldsymbol{\sigma}) &= \operatorname{tr}\boldsymbol{\sigma}, \quad J_2(\boldsymbol{\sigma}) = \frac{1}{2}[(\operatorname{tr}\boldsymbol{\sigma})^2 - \operatorname{tr}\boldsymbol{\sigma}^2], \\ J_3(\boldsymbol{\sigma}) &= \det\boldsymbol{\sigma} = \frac{1}{6}(\operatorname{tr}\boldsymbol{\sigma})^3 - \frac{1}{2}\operatorname{tr}\boldsymbol{\sigma}\operatorname{tr}\boldsymbol{\sigma}^2 + \frac{1}{3}\operatorname{tr}\boldsymbol{\sigma}^3 \end{aligned} \tag{2.2.2}$$

one can write

$$W(\boldsymbol{\sigma}) = W(J_1, J_2, J_3)$$

Any symmetric second rank tensor can be uniquely decomposed into the spherical part and the deviatoric part. For the stress tensor this decomposition can be written down as follows

$$\boldsymbol{\sigma} = \sigma_m\boldsymbol{I} + \boldsymbol{s}, \quad \operatorname{tr}\boldsymbol{s} = 0 \quad \Rightarrow \quad \sigma_m = \frac{1}{3}\operatorname{tr}\boldsymbol{\sigma},$$

where $\boldsymbol{s}$ is the stress deviator and σ_m is the mean stress. With the principal invariants of the stress deviator

$$J_{2D} = -\frac{1}{2}\operatorname{tr}\boldsymbol{s}^2 = -\frac{1}{2}\boldsymbol{s}\cdot\cdot\,\boldsymbol{s}, \quad J_{3D} = \frac{1}{3}\operatorname{tr}\boldsymbol{s}^3 = \frac{1}{3}(\boldsymbol{s}\cdot\boldsymbol{s})\cdot\cdot\,\boldsymbol{s}$$

the potential takes the form

$$W = W(J_1, J_{2D}, J_{3D}),$$

Applying the rule for the derivative of a scalar valued function with respect to a second rank tensor (see Sect. B.4) and (2.1.6) one can obtain

$$\dot{\boldsymbol{\varepsilon}}^{cr} = \frac{\partial W}{\partial J_1}\boldsymbol{I} - \frac{\partial W}{\partial J_{2D}}\boldsymbol{s} + \frac{\partial W}{\partial J_{3D}}\left(\boldsymbol{s}^2 - \frac{1}{3}\operatorname{tr}\boldsymbol{s}^2\boldsymbol{I}\right) \tag{2.2.3}$$

In the classical creep theory it is assumed that the inelastic deformation does not produce a significant change in volume. The spherical part of the creep rate tensor is neglected, i.e $\operatorname{tr}\dot{\boldsymbol{\varepsilon}}^{cr} = 0$. Setting the trace of (2.2.3) to zero results in

$$\operatorname{tr}\dot{\boldsymbol{\varepsilon}}^{cr} = 3\frac{\partial W}{\partial J_1} = 0 \quad \Rightarrow \quad W = W(J_{2D}, J_{3D})$$

From this follows that the creep behavior is not sensitive to the hydrostatic stress state $\boldsymbol{\sigma} = -p\boldsymbol{I}$, where $p > 0$ is the hydrostatic pressure. The creep equation (2.2.3) can be formulated as

$$\dot{\boldsymbol{\varepsilon}}^{cr} = -\frac{\partial W}{\partial J_{2D}}\boldsymbol{s} + \frac{\partial W}{\partial J_{3D}}\left(\boldsymbol{s}^2 - \frac{1}{3}\mathrm{tr}\,\boldsymbol{s}^2\boldsymbol{I}\right) \qquad (2.2.4)$$

The last term in the right-hand side of (2.2.4) is non-linear with respect to the stress deviator $\boldsymbol{s}$. Equations of this type are called tensorial non-linear equations, e.g. [39, 62, 208, 272]. They allow to consider some non-classical or second order effects of the material behavior [39, 69]. As an example let us consider the pure shear stress state $\boldsymbol{s} = \tau(\boldsymbol{m} \otimes \boldsymbol{n} + \boldsymbol{n} \otimes \boldsymbol{m})$, where τ is the magnitude of the shear stress and $\boldsymbol{m}$ and $\boldsymbol{n}$ are orthogonal unit vectors. From (2.2.4) follows

$$\dot{\boldsymbol{\varepsilon}}^{cr} = -\frac{\partial W}{\partial J_{2D}}\tau(\boldsymbol{m} \otimes \boldsymbol{n} + \boldsymbol{n} \otimes \boldsymbol{m}) + \frac{\partial W}{\partial J_{3D}}\tau^2\left(\frac{1}{3}\boldsymbol{I} - \boldsymbol{p} \otimes \boldsymbol{p}\right),$$

where the unit vector $\boldsymbol{p}$ is orthogonal to the plane spanned on $\boldsymbol{m}$ and $\boldsymbol{n}$. We observe that the pure shear load leads to shear creep rate, and additionally to the axial creep rates (Poynting-Swift effect). Within the engineering creep mechanics such effects are usually neglected.

The assumption that the potential is a function of the second invariant of the stress deviator only, i.e.

$$W = W(J_{2D})$$

leads to the classical von Mises type potential [328]. In applications it is convenient to introduce the equivalent stress which allows to compare the creep behavior under different stress states including the uni-axial tension. The von Mises equivalent stress is defined as follows

$$\sigma_{vM} = \sqrt{\frac{3}{2}\boldsymbol{s} \cdot\cdot\, \boldsymbol{s}} = \sqrt{-3J_{2D}}, \qquad (2.2.5)$$

where the factor $3/2$ is used for convenience (in the case of the uni-axial tension with the stress σ the above expression provides $\sigma_{vM} = \sigma$). With $W = W(\sigma_{vM}(\boldsymbol{\sigma}))$ the flow rule (2.1.6) results in

$$\dot{\boldsymbol{\varepsilon}}^{cr} = \frac{\partial W(\sigma_{vM})}{\partial \sigma_{vM}}\frac{\partial \sigma_{vM}}{\partial \boldsymbol{\sigma}} = \frac{\partial W(\sigma_{vM})}{\partial \sigma_{vM}}\frac{3}{2}\frac{\boldsymbol{s}}{\sigma_{vM}} \qquad (2.2.6)$$

The second invariant of $\dot{\boldsymbol{\varepsilon}}^{cr}$ can be calculated as follows

$$\dot{\boldsymbol{\varepsilon}}^{cr} \cdot\cdot\, \dot{\boldsymbol{\varepsilon}}^{cr} = \frac{3}{2}\left[\frac{\partial W(\sigma_{vM})}{\partial \sigma_{vM}}\right]^2$$

Introducing the notation $\dot{\varepsilon}_{vM}^2 = \frac{2}{3}\dot{\boldsymbol{\varepsilon}}^{cr} \cdot\cdot\, \dot{\boldsymbol{\varepsilon}}^{cr}$ and taking into account that

$$P = \frac{\partial W(\sigma_{vM})}{\partial \sigma_{vM}}\sigma_{vM} \geq 0$$

one can write

$$\dot{\boldsymbol{\varepsilon}}^{cr} = \frac{3}{2}\dot{\varepsilon}_{vM}\frac{\boldsymbol{s}}{\sigma_{vM}}, \quad \dot{\varepsilon}_{vM} = \frac{\partial W(\sigma_{vM})}{\partial \sigma_{vM}} \tag{2.2.7}$$

The constitutive equation of steady-state creep (2.2.7) was proposed by Odqvist [243]. Experimental verifications of this equation can be found, for example, in [302] for steel 45, in [235] for titanium alloy Ti-6Al-4V and in [252] for alloys Al-Si, Fe-Co-V and XC 48. In these works tubular specimens were loaded by tension force and torque leading to the plane stress state $\boldsymbol{\sigma} = \sigma \boldsymbol{n} \otimes \boldsymbol{n} + \tau(\boldsymbol{n} \otimes \boldsymbol{m} + \boldsymbol{m} \otimes \boldsymbol{n})$, where σ and τ are the magnitudes of the normal and shear stresses (see Sect. 1.1.2). Surfaces $\sigma_{vM}^2 = \sigma^2 + 3\tau^2 = const$ corresponding to the same steady state values of $\dot{\varepsilon}_{vM}$ were recorded. Assuming the Norton-Bailey type potential (2.1.8), from (2.2.7) it follows

$$\dot{\boldsymbol{\varepsilon}}^{cr} = \frac{3}{2} a \sigma_{vM}^{n-1} \boldsymbol{s} \tag{2.2.8}$$

This model is widely used in estimations of steady-state creep in structures, e.g. [80, 83, 243, 257, 272].

2.2.1.2 Creep Potentials with Three Invariants of the Stress Tensor. In some cases, deviations from the von Mises type equivalent stress were found in experiments. For example, different secondary creep rates under tensile and compressive loading were observed in [201] for Zircaloy-2, in [109] for aluminium alloy ALC101 and in [309], p. 118 for the nickel-based alloy René 95. One way to consider such effects is to construct the creep potential as a function of three invariants of the stress tensor. Below we discuss a generalized creep potential, proposed in [12]. This potential leads to tensorial non-linear constitutive equations and allows to predict the stress state dependent creep behavior and second order effects. The 6 unknown parameters in this law can be identified by some basic tests. Creep potentials formulated in terms of three invariants of the stress tensor are termed non-classical [12].

By analogy to the classical creep equations, the dependence on the stress tensor is defined by means of the equivalent stress σ_{eq}. Various equivalent stress expressions have been proposed in the literature for the formulation of yield or failure criteria, e.g. [30]. In the case of creep, different equivalent stress expressions are summarized in [163]. In [12] the following equivalent stress is proposed

$$\sigma_{eq} = \alpha\sigma_1 + \beta\sigma_2 + \gamma\sigma_3 \tag{2.2.9}$$

with the linear, the quadratic and the cubic invariants

$$\sigma_1 = \mu_1 I_1, \quad \sigma_2^2 = \mu_2 I_1^2 + \mu_3 I_2, \quad \sigma_3^3 = \mu_4 I_1^3 + \mu_5 I_1 I_2 + \mu_6 I_3, \tag{2.2.10}$$

where $I_i = \operatorname{tr} \boldsymbol{\sigma}^i$ $(i = 1,2,3)$ are basic invariants of the stress tensor (see Sect. C.2), μ_j $(j = 1,\ldots,6)$ are parameters, which depend on the material properties. α, β, γ are numerical coefficients for weighting the influence of the different parts in the equivalent stress expression (2.2.9). Such a weighting is usual in phenomenological modelling of material behavior. For example, in [135] similar coefficients are introduced for characterizing different failure modes.

The von Mises equivalent stress (2.2.5) can be obtained from (2.2.9) by setting $\alpha = \gamma = 0$, $\beta = 1$ and $\mu_3 = 1.5$, $\mu_2 = -0.5$. In what follows we set $\beta = 1$ and the equivalent stress takes the form

$$\sigma_{eq} = \alpha\sigma_1 + \sigma_2 + \gamma\sigma_3 \tag{2.2.11}$$

It can be verified that the equivalent stress (2.2.11) satisfies (2.1.9).

The flow rule (2.1.6) allows to formulate the constitutive equation for the creep rate tensor

$$\dot{\boldsymbol{\varepsilon}}^{cr} = \frac{\partial W(\sigma_{eq})}{\partial \sigma_{eq}} \frac{\partial \sigma_{eq}}{\partial \boldsymbol{\sigma}} = \frac{\partial W(\sigma_{eq})}{\partial \sigma_{eq}} \left(\alpha \frac{\partial \sigma_1}{\partial \boldsymbol{\sigma}} + \frac{\partial \sigma_2}{\partial \boldsymbol{\sigma}} + \gamma \frac{\partial \sigma_3}{\partial \boldsymbol{\sigma}} \right) \tag{2.2.12}$$

Taking into account the relations between the invariants σ_i and the basic invariants I_i and using the rules for the derivatives of the invariants (see Sect. B.4), we obtain

$$\begin{aligned} \frac{\partial \sigma_1}{\partial \boldsymbol{\sigma}} &= \mu_1 \boldsymbol{I}, \quad \frac{\partial \sigma_2}{\partial \boldsymbol{\sigma}} = \frac{\mu_2 I_1 \boldsymbol{I} + \mu_3 \boldsymbol{\sigma}}{\sigma_2}, \\ \frac{\partial \sigma_3}{\partial \boldsymbol{\sigma}} &= \frac{\mu_4 I_1^2 \boldsymbol{I} + \frac{\mu_5}{3} I_2 \boldsymbol{I} + \frac{2}{3}\mu_5 I_1 \boldsymbol{\sigma} + \mu_6 \boldsymbol{\sigma} \cdot \boldsymbol{\sigma}}{\sigma_3^2} \end{aligned} \tag{2.2.13}$$

As a result, the creep constitutive equation can be formulated as follows

$$\dot{\boldsymbol{\varepsilon}}^{cr} = \frac{\partial W(\sigma_{eq})}{\partial \sigma_{eq}} \left[\alpha\mu_1 \boldsymbol{I} + \frac{\mu_2 I_1 \boldsymbol{I} + \mu_3 \boldsymbol{\sigma}}{\sigma_2} + \gamma \frac{\left(\mu_4 I_1^2 + \frac{\mu_5}{3} I_2\right) \boldsymbol{I} + \frac{2}{3}\mu_5 I_1 \boldsymbol{\sigma} + \mu_6 \boldsymbol{\sigma} \cdot \boldsymbol{\sigma}}{\sigma_3^2} \right] \tag{2.2.14}$$

Introducing the notation

$$\dot{\varepsilon}_{eq}^{cr} \equiv \frac{\partial W(\sigma_{eq})}{\partial \sigma_{eq}}$$

the constitutive equation takes the form

$$\dot{\boldsymbol{\varepsilon}}^{cr} = \dot{\varepsilon}_{eq}^{cr} \left[\alpha\mu_1 \boldsymbol{I} + \frac{\mu_2 I_1 \boldsymbol{I} + \mu_3 \boldsymbol{\sigma}}{\sigma_2} + \gamma \frac{\left(\mu_4 I_1^2 + \frac{\mu_5}{3} I_2\right) \boldsymbol{I} + \frac{2}{3}\mu_5 I_1 \boldsymbol{\sigma} + \mu_6 \boldsymbol{\sigma} \cdot \boldsymbol{\sigma}}{\sigma_3^2} \right] \tag{2.2.15}$$

Equation (2.2.15) is non-linear with respect to the stress tensor. Therefore, second order effects, e.g. [39, 60, 320] are included in the material behavior description. In addition, the volumetric creep rate can be calculated from (2.2.15) as follows

$$\dot{\varepsilon}_V^{cr} = \dot{\varepsilon}_{eq}^{cr} \left[3\alpha\mu_1 + \frac{(3\mu_2 + \mu_3) I_1}{\sigma_2} + \gamma \frac{(9\mu_4 + 2\mu_5) I_1^2 + 3(\mu_5 + \mu_6) I_2}{3\sigma_3^2} \right] \tag{2.2.16}$$

The volumetric creep rate is different from 0, i.e. the compressibility or dilatation can be considered.

The derived creep equation has the form (2.1.11) of the general relation between two coaxial tensors. The comparison of (2.1.11) and (2.2.15) provides

$$
\begin{aligned}
H_0 &= \dot{\varepsilon}^{cr}_{eq}\left(\alpha\mu_1 + \frac{\mu_2 I_1}{\sigma_2} + \gamma\frac{3\mu_4 I_1^2 + \mu_5 I_2}{3\sigma_3^2}\right),\\
H_1 &= \dot{\varepsilon}^{cr}_{eq}\left(\frac{\mu_3}{\sigma_2} + \gamma\frac{2\mu_5 I_1}{3\sigma_3^2}\right),\\
H_2 &= \dot{\varepsilon}^{cr}_{eq}\gamma\frac{\mu_6}{\sigma_3^2}
\end{aligned}
\qquad (2.2.17)
$$

In [12] the power law function of the equivalent stress (2.1.5) is applied to model creep behavior of several materials. Four independent creep tests are required to identify the material constants. The stress states realized in tests should include uni-axial tension, uni-axial compression, torsion and hydrostatic pressure. Let us note, that experimental data which allows to identify the full set of material constants in (2.2.15) are usually not available. In applications one may consider the following special cases of (2.2.15) with reduced number of material constants.

The classical creep equation based on the von Mises equivalent stress can be derived assuming the following values of material constants

$$
\alpha = \gamma = 0, \quad \mu_2 = -1/2, \quad \mu_3 = 3/2, \qquad (2.2.18)
$$

$$
\sigma_{eq} = \sigma_2 = \sqrt{-\frac{1}{2}I_1^2 + \frac{3}{2}I_2} = \sqrt{\frac{3}{2}\boldsymbol{s}\cdot\cdot\boldsymbol{s}} = \sigma_{vM} \qquad (2.2.19)
$$

The creep rate tensor takes the form

$$
\dot{\boldsymbol{\varepsilon}}^{cr} = \dot{\varepsilon}^{cr}_{eq}\left(\sqrt{\frac{3}{2}\boldsymbol{s}\cdot\cdot\boldsymbol{s}}\right)\frac{3\boldsymbol{\sigma} - I_1\boldsymbol{I}}{2\sqrt{\frac{3}{2}\boldsymbol{s}\cdot\cdot\boldsymbol{s}}} = \frac{3}{2}\frac{\dot{\varepsilon}^{cr}_{eq}(\sigma_{vM})}{\sigma_{vM}}\boldsymbol{s} \qquad (2.2.20)
$$

Assuming identical behavior in tension and compression and neglecting second order effects from $\alpha = \gamma = 0$, the following equivalent stress can be obtained

$$
\sigma_{eq} = \sigma_2 = \sqrt{\mu_2 I_1^2 + \mu_3 I_2} \qquad (2.2.21)
$$

The corresponding creep constitutive equation takes the form

$$
\dot{\boldsymbol{\varepsilon}}^{cr} = \dot{\varepsilon}^{cr}_{eq}(\sigma_2)\frac{\mu_2 I_1\boldsymbol{I} + \mu_3\boldsymbol{\sigma}}{\sigma_2} \qquad (2.2.22)
$$

The parameters μ_2 and μ_3 can be determined from uni-axial tension and torsion tests. Based on the experimental data presented in [170, 171] for technical pure copper M1E (Cu 99,9%) at $T = 573$ K the parameters μ_2 and μ_3 are identified in [27].

Neglecting the influence of the third invariant ($\gamma = 0$), the creep rate tensor can be expressed as follows

$$\dot{\boldsymbol{\varepsilon}}^{cr} = \dot{\varepsilon}^{cr}_{eq}(\sigma_{eq}) \left(\alpha\mu_1 \boldsymbol{I} + \frac{\mu_2 I_1 \boldsymbol{I} + \mu_3 \boldsymbol{\sigma}}{\sigma_2} \right) \tag{2.2.23}$$

The above equation describes different behavior in tension and compression, and includes the volumetric creep rate. Three independent tests, e.g. tension, compression and torsion are required to identify the material constants μ_1, μ_2 and μ_3.

With the quadratic invariant and the reduced cubic invariant several special cases with three material constants can be considered. Setting ($\alpha\mu_1 = \mu_4 = \mu_5 = 0$) the tensorial non-linear equation can be obtained

$$\dot{\boldsymbol{\varepsilon}}^{cr} = \dot{\varepsilon}^{cr}_{eq}(\sigma_{eq}) \left(\frac{\mu_2 I_1 \boldsymbol{I} + \mu_3 \boldsymbol{\sigma}}{\sigma_2} + \gamma \frac{\mu_6 \boldsymbol{\sigma} \cdot \boldsymbol{\sigma}}{\sigma_3^2} \right) \tag{2.2.24}$$

With $\alpha\mu_1 = \mu_4 = \mu_6 = 0$ the creep rate tensor takes the form

$$\dot{\boldsymbol{\varepsilon}}^{cr} = \dot{\varepsilon}^{cr}_{eq}(\sigma_{eq}) \left(\frac{\mu_2 I_1 \boldsymbol{I} + \mu_3 \boldsymbol{\sigma}}{\sigma_2} + \gamma \frac{\mu_5 (I_2 \boldsymbol{I} + 2 I_1 \boldsymbol{\sigma})}{\sigma_3^2} \right) \tag{2.2.25}$$

The material constants in (2.2.23), (2.2.24) and (2.2.25) were identified in [4, 31] according to data from multi-axial creep tests for plastics (PVC) at room temperature [193] and aluminium alloy AK4-1T at 473 K [97, 128, 301]. Furthermore, simulations have been performed in [4, 31] to compare Eqs (2.2.23), (2.2.24) and (2.2.25) as they characterize creep behavior under different loading conditions. The conclusion was made that cubic invariants applied in (2.2.24) and (2.2.25) do not deliver any significant improvement in the material behavior description.

2.2.2 Creep of Initially Anisotropic Materials

Anisotropic creep behavior and anisotropic creep modeling are subjects which are rarely discussed in the classical monographs and textbooks on creep mechanics (only in some books one may found the flow potentials introduced by von Mises [328] and Hill [141]). The reason for this is that the experimental data from creep tests usually show large scatter within the range of 20% or even more. Therefore, it was often difficult to recognize whether the difference in creep curves measured for different specimens (cut from the same material in different directions) is the result of the anisotropy. Therefore, it was no use for anisotropic models with higher order complexity, since the identification of material constants was difficult or even impossible. In the last two decades the importance in modeling anisotropic creep behavior of materials and structures is discussed in many publications. In [51, 206, 265, 266, 267, 268] experimental results of creep of superalloys SRR99 and CMSX-4 are reported, which demonstrate significant anisotropy of creep behavior for different orientations of specimens with respect to the crystallographic axes. In [144] experimental creep curves of a 9CrMoNbV weld metal are presented. They show significant difference for specimens cut in longitudinal (welding) direction and transverse directions. Another example is a material reinforced by fibers, showing quite different creep behavior in direction of fibers and in the transverse direction, e.g. [280, 281].

Within the creep mechanics one usually distinguishes between two kinds of anisotropy: the initial anisotropy and the deformation or damage induced anisotropy. In what follows the first case will be introduced. The second case will be discussed in Sects. 2.3.2 and 2.4.2.

The modeling of anisotropic behavior starts with the concepts of material symmetry, physical symmetry, symmetry transformation and symmetry group, e.g. [340]. The material symmetry group is related to the symmetries of the materials microstructure, e.g. the crystal symmetries, the symmetries due to the arrangement of fibers in a fiber-reinforced materials, etc. The symmetry transformations are described by means of orthogonal tensors. Two important of them are

– the reflection

$$\boldsymbol{Q}(\boldsymbol{n}) = \boldsymbol{I} - 2\boldsymbol{n} \otimes \boldsymbol{n}, \tag{2.2.26}$$

where $\boldsymbol{n}$ is the unit normal to the mirror plane,

– the rotation about a fixed axis

$$\boldsymbol{Q}(\varphi \boldsymbol{m}) = \boldsymbol{m} \otimes \boldsymbol{m} + \cos\varphi(\boldsymbol{I} - \boldsymbol{m} \otimes \boldsymbol{m}) + \sin\varphi \boldsymbol{m} \times \boldsymbol{I}, \tag{2.2.27}$$

where $\boldsymbol{m}$ is the axis of rotation and φ is the angle of rotation.

Any arbitrary rotation of a rigid body can be described as a composition of three rotations (2.2.27) about three fixed axes [342]. Any symmetry transformation can be represented by means of rotations and reflections, i.e. the tensors of the type (2.2.26) and (2.2.27). The notion of the symmetry group as a set of symmetry transformations was introduced in [237]. The symmetry groups of polar and axial tensors are discussed in [341]. According to [321], p. 82 a "simple solid" is called aelotropic or anisotropic, if its symmetry group is a proper subgroup of the orthogonal group.

The concept of the "physical symmetry group" is related to the symmetries of the material behavior, e.g. linear elasticity, thermal expansion, plasticity, creep, etc. It can only be established based on experimental observations. Physical symmetries must be considered in the formulation of constitutive equations and constitutive functions. As an example let us consider the symmetry group of the fourth rank elasticity tensor ${}^{(4)}\boldsymbol{C} = C^{ijkl}\boldsymbol{e}_i \otimes \boldsymbol{e}_j \otimes \boldsymbol{e}_k \otimes \boldsymbol{e}_l$ as the set of orthogonal tensors $\boldsymbol{Q}$ satisfying the equation, e.g. [28, 341],

$${}^{(4)}\boldsymbol{C}' = C^{ijkl}\boldsymbol{Q} \cdot \boldsymbol{e}_i \otimes \boldsymbol{Q} \cdot \boldsymbol{e}_j \otimes \boldsymbol{Q} \cdot \boldsymbol{e}_k \otimes \boldsymbol{Q} \cdot \boldsymbol{e}_l = {}^{(4)}\boldsymbol{C} \tag{2.2.28}$$

The physical symmetries or the set of orthogonal solutions of (2.2.28) can be found only if all the 21 coordinates of the elasticity tensor ${}^{(4)}\boldsymbol{C}$ for a selected basis are identified from tests. Vice versa, if the physical symmetry group is known then one can find the general structure of the elasticity tensor based on (2.2.28). Clearly, neither the elasticity tensor nor the physical symmetry group of the linear elastic behavior can be exactly found from tests. Establishment of physical symmetries of creep behavior is rather complicated due to relatively large scatter of experimental data. However, one can relate physical symmetries to the known symmetries of materials microstructure. According to the Neumann principle widely used in different branches of physics and continuum mechanics, e.g. [28, 239, 341]

> *The symmetry group of the reason belongs to the symmetry group of the consequence.*

Considering the material symmetries as one of the "reasons" and the physical symmetries as a "consequence" one can apply the following statement [340]

> *For a material element and for any of its physical properties every material symmetry transformation of the material element is a physical symmetry transformation of the physical property.*

In many cases the material symmetry elements are evident from the arrangement of the materials microstructure as a consequence of manufacturing conditions, for example. The above principle states that the physical behavior, e.g. the steady-state creep, contains all elements of the material symmetry. The physical symmetry group usually possesses more elements than the material symmetry group, e.g. [239].

2.2.2.1 Classical Creep Equations. Here we discuss steady-state creep equations based on the flow rule (2.1.6) and assumption that the creep potential has a quadratic form with respect to the invariants of the stress tensor. These invariants must be established according to the assumed symmetry elements of the creep behavior. The assumption of the quadratic form of the flow potential originates from the von Mises work on plasticity of crystals [328]. Therefore, the equations presented below may be termed as von Mises type equations.

Transverse Isotropy. In this case the potential $W(\boldsymbol{\sigma})$ must satisfy the following restriction

$$W(\boldsymbol{Q}\cdot\boldsymbol{\sigma}\cdot\boldsymbol{Q}^T) = W(\boldsymbol{\sigma}), \quad \boldsymbol{Q}(\varphi\boldsymbol{m}) = \boldsymbol{m}\otimes\boldsymbol{m} + \cos\varphi(\boldsymbol{I}-\boldsymbol{m}\otimes\boldsymbol{m}) + \sin\varphi\boldsymbol{m}\times\boldsymbol{I} \qquad (2.2.29)$$

In (2.2.29) $\boldsymbol{Q}(\varphi\boldsymbol{m})$ is the assumed element of the symmetry group, whereby $\boldsymbol{m}$ is a constant unit vector and φ is the arbitrary angle of rotation about $\boldsymbol{m}$. From the restriction (2.2.29) follows that the potential W must satisfy the following partial differential equation (see Sect. C.3)

$$(\boldsymbol{m}\times\boldsymbol{\sigma} - \boldsymbol{\sigma}\times\boldsymbol{m})\cdot\cdot\left(\frac{\partial W}{\partial\boldsymbol{\sigma}}\right)^T = 0 \qquad (2.2.30)$$

The set of integrals of this equation represent the set of functionally independent scalar valued arguments of the potential W with respect to the symmetry transformation (2.2.29). The characteristic system of (2.2.30) is the system of ordinary differential equations

$$\frac{d\boldsymbol{\sigma}}{ds} = (\boldsymbol{m} \times \boldsymbol{\sigma} - \boldsymbol{\sigma} \times \boldsymbol{m}) \tag{2.2.31}$$

Any system of n linear ordinary differential equations has not more than $n-1$ functionally independent integrals [95]. Since $\boldsymbol{\sigma}$ is symmetric, (2.2.31) is a system of six ordinary differential equations and has not more than five functionally independent integrals. The lists of these integrals are presented by (C.3.10) and (C.3.21). Within the classical von Mises type theory second order effects are neglected. Therefore, we have to neglect the arguments which are cubic with respect to the stress tensor. In this case the difference between various kinds of transverse isotropy considered in Sect. C.3 vanishes. It is possible to use different lists of of scalar arguments. The linear and quadratic arguments from (C.3.10) are

$$\operatorname{tr}\boldsymbol{\sigma}, \quad \operatorname{tr}\boldsymbol{\sigma}^2, \quad \boldsymbol{m}\cdot\boldsymbol{\sigma}\cdot\boldsymbol{m}, \quad \boldsymbol{m}\cdot\boldsymbol{\sigma}^2\cdot\boldsymbol{m} \tag{2.2.32}$$

Instead of (2.2.32) one can use other arguments, for example [280],

$$\begin{aligned}
&\operatorname{tr}\boldsymbol{\sigma}, \quad \operatorname{tr}\boldsymbol{s}^2 = \operatorname{tr}\boldsymbol{\sigma}^2 - \frac{1}{3}(\operatorname{tr}\boldsymbol{\sigma})^2,\\
&\boldsymbol{m}\cdot\boldsymbol{s}\cdot\boldsymbol{m} = \boldsymbol{m}\cdot\boldsymbol{\sigma}\cdot\boldsymbol{m} - \frac{1}{3}\operatorname{tr}\boldsymbol{\sigma},\\
&\boldsymbol{m}\cdot\boldsymbol{s}^2\cdot\boldsymbol{m} = \boldsymbol{m}\cdot\boldsymbol{\sigma}^2\cdot\boldsymbol{m} - \frac{2}{3}\boldsymbol{m}\cdot\boldsymbol{s}\cdot\boldsymbol{m}\operatorname{tr}\boldsymbol{\sigma} - \frac{1}{9}(\operatorname{tr}\boldsymbol{\sigma})^2
\end{aligned} \tag{2.2.33}$$

In what follows we prefer another set of invariants which can be related to (2.2.32) but has a more clear mechanical interpretation. Let us decompose the stress tensor as follows

$$\boldsymbol{\sigma} = \sigma_{mm}\boldsymbol{m}\otimes\boldsymbol{m} + \boldsymbol{\sigma}_p + \boldsymbol{\tau}_m\otimes\boldsymbol{m} + \boldsymbol{m}\otimes\boldsymbol{\tau}_m \tag{2.2.34}$$

with the projections

$$\begin{aligned}
\sigma_{mm} &= \boldsymbol{m}\cdot\boldsymbol{\sigma}\cdot\boldsymbol{m},\\
\boldsymbol{\sigma}_p &= (\boldsymbol{I} - \boldsymbol{m}\otimes\boldsymbol{m})\cdot\boldsymbol{\sigma}\cdot(\boldsymbol{I} - \boldsymbol{m}\otimes\boldsymbol{m}),\\
\boldsymbol{\tau}_m &= \boldsymbol{m}\cdot\boldsymbol{\sigma}\cdot(\boldsymbol{I} - \boldsymbol{m}\otimes\boldsymbol{m})
\end{aligned} \tag{2.2.35}$$

The meaning of the decomposition (2.2.34) is obvious. σ_{mm} is the normal stress acting in the plane with the unit normal $\boldsymbol{m}$, $\boldsymbol{\sigma}_p$ stands for the "plane" part of the stress tensor representing the stress state in the isotropy plane. $\boldsymbol{\tau}_m$ is the shear stress vector in the plane with the unit normal $\boldsymbol{m}$. For the orthonormal basis $\boldsymbol{k}$, $\boldsymbol{l}$ and $\boldsymbol{m}$ the projections are (see Fig. 2.1)

$$\begin{aligned}
\boldsymbol{\tau}_m &= \tau_{mk}\boldsymbol{k} + \tau_{ml}\boldsymbol{l},\\
\boldsymbol{\sigma}_p &= \sigma_{kk}\boldsymbol{k}\otimes\boldsymbol{k} + \sigma_{ll}\boldsymbol{l}\otimes\boldsymbol{l} + \tau_{kl}(\boldsymbol{k}\otimes\boldsymbol{l} + \boldsymbol{l}\otimes\boldsymbol{k})
\end{aligned}$$

The plane part of the stress tensor can be further decomposed as follows

$$\boldsymbol{\sigma}_p = \boldsymbol{s}_p + \frac{1}{2}\operatorname{tr}\boldsymbol{\sigma}_p(\boldsymbol{I} - \boldsymbol{m}\otimes\boldsymbol{m}), \quad \operatorname{tr}\boldsymbol{s}_p = 0 \tag{2.2.36}$$

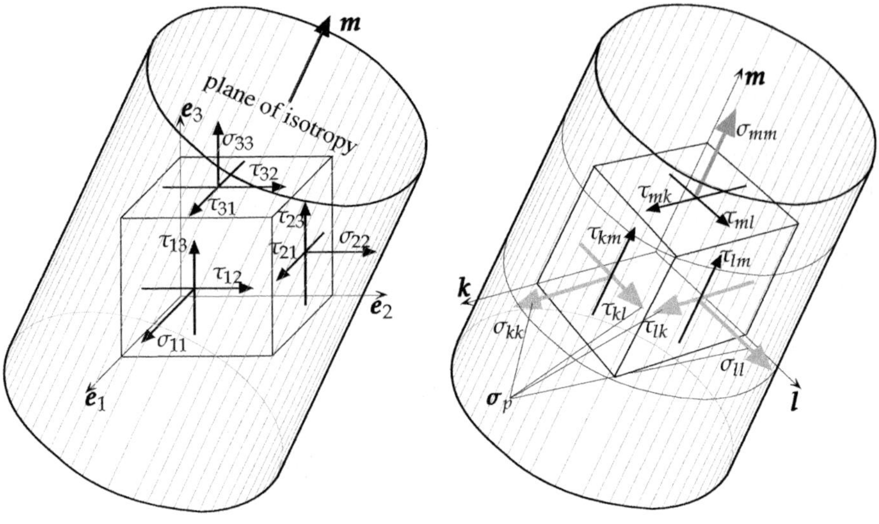

Fig. 2.1 Stress state in a transversely isotropic medium and corresponding projections σ_{mm}, $\boldsymbol{\sigma}_p$ and $\boldsymbol{\tau}_m$

Now we can introduce the following set of transversely isotropic invariants

$$
\begin{aligned}
I_{1m} &= \sigma_{mm} = \boldsymbol{m} \cdot \boldsymbol{\sigma} \cdot \boldsymbol{m}, \\
I_{2m} &= \operatorname{tr} \boldsymbol{\sigma}_p = \operatorname{tr} \boldsymbol{\sigma} - \boldsymbol{m} \cdot \boldsymbol{\sigma} \cdot \boldsymbol{m}, \\
I_{3m} &= \frac{1}{2} \operatorname{tr} \boldsymbol{s}_p^2 = \frac{1}{2} \operatorname{tr} \boldsymbol{\sigma}_p^2 - \frac{1}{4} (\operatorname{tr} \boldsymbol{\sigma}_p)^2 \\
&= \frac{1}{2} \left(\operatorname{tr} \boldsymbol{\sigma}^2 + (\boldsymbol{m} \cdot \boldsymbol{\sigma} \cdot \boldsymbol{m})^2 \right) - \boldsymbol{m} \cdot \boldsymbol{\sigma}^2 \cdot \boldsymbol{m} - \frac{1}{4} (\operatorname{tr} \boldsymbol{\sigma} - \boldsymbol{m} \cdot \boldsymbol{\sigma} \cdot \boldsymbol{m})^2, \\
I_{4m} &= \boldsymbol{\tau}_m \cdot \boldsymbol{\tau}_m = \boldsymbol{m} \cdot \boldsymbol{\sigma}^2 \cdot \boldsymbol{m} - (\boldsymbol{m} \cdot \boldsymbol{\sigma} \cdot \boldsymbol{m})^2 = (\boldsymbol{m} \times \boldsymbol{\sigma} \cdot \boldsymbol{m}) \cdot (\boldsymbol{m} \times \boldsymbol{\sigma} \cdot \boldsymbol{m})
\end{aligned}
\tag{2.2.37}
$$

In the above list I_{2m} and I_{3m} are two invariants of $\boldsymbol{\sigma}_p$ and $I_{4m} = \boldsymbol{\tau}_m^2 = \boldsymbol{\tau}_m \cdot \boldsymbol{\tau}_m$ is the square of the length of the shear stress vector acting in the plane with the unit normal $\boldsymbol{m}$. It is shown in Sect. C.3 that the above invariants are integrals of (2.2.31).

Taking into account the relations

$$
\frac{\partial I_{1m}}{\partial \boldsymbol{\sigma}} = \boldsymbol{m} \otimes \boldsymbol{m}, \quad \frac{\partial I_{2m}}{\partial \boldsymbol{\sigma}} = \boldsymbol{I} - \boldsymbol{m} \otimes \boldsymbol{m},
$$
$$
\frac{\partial I_{3m}}{\partial \boldsymbol{\sigma}} = \boldsymbol{s}_p, \quad \frac{\partial I_{4m}}{\partial \boldsymbol{\sigma}} = \boldsymbol{\tau}_m \otimes \boldsymbol{m} + \boldsymbol{m} \otimes \boldsymbol{\tau}_m
$$

and the flow rule (2.1.6) we obtain the following creep equation

$$
\begin{aligned}
\dot{\boldsymbol{\varepsilon}}^{cr} &= \frac{\partial W}{\partial I_{1m}} \boldsymbol{m} \otimes \boldsymbol{m} + \frac{\partial W}{\partial I_{2m}} (\boldsymbol{I} - \boldsymbol{m} \otimes \boldsymbol{m}) + \frac{\partial W}{\partial I_{3m}} \boldsymbol{s}_p \\
&+ \frac{\partial W}{\partial I_{4m}} (\boldsymbol{\tau}_m \otimes \boldsymbol{m} + \boldsymbol{m} \otimes \boldsymbol{\tau}_m)
\end{aligned}
\tag{2.2.38}
$$

The next assumption of the classical theory is the zero volumetric creep rate. Taking the trace of (2.2.38) we obtain

$$\operatorname{tr} \dot{\boldsymbol{\varepsilon}}^{cr} = \frac{\partial W}{\partial I_{1m}} + 2\frac{\partial W}{\partial I_{2m}} = 0 \quad \Rightarrow \quad W = W(I_{1m} - \frac{1}{2}I_{2m}, I_{3m}, I_{4m}) \tag{2.2.39}$$

Introducing the notation

$$J_m \equiv I_{1m} - \frac{1}{2}I_{2m} = \boldsymbol{m} \cdot \boldsymbol{\sigma} \cdot \boldsymbol{m} - \frac{1}{2}\operatorname{tr} \boldsymbol{\sigma}_p$$

the creep equation (2.2.38) takes the form

$$\dot{\boldsymbol{\varepsilon}}^{cr} = \frac{1}{2}\frac{\partial W}{\partial J_m}(3\boldsymbol{m} \otimes \boldsymbol{m} - \boldsymbol{I}) + \frac{\partial W}{\partial I_{3m}}\boldsymbol{s}_p + \frac{\partial W}{\partial I_{4m}}(\boldsymbol{\tau_m} \otimes \boldsymbol{m} + \boldsymbol{m} \otimes \boldsymbol{\tau_m}) \tag{2.2.40}$$

By analogy to the isotropic case we formulate the equivalent stress as follows

$$\begin{aligned} \sigma_{eq}^2 &= \alpha_1 J_m^2 + 3\alpha_2 I_{3m} + 3\alpha_3 I_{4m} \\ &= \alpha_1 \left(\boldsymbol{m} \cdot \boldsymbol{\sigma} \cdot \boldsymbol{m} - \frac{1}{2}\operatorname{tr} \boldsymbol{\sigma}_p\right)^2 + \frac{3}{2}\alpha_2 \operatorname{tr} \boldsymbol{s}_p^2 + 3\alpha_3 \tau_{\boldsymbol{m}}^2 \end{aligned} \tag{2.2.41}$$

The positive definiteness of the quadratic form (2.2.41) is provided by the conditions $\alpha_i > 0, i = 1, 2, 3$. The deviatoric part $\boldsymbol{s}$ of the stress tensor and its second invariant can be computed by

$$\begin{aligned} \boldsymbol{s} &= J_m \left(\boldsymbol{m} \otimes \boldsymbol{m} - \frac{1}{3}\boldsymbol{I}\right) + \boldsymbol{s}_p + \boldsymbol{\tau}_m \otimes \boldsymbol{m} + \boldsymbol{m} \otimes \boldsymbol{\tau}_m, \\ \operatorname{tr} \boldsymbol{s}^2 &= \frac{2}{3}J_m^2 + \operatorname{tr} \boldsymbol{s}_p^2 + 2\tau_{\boldsymbol{m}}^2 \end{aligned}$$

Consequently, the von Mises equivalent stress (2.2.5) follows from (2.2.41) by setting $\alpha_1 = \alpha_2 = \alpha_3 = 1$.

The advantage of the introduced invariants over (2.2.32) or (2.2.33) is that they can be specified independently from each other. For example, set the second invariant in (2.2.32) to zero, i.e. $\operatorname{tr} \boldsymbol{\sigma}^2 = \boldsymbol{\sigma} \cdot\cdot \boldsymbol{\sigma} = 0$. From this follows that $\boldsymbol{\sigma} = \boldsymbol{0}$ and consequently all other invariants listed in (2.2.32) are simultaneously equal to zero. In addition, the introduced invariants can be related to typical stress states which should be realized in creep tests for the identification of constitutive functions and material constants. With the equivalent stress (2.2.41) the creep equation (2.2.40) can be rewritten as follows

$$\dot{\boldsymbol{\varepsilon}}^{cr} = \frac{3}{2\sigma_{eq}}\frac{\partial W}{\partial \sigma_{eq}}\left[\alpha_1 J_m \left(\boldsymbol{m} \otimes \boldsymbol{m} - \frac{1}{3}\boldsymbol{I}\right) + \alpha_2 \boldsymbol{s}_p + \alpha_3(\boldsymbol{\tau}_m \otimes \boldsymbol{m} + \boldsymbol{m} \otimes \boldsymbol{\tau}_m)\right] \tag{2.2.42}$$

With the notation $\dot{\varepsilon}_{eq}^{cr} \equiv \frac{\partial W}{\partial \sigma_{eq}}$ (2.2.42) takes the form

$$\dot{\boldsymbol{\varepsilon}}^{cr} = \frac{3}{2}\frac{\dot{\varepsilon}_{eq}^{cr}}{\sigma_{eq}}\left[\alpha_1 J_m \left(\boldsymbol{m} \otimes \boldsymbol{m} - \frac{1}{3}\boldsymbol{I}\right) + \alpha_2 \boldsymbol{s}_p + \alpha_3(\boldsymbol{\tau}_m \otimes \boldsymbol{m} + \boldsymbol{m} \otimes \boldsymbol{\tau}_m)\right] \tag{2.2.43}$$

Let us introduce the following parts of the creep rate tensor

$$\begin{aligned}
\dot{\varepsilon}^{cr}_{mm} &\equiv \boldsymbol{m} \cdot \dot{\boldsymbol{\varepsilon}}^{cr} \cdot \boldsymbol{m}, \\
\dot{\boldsymbol{\varepsilon}}^{cr}_{p} &\equiv (\boldsymbol{I} - \boldsymbol{m} \otimes \boldsymbol{m}) \cdot \dot{\boldsymbol{\varepsilon}}^{cr} \cdot (\boldsymbol{I} - \boldsymbol{m} \otimes \boldsymbol{m}), \\
\dot{\boldsymbol{e}}^{cr}_{p} &\equiv \dot{\boldsymbol{\varepsilon}}^{cr}_{p} - \frac{1}{2}\dot{\varepsilon}^{cr}_{mm}(\boldsymbol{I} - \boldsymbol{m} \otimes \boldsymbol{m}), \\
\dot{\boldsymbol{\gamma}}^{cr}_{m} &\equiv \boldsymbol{m} \cdot \dot{\boldsymbol{\varepsilon}}^{cr} \cdot (\boldsymbol{I} - \boldsymbol{m} \otimes \boldsymbol{m})
\end{aligned} \tag{2.2.44}$$

From (2.2.42) we obtain

$$\dot{\varepsilon}^{cr}_{mm} = \alpha_1 \frac{\dot{\varepsilon}^{cr}_{eq}}{\sigma_{eq}} J_m, \qquad \dot{\boldsymbol{e}}^{cr}_{p} = \frac{3}{2}\alpha_2 \frac{\dot{\varepsilon}^{cr}_{eq}}{\sigma_{eq}} \boldsymbol{s}_p, \qquad \dot{\boldsymbol{\gamma}}^{cr}_{m} = \frac{3}{2}\alpha_3 \frac{\dot{\varepsilon}^{cr}_{eq}}{\sigma_{eq}} \boldsymbol{\tau}_m \tag{2.2.45}$$

Similarly to the isotropic case the equivalent creep rate can be calculated as follows

$$\dot{\varepsilon}^{cr}_{eq} = \sqrt{\frac{1}{\alpha_1}(\dot{\varepsilon}^{cr}_{mm})^2 + \frac{2}{3}\frac{1}{\alpha_2}\dot{\boldsymbol{e}}^{cr}_{p} \cdot\cdot\, \dot{\boldsymbol{e}}^{cr}_{p} + \frac{4}{3}\frac{1}{\alpha_3}\dot{\boldsymbol{\gamma}}^{cr}_{m} \cdot \dot{\boldsymbol{\gamma}}^{cr}_{m}} \tag{2.2.46}$$

The equivalent creep rate (2.2.46) is useful for the verification of the creep potential hypothesis and the assumed quadratic form of the equivalent stress with respect to the transversely isotropic invariants of the stress tensor. The introduced creep equation contains three material constants α_i and the equivalent creep rate $\dot{\varepsilon}^{cr}_{eq}$.

The assumptions of transverse isotropy and the quadratic form of the equivalent stress are widely used in models of elasticity, plasticity, creep and failure of fiber reinforced composites, e.g. [10, 77, 280, 281, 286, 306], and directionally solidified superalloys [46, 219]. The proposed equations will be applied in Sect. 3.2 to the description of anisotropic creep in a multi-pass weld metal.

Orthotropic Symmetry. In this case the potential $W(\boldsymbol{\sigma})$ must satisfy the following restriction

$$W(\boldsymbol{Q}_i \cdot \boldsymbol{\sigma} \cdot \boldsymbol{Q}_i^T) = W(\boldsymbol{\sigma}), \quad \boldsymbol{Q}_i = \boldsymbol{I} - 2\boldsymbol{n}_i \otimes \boldsymbol{n}_i, \quad i = 1,2,3 \tag{2.2.47}$$

In (2.2.47) $\boldsymbol{Q}_i$ denote the assumed symmetry elements - three reflections with respect to the planes with unit normals $\pm\boldsymbol{n}_i$, Fig. 2.2. The unit vectors $\pm\boldsymbol{n}_1, \pm\boldsymbol{n}_2, \pm\boldsymbol{n}_3$ are assumed to be orthogonal, i.e. $\boldsymbol{n}_i \cdot \boldsymbol{n}_j = 0,\ i \neq j$. In Sect. C.4 the sets of scalar arguments which satisfy the above restrictions are presented. As in the previous paragraph we assume the quadratic form of the potential with respect to the stress tensor. One can use different sets of scalar arguments of the stress tensor satisfying (2.2.47), see for example [76],

$$\begin{aligned}
&\boldsymbol{n}_1 \cdot \boldsymbol{\sigma} \cdot \boldsymbol{n}_1, \quad \boldsymbol{n}_2 \cdot \boldsymbol{\sigma} \cdot \boldsymbol{n}_2, \quad \boldsymbol{n}_3 \cdot \boldsymbol{\sigma} \cdot \boldsymbol{n}_3, \\
&\boldsymbol{n}_1 \cdot \boldsymbol{\sigma}^2 \cdot \boldsymbol{n}_1, \quad \boldsymbol{n}_2 \cdot \boldsymbol{\sigma}^2 \cdot \boldsymbol{n}_2, \quad \boldsymbol{n}_3 \cdot \boldsymbol{\sigma}^2 \cdot \boldsymbol{n}_3
\end{aligned}$$

Figure 2.2 shows the components of the stress tensor in a Cartesian frame $\boldsymbol{e}_i$, three planes of symmetry characterized by the unit vectors $\pm\boldsymbol{n}_i$ and components of the

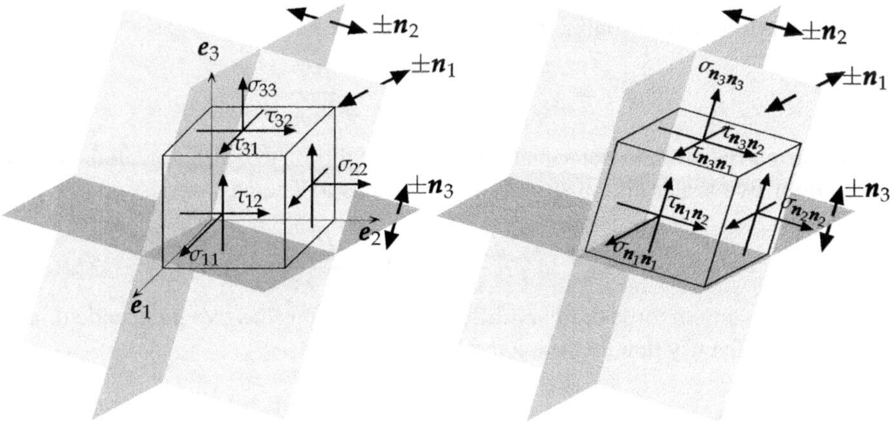

Fig. 2.2 Stress state in an orthotropic medium and corresponding projections $\sigma_{n_in_i}$, $\tau_{n_in_j}$

stress tensor with respect to the planes of symmetry. The stress tensor can be represented as follows

$$
\begin{aligned}
\boldsymbol{\sigma} &= \sigma_{n_1n_1}\boldsymbol{n}_1\otimes\boldsymbol{n}_1+\sigma_{n_2n_2}\boldsymbol{n}_2\otimes\boldsymbol{n}_2+\sigma_{n_3n_3}\boldsymbol{n}_3\otimes\boldsymbol{n}_3\\
&+ \tau_{n_1n_2}(\boldsymbol{n}_1\otimes\boldsymbol{n}_2+\boldsymbol{n}_2\otimes\boldsymbol{n}_1)+\tau_{n_1n_3}(\boldsymbol{n}_1\otimes\boldsymbol{n}_3+\boldsymbol{n}_3\otimes\boldsymbol{n}_1)\\
&+ \tau_{n_2n_3}(\boldsymbol{n}_2\otimes\boldsymbol{n}_3+\boldsymbol{n}_3\otimes\boldsymbol{n}_2)
\end{aligned}
$$

with

$$
\begin{aligned}
\sigma_{n_1n_1} &= \boldsymbol{n}_1\cdot\boldsymbol{\sigma}\cdot\boldsymbol{n}_1, \quad \sigma_{n_2n_2}=\boldsymbol{n}_2\cdot\boldsymbol{\sigma}\cdot\boldsymbol{n}_2, \quad \sigma_{n_3n_3}=\boldsymbol{n}_3\cdot\boldsymbol{\sigma}\cdot\boldsymbol{n}_3,\\
\tau_{n_1n_2} &= \boldsymbol{n}_1\cdot\boldsymbol{\sigma}\cdot\boldsymbol{n}_2, \quad \tau_{n_1n_3}=\boldsymbol{n}_1\cdot\boldsymbol{\sigma}\cdot\boldsymbol{n}_3, \quad \tau_{n_2n_3}=\boldsymbol{n}_2\cdot\boldsymbol{\sigma}\cdot\boldsymbol{n}_3
\end{aligned}
$$

According to Sect. C.4 we use the following orthotropic invariants of the stress tensor

$$
\begin{aligned}
I_{n_1n_1} &= \sigma_{n_1n_1}, \quad I_{n_2n_2}=\sigma_{n_2n_2}, \quad I_{n_3n_3}=\sigma_{n_3n_3},\\
I_{n_1n_2} &= \tau^2_{n_1n_2}, \quad I_{n_1n_3}=\tau^2_{n_1n_3}, \quad I_{n_2n_3}=\tau^2_{n_2n_3}
\end{aligned}
\tag{2.2.48}
$$

Assuming that the creep potential is a function of six arguments introduced, the flow rule (2.1.6) leads to the following creep equation

$$
\begin{aligned}
\dot{\boldsymbol{\varepsilon}}^{cr} &= \frac{\partial W}{\partial I_{n_1n_1}}\boldsymbol{n}_1\otimes\boldsymbol{n}_1+\frac{\partial W}{\partial I_{n_2n_2}}\boldsymbol{n}_2\otimes\boldsymbol{n}_2+\frac{\partial W}{\partial I_{n_3n_3}}\boldsymbol{n}_3\otimes\boldsymbol{n}_3\\
&+ \frac{\partial W}{\partial I_{n_1n_2}}\boldsymbol{n}_1\cdot\boldsymbol{\sigma}\cdot\boldsymbol{n}_2(\boldsymbol{n}_1\otimes\boldsymbol{n}_2+\boldsymbol{n}_2\otimes\boldsymbol{n}_1)\\
&+ \frac{\partial W}{\partial I_{n_1n_3}}\boldsymbol{n}_1\cdot\boldsymbol{\sigma}\cdot\boldsymbol{n}_3(\boldsymbol{n}_1\otimes\boldsymbol{n}_3+\boldsymbol{n}_3\otimes\boldsymbol{n}_1)\\
&+ \frac{\partial W}{\partial I_{n_2n_3}}\boldsymbol{n}_2\cdot\boldsymbol{\sigma}\cdot\boldsymbol{n}_3(\boldsymbol{n}_2\otimes\boldsymbol{n}_3+\boldsymbol{n}_3\otimes\boldsymbol{n}_2)
\end{aligned}
\tag{2.2.49}
$$

The assumption of zero volumetric creep rate leads to

$$\operatorname{tr}\dot{\boldsymbol{\varepsilon}}^{cr} = \frac{\partial W}{\partial I_{\boldsymbol{n}_1\boldsymbol{n}_1}} + \frac{\partial W}{\partial I_{\boldsymbol{n}_2\boldsymbol{n}_2}} + \frac{\partial W}{\partial I_{\boldsymbol{n}_3\boldsymbol{n}_3}} = 0 \tag{2.2.50}$$

From the partial differential equation (2.2.50) follows that the potential W is a function of five scalar arguments of the stress tensor. The characteristic system of (2.2.50) is

$$\frac{dI_{\boldsymbol{n}_1\boldsymbol{n}_1}}{ds} = 1, \quad \frac{dI_{\boldsymbol{n}_2\boldsymbol{n}_2}}{ds} = 1, \quad \frac{dI_{\boldsymbol{n}_3\boldsymbol{n}_3}}{ds} = 1 \tag{2.2.51}$$

The above system of three ordinary differential equations has two independent integrals. One can verify that the following invariants

$$J_1 = \frac{1}{2}(I_{\boldsymbol{n}_2\boldsymbol{n}_2} - I_{\boldsymbol{n}_3\boldsymbol{n}_3}), \quad J_2 = \frac{1}{2}(I_{\boldsymbol{n}_3\boldsymbol{n}_3} - I_{\boldsymbol{n}_1\boldsymbol{n}_1}), \quad J_3 = \frac{1}{2}(I_{\boldsymbol{n}_1\boldsymbol{n}_1} - I_{\boldsymbol{n}_2\boldsymbol{n}_2}) \tag{2.2.52}$$

are integrals of (2.2.51). Only two of them are independent due to the relation $J_1 + J_2 + J_3 = 0$. If the principal directions of the stress tensor coincide with the directions $\boldsymbol{n}_i$ then $\tau_{\boldsymbol{n}_i\boldsymbol{n}_j} = 0, i \neq j$ and the above invariants represent the principal shear stresses. An alternative set of integrals of (2.2.51) is

$$\tilde{J}_1 = I_{\boldsymbol{n}_1\boldsymbol{n}_1} - \frac{1}{3}\operatorname{tr}\boldsymbol{\sigma}, \quad \tilde{J}_2 = I_{\boldsymbol{n}_2\boldsymbol{n}_2} - \frac{1}{3}\operatorname{tr}\boldsymbol{\sigma}, \quad \tilde{J}_3 = I_{\boldsymbol{n}_3\boldsymbol{n}_3} - \frac{1}{3}\operatorname{tr}\boldsymbol{\sigma} \tag{2.2.53}$$

If the principal directions of the stress tensor coincide with $\boldsymbol{n}_i$ then the above invariants are the principal values of the stress deviator. For the formulation of the creep potential in terms of invariants the relation $\tilde{J}_1 + \tilde{J}_2 + \tilde{J}_3 = 0$ must be taken into account.

In what follows we apply the invariants (2.2.52). The equivalent stress can be formulated as follows

$$\begin{aligned} \sigma_{eq}^2 &= 2\beta_1 J_1^2 + 2\beta_2 J_2^2 + 2\beta_3 J_3^2 \\ &+ 3\beta_{12} I_{\boldsymbol{n}_1\boldsymbol{n}_2} + 3\beta_{13} I_{\boldsymbol{n}_1\boldsymbol{n}_3} + 3\beta_{23} I_{\boldsymbol{n}_2\boldsymbol{n}_3} \end{aligned} \tag{2.2.54}$$

The von Mises equivalent stress (2.2.5) follows from (2.2.54) by setting $\beta_1 = \beta_2 = \beta_3 = \beta_{12} = \beta_{13} = \beta_{23} = 1$. Applying the flow rule (2.1.6) we obtain the following creep equation

$$\begin{aligned} \dot{\boldsymbol{\varepsilon}}^{cr} = \frac{\dot{\varepsilon}_{eq}^{cr}}{\sigma_{eq}} \Big[& \beta_1 J_1 (\boldsymbol{n}_2 \otimes \boldsymbol{n}_2 - \boldsymbol{n}_3 \otimes \boldsymbol{n}_3) \\ & + \beta_2 J_2 (\boldsymbol{n}_3 \otimes \boldsymbol{n}_3 - \boldsymbol{n}_1 \otimes \boldsymbol{n}_1) \\ & + \beta_3 J_3 (\boldsymbol{n}_1 \otimes \boldsymbol{n}_1 - \boldsymbol{n}_2 \otimes \boldsymbol{n}_2) \\ & + \frac{3}{2}\beta_{12}\tau_{\boldsymbol{n}_1\boldsymbol{n}_2} (\boldsymbol{n}_1 \otimes \boldsymbol{n}_2 + \boldsymbol{n}_2 \otimes \boldsymbol{n}_1) \\ & + \frac{3}{2}\beta_{13}\tau_{\boldsymbol{n}_1\boldsymbol{n}_3} (\boldsymbol{n}_1 \otimes \boldsymbol{n}_3 + \boldsymbol{n}_3 \otimes \boldsymbol{n}_1) \\ & + \frac{3}{2}\beta_{23}\tau_{\boldsymbol{n}_2\boldsymbol{n}_3} (\boldsymbol{n}_2 \otimes \boldsymbol{n}_3 + \boldsymbol{n}_3 \otimes \boldsymbol{n}_2) \Big] \end{aligned} \tag{2.2.55}$$

The equivalent stress and the creep equation includes six independent material constants. Therefore six independent homogeneous stress states should be realized in order to identify the whole set of constants. In addition, the dependence of the creep rate on the equivalent stress must be fitted from the results of uni-axial creep tests for different constant stress values. For example, if the power law stress function provides a satisfactory description of steady-state creep then the constant n must be additionally identified.

An example of orthotropic creep is discussed in [167] for the aluminium alloy D16AT. Plane specimens were removed from rolled sheet along three directions: the rolling direction, the transverse direction as well as under the angle of 45° to the rolling direction. Uni-axial creep tests were performed at 273°C and 300°C within the stress range 63-90 MPa. The results have shown that at 273°C creep curves depend on the loading direction while at 300°C the creep behavior is isotropic.

Other cases. The previous models are based on the assumption of the quadratic form of the creep potential with respect to the stress tensor. The most general quadratic form can be formulated as follows

$$\sigma_{eq}^2 = \frac{1}{2}\boldsymbol{\sigma} \cdot\cdot {}^{(4)}\boldsymbol{B} \cdot\cdot \boldsymbol{\sigma}, \tag{2.2.56}$$

where σ_{eq} plays the role of the equivalent stress. The fourth rank tensor ${}^{(4)}\boldsymbol{B}$ must satisfy the following restrictions

$$\begin{gathered}\boldsymbol{a} \cdot\cdot {}^{(4)}\boldsymbol{B} \cdot\cdot \boldsymbol{a} \geq 0, \quad \boldsymbol{a} \cdot\cdot {}^{(4)}\boldsymbol{B} = {}^{(4)}\boldsymbol{B} \cdot\cdot \boldsymbol{a}, \quad \boldsymbol{c} \cdot\cdot {}^{(4)}\boldsymbol{B} = \boldsymbol{0}, \\ \forall \boldsymbol{a}, \boldsymbol{c} \text{ with } \boldsymbol{a} = \boldsymbol{a}^T, \ \boldsymbol{c} = -\boldsymbol{c}^T,\end{gathered} \tag{2.2.57}$$

where $\boldsymbol{a}$ and $\boldsymbol{c}$ are second rank tensors. Additional restrictions follow from the assumed symmetries of the steady-state creep behavior. For example, if the orthogonal tensor $\boldsymbol{Q}$ stands for a symmetry element, the structure of the tensor ${}^{(4)}\boldsymbol{B}$ can be established from the following equation

$${}^{(4)}\boldsymbol{B}' = B^{ijkl}\boldsymbol{Q} \cdot \boldsymbol{e}_i \otimes \boldsymbol{Q} \cdot \boldsymbol{e}_j \otimes \boldsymbol{Q} \cdot \boldsymbol{e}_k \otimes \boldsymbol{Q} \cdot \boldsymbol{e}_l = {}^{(4)}\boldsymbol{B}, \tag{2.2.58}$$

where $\boldsymbol{e}_i$, $i = 1, 2, 3$ are basis vectors.

The flow rule (2.1.6) provides the following generalized anisotropic creep equation

$$\dot{\boldsymbol{\varepsilon}}^{cr} = \frac{\dot{\varepsilon}_{eq}^{cr}}{2\sigma_{eq}} {}^{(4)}\boldsymbol{B} \cdot\cdot \boldsymbol{\sigma}, \quad \dot{\varepsilon}_{eq}^{cr} \equiv \frac{\partial W}{\partial \sigma_{eq}} \tag{2.2.59}$$

The fourth rank tensors satisfying the restrictions (2.2.57) are well-known from the theory of linear elasticity. They are used to represent elastic material properties in the generalized Hooke's law. The components of these tensors in a Cartesian coordinate system are given in the matrix notation in many textbooks on linear elasticity as well as in books and monographs on composite materials, e.g. [9, 10, 32, 125, 263, 317]. Furthermore, different coordinate free representations of fourth rank tensors of this type are discussed in the literature. For a review we

Table 2.1 Basic symmetry classes and number of independent coordinates of the tensor ${}^{(4)}\boldsymbol{B}$

Symmetry class	Number of independent coordinates of ${}^{(4)}\boldsymbol{B}$
triclinic symmetry	21
monoclinic symmetry	13
orthotropic or rhombic symmetry	9
trigonal symmetry	6
tetragonal symmetry	6
transverse isotropy or hexagonal symmetry	5
cubic symmetry	3
isotropic symmetry	2

refer to [79]. One of these representations - the projector representation is applied in [51, 52, 206] to constitutive modeling of creep in single crystal alloys under assumption of the cubic symmetry.

Let us recall that (2.2.59) is the consequence of the creep potential hypothesis and the quadratic form of the equivalent stress with respect to the stress tensor. Similarly to the case of linear elasticity [317] one can prove that only eight basic symmetry classes are relevant according to these assumptions. The basic symmetry classes and the corresponding number of independent coordinates of the tensor ${}^{(4)}\boldsymbol{B}$ are listed in Table 2.1. The number of independent coordinates indicates the number of material constants which should be identified from creep tests. This number can be reduced if the volume constancy is additionally assumed. For example, in the cases of transverse isotropy and orthotropic symmetry the number of independent coordinates of $\boldsymbol{B}$ reduces to 3 and 5, respectively (see previous paragraphs).

2.2.2.2 Non-Classical Creep Equations. Non-classical effects are the dependence of secondary creep rate on the kind of loading and second order effects, see Sect. 2.2.1. Examples of such behavior are different creep rates under tensile and compressive stress or the effect of reversal of the shear stress. The last case is observed in creep tests on tubular specimens under applied torque. The change of the direction of the applied torque leads to different values of the shear strain rate. The effect of shear stress reversal is usually explained to be the result of the anisotropy induced by the deformation process (e.g. anisotropic hardening) or anisotropy induced by damage evolution. Phenomenological models of induced anisotropy will be introduced in Sects. 2.3.2 and 2.4. Here we consider the case of initial anisotropy without discussion of histories of the deformation, damage or manufacturing processes. Nevertheless, a phenomenological model of anisotropic creep should be able to reflect the above mentioned effects since they are observed experimentally. In order to describe non-classical effects the quadratic form of the creep potential should be replaced by a more general form including all invariants of the stress tensor for the assumed symmetry group. In this case the number of material constants rapidly increases. Furthermore, the identification and verification of the model requires creep tests under combined multi-axial stress states. In what follows we limit ourselves to some remarks regarding the general structure of constitutive equations and kinds of tests for the identification.

Transverse Isotropy. The creep potential must satisfy the restriction (2.2.29) leading to the partial differential equation (2.2.30). The integrals represent the set of functionally independent arguments of the creep potential. They are listed in Sect. C.3 for two transverse isotropy groups. The first group is formed by all the rotations about a given axis $\boldsymbol{m}$, i.e

$$\boldsymbol{Q}(\psi\boldsymbol{m}) = \boldsymbol{m} \otimes \boldsymbol{m} + \cos\psi(\boldsymbol{I} - \boldsymbol{m} \otimes \boldsymbol{m}) + \sin\psi\boldsymbol{m} \times \boldsymbol{I}$$

The second group additionally includes rotations on the angle π about any axis orthogonal to $\boldsymbol{m}$, i.e.

$$\boldsymbol{Q}_1 = \boldsymbol{Q}(\pi\boldsymbol{p}) = 2\boldsymbol{p} \otimes \boldsymbol{p} - \boldsymbol{I}, \quad \det\boldsymbol{Q} = 1, \quad \boldsymbol{p} \cdot \boldsymbol{m} = 0$$

Let us note that there is an essential difference in these two groups since the creep potential depends on different non-quadratic arguments of the stress tensor. Here we limit our considerations to the second case which is widely discussed in the literature on anisotropic elasticity, plasticity and creep [62, 76, 87, 286, 293], where the following invariants are applied [2]

$$\mathrm{tr}\,\boldsymbol{\sigma}, \quad \mathrm{tr}\,\boldsymbol{\sigma}^2, \quad \mathrm{tr}\,\boldsymbol{\sigma}^3, \quad \boldsymbol{m} \cdot \boldsymbol{\sigma} \cdot \boldsymbol{m}, \quad \boldsymbol{m} \cdot \boldsymbol{\sigma}^2 \cdot \boldsymbol{m} \tag{2.2.60}$$

To be consistent with derivations in Sect. 2.2.2.1 let us use the decomposition of the stress tensor (2.2.34) leading to the following set of invariants

$$\begin{aligned}
I_{1m} &= \sigma_{mm} = \boldsymbol{m} \cdot \boldsymbol{\sigma} \cdot \boldsymbol{m}, \\
I_{2m} &= \mathrm{tr}\,\boldsymbol{\sigma}_p = \mathrm{tr}\,\boldsymbol{\sigma} - \boldsymbol{m} \cdot \boldsymbol{\sigma} \cdot \boldsymbol{m}, \\
I_{3m} &= \frac{1}{2}\mathrm{tr}\,\mathbf{s}_p^2 = \frac{1}{2}\mathrm{tr}\,\boldsymbol{\sigma}_p^2 - \frac{1}{4}(\mathrm{tr}\,\boldsymbol{\sigma}_p)^2 \\
&= \frac{1}{2}\left[\mathrm{tr}\,\boldsymbol{\sigma}^2 + (\boldsymbol{m} \cdot \boldsymbol{\sigma} \cdot \boldsymbol{m})^2\right] - \boldsymbol{m} \cdot \boldsymbol{\sigma}^2 \cdot \boldsymbol{m} - \frac{1}{4}(\mathrm{tr}\,\boldsymbol{\sigma} - \boldsymbol{m} \cdot \boldsymbol{\sigma} \cdot \boldsymbol{m})^2, \\
I_{4m} &= \boldsymbol{\tau}_m \cdot \boldsymbol{\tau}_m = \boldsymbol{m} \cdot \boldsymbol{\sigma}^2 \cdot \boldsymbol{m} - (\boldsymbol{m} \cdot \boldsymbol{\sigma} \cdot \boldsymbol{m})^2 = (\boldsymbol{m} \times \boldsymbol{\sigma} \cdot \boldsymbol{m}) \cdot (\boldsymbol{m} \times \boldsymbol{\sigma} \cdot \boldsymbol{m}) \\
I_{5m} &= \boldsymbol{\tau}_m \cdot \mathbf{s}_p \cdot \boldsymbol{\tau}_m = \boldsymbol{m} \cdot \boldsymbol{\sigma}^3 \cdot \boldsymbol{m} - 2(\boldsymbol{m} \cdot \boldsymbol{\sigma} \cdot \boldsymbol{m})(\boldsymbol{m} \cdot \boldsymbol{\sigma}^2 \cdot \boldsymbol{m}) \\
&\quad + (\boldsymbol{m} \cdot \boldsymbol{\sigma} \cdot \boldsymbol{m})^3 - \frac{1}{2}(\mathrm{tr}\,\boldsymbol{\sigma} - \boldsymbol{m} \cdot \boldsymbol{\sigma} \cdot \boldsymbol{m})\left[\boldsymbol{m} \cdot \boldsymbol{\sigma}^2 \cdot \boldsymbol{m} - (\boldsymbol{m} \cdot \boldsymbol{\sigma} \cdot \boldsymbol{m})^2\right]
\end{aligned} \tag{2.2.61}$$

The meaning of the first four invariants is explained in in Sect. 2.2.2.1. The last cubic invariant is introduced instead $\mathrm{tr}\,\boldsymbol{\sigma}^3$. One can prove the following relation

$$\mathrm{tr}\,\boldsymbol{\sigma}^3 = I_{1m}^3 + 3I_{1m}I_{4m} + 3I_{2m}I_{3m} + \frac{3}{2}I_{2m}I_{4m} + \frac{1}{2}I_{2m}^3 + 3I_{5m}$$

[2] For the description of elastic material behavior instead of $\boldsymbol{\sigma}$ a strain tensor, e.g. the Cauchy-Green strain tensor is introduced. The five transversely isotropic invariants are the arguments of the strain energy density function.

Assuming that the creep potential W is a function of five scalar arguments (2.2.61) and applying the flow rule (2.1.6) we obtain the following creep equation

$$\dot{\boldsymbol{\varepsilon}}^{cr} = h_1 \boldsymbol{m} \otimes \boldsymbol{m} + (h_2 - \frac{1}{2} h_5 I_{4m})(\boldsymbol{I} - \boldsymbol{m} \otimes \boldsymbol{m}) + h_3 \boldsymbol{\sigma}_p + h_4 (\boldsymbol{\tau}_m \otimes \boldsymbol{m} + \boldsymbol{m} \otimes \boldsymbol{\tau}_m)$$
$$+ h_5 \left(\boldsymbol{\tau}_m \otimes \boldsymbol{\tau}_m + \boldsymbol{m} \otimes \boldsymbol{\sigma}_p \cdot \boldsymbol{\tau}_m + \boldsymbol{\tau}_m \cdot \boldsymbol{\sigma}_p \otimes \boldsymbol{m} \right), \tag{2.2.62}$$

where

$$h_i = \frac{\partial W}{\partial I_{im}}, \quad i = 1, 2, \ldots, 5$$

The last term in the right-hand side of (2.2.62) describes second order effects. The meaning of these effects is obvious. In the case of non-zero "transverse shear stress" vector

$$\boldsymbol{\tau}_m = \boldsymbol{m} \cdot \boldsymbol{\sigma} \cdot (\boldsymbol{I} - \boldsymbol{m} \otimes \boldsymbol{m})$$

the elongation in the direction of $\boldsymbol{\tau}_m$ can be considered. The vector $\boldsymbol{\varsigma}_m = \boldsymbol{s}_p \cdot \boldsymbol{\tau}_m$ belongs to the isotropy plane, i.e. $\boldsymbol{\varsigma}_m \cdot \boldsymbol{m} = 0$. In the case that $\boldsymbol{\varsigma}_m \neq \boldsymbol{0}$ (2.2.62) describes an additional "transverse shear strain rate" effect.

In order to formulate the creep constitutive equation one should specify an expression for the equivalent stress as a function of the introduced invariants. As an example we present the equivalent stress by use of polynomials of the type (2.2.9) and (2.2.10)

$$\sigma_{eq} = \alpha \sigma_1 + \sigma_2 + \gamma \sigma_3, \tag{2.2.63}$$

with

$$\begin{aligned}
\sigma_1 &= \mu_{11} I_{1m} + \mu_{12} I_{2m}, \\
\sigma_2^2 &= \mu_{21} I_{1m}^2 + \mu_{22} I_{1m} I_{2m} + \mu_{23} I_{2m}^2 + \mu_{24} I_{3m} + \mu_{25} I_{4m}, \\
\sigma_3^3 &= \mu_{31} I_{1m}^3 + \mu_{32} I_{1m}^2 I_{2m} + \mu_{33} I_{1m} I_{2m}^2 + \mu_{34} I_{2m}^3 + \mu_{35} I_{1m} I_{3m} \\
&+ \mu_{36} I_{2m} I_{3m} + \mu_{37} I_{1m} I_{4m} + \mu_{38} I_{2m} I_{4m} + \mu_{39} I_{5m}
\end{aligned} \tag{2.2.64}$$

The equivalent stress (2.2.63) includes 16 material constants μ_{ij} and two weighting factors α and γ. The identification of all material constants requires different independent creep tests under multi-axial stress states. For example, in order to find the constant μ_{39} creep tests under stress states with nonzero cubic invariant I_{5m} should be carried out. An example is the tension in the isotropy plane combined with the transverse shear stress leading to the stress state of the type $\boldsymbol{\sigma} = \sigma_0 \boldsymbol{n}_1 \otimes \boldsymbol{n}_1 + \tau_0 (\boldsymbol{n}_1 \otimes \boldsymbol{m} + \boldsymbol{m} \otimes \boldsymbol{n}_1)$, where $\sigma_0 > 0$ and $\tau_0 > 0$ are the magnitudes of the applied stresses, $\boldsymbol{n}_1$ is the direction of tension and $\boldsymbol{n}_1 \cdot \boldsymbol{m} = 0$. In this case

$$\boldsymbol{s}_p = \frac{1}{2} \sigma_0 (\boldsymbol{n}_1 \otimes \boldsymbol{n}_1 - \boldsymbol{n}_2 \otimes \boldsymbol{n}_2), \quad \boldsymbol{n}_1 \cdot \boldsymbol{n}_2 = 0, \qquad \boldsymbol{\tau}_m = \tau_0 \boldsymbol{n}_1, \quad I_{5m} = \frac{1}{2} \sigma_0 \tau_0^2$$

By analogy to the non-classical models of isotropic creep discussed in Sect. 2.2.1 different special cases can be introduced. Setting $\gamma = 0$ in (2.2.64), second order effects will be neglected. The resulting constitutive model takes into account different behavior under tension and compression. To find the constants μ_{11} and μ_{12} creep tests under tension (compression) along the direction $\boldsymbol{m}$ as well as tension (compression) along any direction in the isotropy plane should be carried out. Setting $\alpha = 0$ the model with the quadratic form of the creep potential with 5 constants can be obtained. The assumption of the zero volumetric creep rate will lead to the model discussed in Sect. 2.2.2.1.

Second order effects of anisotropic creep were discussed by Betten [56, 62]. He found disagreements between creep equations based on the theory of isotropic functions and the creep equation of the type (2.2.62) according to the potential hypothesis and the flow rule. The conclusion was made that the potential theory leads to restrictive forms of constitutive equations if compared to the representations of tensor functions.

Let us recall the results following from the algebra of isotropic tensor functions [74]. In the case of transverse isotropy group characterized by the symmetry elements (C.3.13) the statement of the problem is to find the general representation of the isotropic tensor function of the stress tensor $\boldsymbol{\sigma}$ and the dyad $\boldsymbol{m} \otimes \boldsymbol{m}$ (so-called structure tensor). The constitutive equation describing the creep behavior must be found as follows

$$\dot{\boldsymbol{\varepsilon}}^{cr} = \boldsymbol{f}(\boldsymbol{\sigma}, \boldsymbol{m} \otimes \boldsymbol{m}),$$

where $\boldsymbol{f}$ is an isotropic tensor function of two tensor arguments. The general representation of this function is [76]

$$\begin{aligned} \boldsymbol{f}(\boldsymbol{\sigma}, \boldsymbol{m} \otimes \boldsymbol{m}) &= f_1 \boldsymbol{m} \otimes \boldsymbol{m} + f_2(\boldsymbol{I} - \boldsymbol{m} \otimes \boldsymbol{m}) + f_3 \boldsymbol{\sigma} + f_4 \boldsymbol{\sigma}^2 \\ &+ f_5(\boldsymbol{m} \otimes \boldsymbol{m} \cdot \boldsymbol{\sigma} + \boldsymbol{\sigma} \cdot \boldsymbol{m} \otimes \boldsymbol{m}) + f_6(\boldsymbol{m} \otimes \boldsymbol{m} \cdot \boldsymbol{\sigma}^2 + \boldsymbol{\sigma}^2 \cdot \boldsymbol{m} \otimes \boldsymbol{m}), \end{aligned} \tag{2.2.65}$$

where the scalars f_i, $i = 1, \ldots, 6$, depend on the five invariants of the stress tensor (2.2.60). Betten found that the last term in (2.2.65) is missing in the constitutive equation which is based on the potential theory. In order to discuss the meaning of the last term in (2.2.65) let us introduce the identities which follow from the decomposition of the stress tensor by Eqs (2.2.34) and (2.2.36)

$$\begin{aligned} \boldsymbol{\sigma}^2 &= I_{2m} \boldsymbol{s}_p + \left(I_{3m} + \frac{1}{4} I_{2m}^2\right)(\boldsymbol{I} - \boldsymbol{m} \otimes \boldsymbol{m}) + \boldsymbol{m} \otimes \boldsymbol{s}_p \cdot \boldsymbol{\tau}_m + \boldsymbol{\tau}_m \cdot \boldsymbol{s}_p \otimes \boldsymbol{m} \\ &+ \left(I_{1m} + \frac{1}{2} I_{2m}\right)(\boldsymbol{\tau}_m \otimes \boldsymbol{m} + \boldsymbol{m} \otimes \boldsymbol{\tau}_m) + (I_{1m}^2 + I_{4m}) \boldsymbol{m} \otimes \boldsymbol{m} + \boldsymbol{\tau}_m \otimes \boldsymbol{\tau}_m, \\ \boldsymbol{m} \otimes \boldsymbol{m} \cdot \boldsymbol{\sigma} + \boldsymbol{\sigma} \cdot \boldsymbol{m} \otimes \boldsymbol{m} &= \boldsymbol{\tau}_m \otimes \boldsymbol{m} + \boldsymbol{m} \otimes \boldsymbol{\tau}_m + 2 I_{1m} \boldsymbol{m} \otimes \boldsymbol{m}, \\ \boldsymbol{m} \otimes \boldsymbol{m} \cdot \boldsymbol{\sigma}^2 + \boldsymbol{\sigma}^2 \cdot \boldsymbol{m} \otimes \boldsymbol{m} &= \boldsymbol{m} \otimes \boldsymbol{s}_p \cdot \boldsymbol{\tau}_m + \boldsymbol{\tau}_m \cdot \boldsymbol{s}_p \otimes \boldsymbol{m} \\ &+ \left(I_{1m} + \frac{1}{2} I_{2m}\right)(\boldsymbol{\tau}_m \otimes \boldsymbol{m} + \boldsymbol{m} \otimes \boldsymbol{\tau}_m) \\ &+ 2(I_{4m} + I_{1m}^2) \boldsymbol{m} \otimes \boldsymbol{m} \end{aligned} \tag{2.2.66}$$

After inserting (2.2.66), (2.2.34) and (2.2.36) into (2.2.65) we obtain the following creep equation

$$\begin{aligned}\dot{\boldsymbol{\varepsilon}}^{cr} &= g_1 \boldsymbol{m}\otimes\boldsymbol{m} + g_2(\boldsymbol{I} - \boldsymbol{m}\otimes\boldsymbol{m}) + g_3\boldsymbol{s}_p + g_4(\boldsymbol{m}\otimes\boldsymbol{\tau}_m + \boldsymbol{\tau}_m\otimes\boldsymbol{m})\\ &+ g_5(\boldsymbol{m}\otimes\boldsymbol{s}_p\cdot\boldsymbol{\tau}_m + \boldsymbol{\tau}_m\cdot\boldsymbol{s}_p\otimes\boldsymbol{m}) + g_6\boldsymbol{\tau}_m\otimes\boldsymbol{\tau}_m\end{aligned} \tag{2.2.67}$$

with

$$\begin{aligned}g_1 &= f_1 + f_4(I_{1m}^2 + I_{4m}) + 2f_5 I_{1m} + 2f_6(I_{4m} + I_{1m}^2),\\ g_2 &= f_2 + \frac{1}{2}f_3 I_{2m} + f4(I_{3m} + \frac{1}{4}I_{2m}^2),\\ g_3 &= f_3 + I_{2m}f_4,\\ g_4 &= (f_4 + f_6)(I_{1m} + \frac{1}{2}I_{2m}) + f_5,\\ g_5 &= f_4 + f_6,\\ g_6 &= f_4\end{aligned}$$

We observe that Eq. (2.2.67) based on the theory of isotropic tensor functions does not deliver any new second order effect in comparison to (2.2.62). The only difference is that the two last terms in (2.2.67) characterizing the second order effects appear with two different influence functions. The comparison of (2.2.67) with (2.2.62) provides the following conditions for the existence of the potential

$$\begin{aligned}&\frac{\partial W}{\partial I_{1m}} = g_1, \quad \frac{\partial W}{\partial I_{2m}} = g_2 + \frac{1}{2}g_5 I_{4m},\\ &\frac{\partial W}{\partial I_{3m}} = g_3, \quad \frac{\partial W}{\partial I_{4m}} = g_4, \quad \frac{\partial W}{\partial I_{5m}} = g_5, \quad g_6 = g_5\end{aligned}$$

Furthermore, the functions g_i must satisfy the integrability conditions which can be obtained by equating the mixed derivatives of the potential with respect to invariants, i.e.

$$\frac{\partial^2 W}{\partial I_{im}\partial I_{km}} = \frac{\partial^2 W}{\partial I_{km}\partial I_{im}}, \quad i \neq k, \quad i,k = 1,2,\ldots,5$$

Let us note that the models (2.2.62) and (2.2.67) are restricted to the special case of transverse isotropy. In the general case one should analyze the creep potential with the invariants listed in (C.3.21).

Other Cases. Alternatively a phenomenological constitutive equation of anisotropic creep can be formulated with the help of material tensors, e.g. [4]. Introducing three material tensors $\boldsymbol{A}$, ${}^{(4)}\boldsymbol{B}$ and ${}^{(6)}\boldsymbol{C}$ the equivalent stress (2.2.63) can be generalized as follows

$$\sigma_{eq} = \alpha\sigma_1 + \sigma_2 + \gamma\sigma_3 \tag{2.2.68}$$

with

$$\sigma_1 = \boldsymbol{A}\cdot\cdot\boldsymbol{\sigma}, \quad \sigma_2^2 = \boldsymbol{\sigma}\cdot\cdot{}^{(4)}\boldsymbol{B}\cdot\cdot\boldsymbol{\sigma}, \quad \sigma_3^3 = \boldsymbol{\sigma}\cdot\cdot(\boldsymbol{\sigma}\cdot\cdot{}^{(6)}\boldsymbol{C}\cdot\cdot\boldsymbol{\sigma}) \tag{2.2.69}$$

The structure of the material tensors must be established from the following restrictions

$$
\begin{aligned}
\boldsymbol{A}' &= \boldsymbol{Q}\cdot\boldsymbol{A}\cdot\boldsymbol{Q}^T = A^{ij}\boldsymbol{Q}\cdot\boldsymbol{e}_i\otimes\boldsymbol{Q}\cdot\boldsymbol{e}_j = \boldsymbol{A},\\
{}^{(4)}\boldsymbol{B}' &= B^{ijkl}\boldsymbol{Q}\cdot\boldsymbol{e}_i\otimes\boldsymbol{Q}\cdot\boldsymbol{e}_j\otimes\boldsymbol{Q}\cdot\boldsymbol{e}_k\otimes\boldsymbol{Q}\cdot\boldsymbol{e}_l = {}^{(4)}\boldsymbol{B},\\
{}^{(6)}\boldsymbol{C}' &= C^{ijklmn}\boldsymbol{Q}\cdot\boldsymbol{e}_i\otimes\boldsymbol{Q}\cdot\boldsymbol{e}_j\otimes\boldsymbol{Q}\cdot\boldsymbol{e}_k\otimes\boldsymbol{Q}\cdot\boldsymbol{e}_l\otimes\boldsymbol{Q}\cdot\boldsymbol{e}_m\otimes\boldsymbol{Q}\cdot\boldsymbol{e}_n = {}^{(6)}\boldsymbol{C},
\end{aligned}
\tag{2.2.70}
$$

where $\boldsymbol{Q}$ is an element of the physical symmetry group. The creep potential hypothesis and the flow rule (2.1.6) lead to the following creep equation

$$
\dot{\boldsymbol{\varepsilon}}^{cr} = \frac{\partial W}{\partial\sigma_{eq}}\left(\alpha\frac{\partial\sigma_1}{\partial\boldsymbol{\sigma}} + \frac{\partial\sigma_2}{\partial\boldsymbol{\sigma}} + \gamma\frac{\partial\sigma_3}{\partial\boldsymbol{\sigma}}\right)
\tag{2.2.71}
$$

Taking into account the relations

$$
\frac{\partial\sigma_1}{\partial\boldsymbol{\sigma}} = \boldsymbol{A}, \quad \frac{\partial\sigma_2}{\partial\boldsymbol{\sigma}} = \frac{{}^{(4)}\boldsymbol{B}\cdot\cdot\boldsymbol{\sigma}}{\sigma_2}, \quad \frac{\partial\sigma_3}{\partial\boldsymbol{\sigma}} = \frac{\boldsymbol{\sigma}\cdot\cdot{}^{(6)}\boldsymbol{C}\cdot\cdot\boldsymbol{\sigma}}{\sigma_3^2}
\tag{2.2.72}
$$

a generalized anisotropic creep equation can be formulated as follows

$$
\dot{\boldsymbol{\varepsilon}}^{cr} = \dot{\varepsilon}^{cr}_{eq}\left(\alpha\boldsymbol{A} + \frac{{}^{(4)}\boldsymbol{B}\cdot\cdot\boldsymbol{\sigma}}{\sigma_2} + \gamma\frac{\boldsymbol{\sigma}\cdot\cdot{}^{(6)}\boldsymbol{C}\cdot\cdot\boldsymbol{\sigma}}{\sigma_3^2}\right), \quad \dot{\varepsilon}^{cr}_{eq} \equiv \frac{\partial W}{\partial\sigma_{eq}}
\tag{2.2.73}
$$

In [55, 272] the following anisotropic creep equation is proposed

$$
\dot{\boldsymbol{\varepsilon}}^{cr} = \boldsymbol{H} + {}^{(4)}\boldsymbol{M}\cdot\cdot\boldsymbol{\sigma} + \left({}^{(6)}\boldsymbol{L}\cdot\cdot\boldsymbol{\sigma}\right)\cdot\cdot\boldsymbol{\sigma}
\tag{2.2.74}
$$

Comparing the Eqs (2.2.73) and (2.2.74) the material tensors $\boldsymbol{H}$, ${}^{(4)}\boldsymbol{M}$ and ${}^{(6)}\boldsymbol{L}$ can be related to the tensors $\boldsymbol{A}$, ${}^{(4)}\boldsymbol{B}$ and ${}^{(6)}\boldsymbol{C}$.

The tensors $\boldsymbol{A}$, ${}^{(4)}\boldsymbol{B}$ and ${}^{(6)}\boldsymbol{C}$ contain 819 coordinates ($\boldsymbol{A}$ - 9, ${}^{(4)}\boldsymbol{B}$ - 81, ${}^{(6)}\boldsymbol{C}$ - 729). From the symmetry of the stress tensor and the creep rate tensor as well as from the potential hypothesis follows that "only" 83 coordinates are independent ($\boldsymbol{A}$ - 6, ${}^{(4)}\boldsymbol{B}$ - 21, ${}^{(6)}\boldsymbol{C}$ - 56). Further reduction is based on the symmetry considerations. The structure of material tensors and the number of independent coordinates can be obtained by solving (2.2.70).

Another possibility of simplification is the establishing of special cases of (2.2.73). For instance, equations with a reduced number of parameters can be derived as follows

– $\alpha = 1, \gamma = 0$:

$$
\sigma_{eq} = \sigma_1 + \sigma_2, \quad \dot{\boldsymbol{\varepsilon}}^{cr} = \dot{\varepsilon}^{cr}_{eq}\left(\boldsymbol{A} + \frac{{}^{(4)}\boldsymbol{B}\cdot\cdot\boldsymbol{\sigma}}{\sigma_2}\right),
\tag{2.2.75}
$$

– $\alpha = 0, \gamma = 1$:

$$
\sigma_{eq} = \sigma_2 + \sigma_3, \quad \dot{\boldsymbol{\varepsilon}}^{cr} = \dot{\varepsilon}^{cr}_{eq}\left(\frac{{}^{(4)}\boldsymbol{B}\cdot\cdot\boldsymbol{\sigma}}{\sigma_2} + \frac{\boldsymbol{\sigma}\cdot\cdot{}^{(6)}\boldsymbol{C}\cdot\cdot\boldsymbol{\sigma}}{\sigma_3^2}\right),
\tag{2.2.76}
$$

– $\alpha = 0, \gamma = 0$:

$$\sigma_{eq} = \sigma_2, \quad \dot{\boldsymbol{\varepsilon}}^{cr} = \dot{\varepsilon}^{cr}_{eq} \left(\frac{{}^{(4)}\boldsymbol{B} \cdot\cdot \boldsymbol{\sigma}}{\sigma_2} \right) \tag{2.2.77}$$

The last case has been discussed in Sect. 2.2.2.1. Examples of application of constitutive equation (2.2.73) as well as different cases of symmetries are discussed in [4, 12].

2.2.3 Functions of Stress and Temperature

In all constitutive equations discussed in Sects 2.2.1 and 2.2.2 the creep potential or the equivalent creep rate must be specified as functions of the equivalent stress and the temperature, i.e.

$$\dot{\varepsilon}^{cr}_{eq} = \frac{\partial W}{\partial \sigma_{eq}} = f(\sigma_{eq}, T)$$

In [181] the function f is termed to be the constitutive or response function. For the formulation of constitutive functions one may apply theoretical foundations from materials science with regard to mechanisms of creep deformation and related forms of stress and temperature functions. Furthermore, experimental data including families of creep curves obtained from uni-axial creep tests for certain ranges of stress and temperature are required. It is convenient to present these families in a form of minimum creep rate vs. stress and minimum creep rate vs. temperature curves in order to find mechanical properties of the material within the steady-state creep range.

Many empirical functions of stress and temperature which allow to fit experimental data have been proposed in the literature, e.g. [243, 257, 273, 299]. The starting point is the assumption that the creep rate may be descried as a product of two separate functions of stress and temperature

$$\dot{\varepsilon}^{cr}_{eq} = f_\sigma(\sigma_{eq}) f_T(T)$$

The widely used functions of stress are:

– power law

$$f_\sigma(\sigma_{eq}) = \dot{\varepsilon}_0 \left| \frac{\sigma_{eq}}{\sigma_0} \right|^{n-1} \frac{\sigma_{eq}}{\sigma_0} \tag{2.2.78}$$

The power law contains three constants $(\dot{\varepsilon}_0, \sigma_0, n)$ but only two of them are independent. Instead of $\dot{\varepsilon}_0$ and σ_0 one material constant

$$a \equiv \frac{\dot{\varepsilon}_0}{\sigma_0^n}$$

can be introduced.

– power law including the creep limit

$$f_\sigma(\sigma_{eq}) = \dot{\varepsilon}'_0 \left(\frac{\sigma_{eq}}{\sigma'_0} - 1 \right)^{n'}, \quad \sigma_{eq} > \sigma'_0$$

If $\sigma_{eq} \leq \sigma'_0$ the creep rate is equal zero. In this case σ'_0 is the assumed creep limit. Let us note that the experimental identification of its value is difficult, e.g. [273].

– exponential law

$$f_\sigma(\sigma_{eq}) = \dot{\varepsilon}_0 \exp \frac{\sigma_{eq}}{\sigma_0}$$

$\dot{\varepsilon}_0, \sigma_0$ are material constants. The disadvantage of this expression is that it predicts a nonzero creep rate for a zero equivalent stress

$$f_\sigma(0) = \dot{\varepsilon}_0 \neq 0$$

– hyperbolic sine law

$$f_\sigma(\sigma_{eq}) = \dot{\varepsilon}_0 \sinh \frac{\sigma_{eq}}{\sigma_0}$$

For low stress values this function provides the linear dependence on the stress

$$f_\sigma(\sigma_{eq}) \approx \dot{\varepsilon}_0 \frac{\sigma_{eq}}{\sigma_0}$$

Assuming the constant temperature, equations for the equivalent creep rate can be summarized as follows

$$\begin{aligned}
\dot{\varepsilon}^{cr}_{eq} &= a\sigma^n_{eq} && \text{Norton, 1929, Bailey, 1929,} \\
\dot{\varepsilon}^{cr}_{eq} &= b\left(\exp\frac{\sigma_{eq}}{\sigma_0} - 1\right) && \text{Soderberg, 1936,} \\
\dot{\varepsilon}^{cr}_{eq} &= a\sinh\frac{\sigma_{eq}}{\sigma_0} && \text{Prandtl, 1928, Nadai, 1938, McVetty, 1943,} \\
\dot{\varepsilon}^{cr}_{eq} &= a_1\sigma^{n_1}_{eq} + a_2\sigma^{n_2}_{eq} && \text{Johnson et al., 1963,} \\
\dot{\varepsilon}^{cr}_{eq} &= a\left(\sinh\frac{\sigma_{eq}}{\sigma_0}\right)^n && \text{Garofalo, 1965,}
\end{aligned} \tag{2.2.79}$$

where $a, b, a_1, a_2, \sigma_0, n, n_1$ and n_2 are material constants. The dependence on the temperature is usually expressed by the Arrhenius law

$$f_T(T) = \exp[-Q/RT],$$

where Q and R denote the activation energy and the Boltzmann's constant, respectively.

For the use of stress and temperature functions one should take into account that different deformation mechanisms may operate for different specific ranges of stress and temperature. An overview is provided by the deformation mechanisms

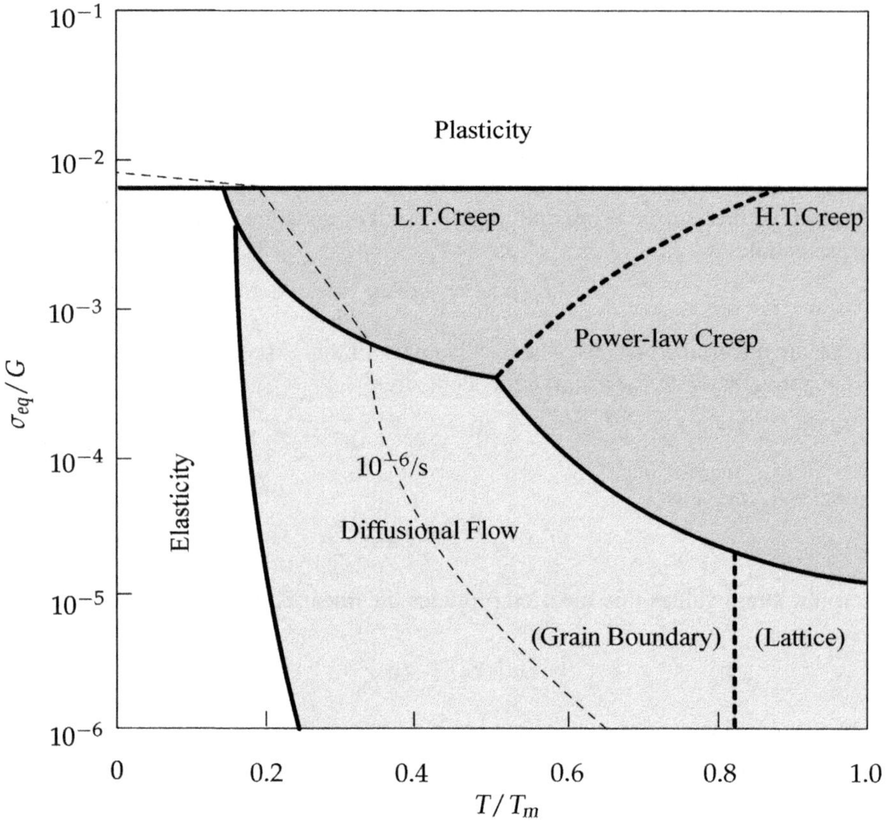

Fig. 2.3 Schematic deformation-mechanism map (L.T.Creep - low temperature creep, H.T. Creep - high temperature creep)

maps proposed by Frost and Ashby [120], Fig. 2.3. Contours of constant strain rates are presented as functions of the normalized equivalent stress σ_{eq}/G and the homologous temperature T/T_m, where G is the shear modulus and T_m is the melting temperature. For a given combination of the stress and the temperature, the map provides the dominant creep mechanism and the strain rate.

Let us briefly discuss different regions on the map, the mechanisms of creep deformation and constitutive functions derived in materials science. For comprehensive reviews one may consult [119, 159, 228]. The origins of the inelastic deformation at the temperature range $0.5 < T/T_m < 0.7$ are transport processes associated with motion and interaction of dislocations and diffusion of vacancies. Here we limit our consideration to the two classes of physical models - dislocation and diffusion creep. Various creep rate equations within the dislocation creep range are based on the Bailey-Orowan recovery hypothesis. An internal barrier stress σ_{int} being opposed to the dislocation movement is assumed. When the plastic strain occurs the internal stress increases as a result of work hardening due to accumulation of deformation and due to increase of the dislocation density. As the material is subjected to the load and temperature over certain time, the internal stress σ_{int} recovers. In the uni-axial case the rate of change of the internal stress is assumed as follows

$$\dot{\sigma}_{int} = h\dot{\varepsilon}^{cr} - r\sigma_{int},$$

where h and r are material properties related to hardening and recovery, respectively. In the steady state $\dot{\sigma}_{int} = 0$ so that

$$\dot{\varepsilon}^{cr} = \frac{r\sigma_{int}}{h}$$

Specifying the values for r, h and σ_{int} various models for the steady-state creep rate have been derived. An example is the following expression (for details of derivation we refer to [119])

$$\dot{\varepsilon}^{cr} \propto \frac{D}{RT}\frac{\sigma^4}{G^3}\exp\left(-\frac{Q}{RT}\right),$$

where D is the diffusion coefficient.

Further models of dislocation creep are discussed under the assumption of the climb-plus-glide deformation mechanism. At high temperatures and moderate stresses, dislocations can climb as well as glide. The glide of dislocations produced by the applied stress is opposed by obstacles. Due to diffusion of vacancies, the dislocations can climb around strengthening particles. The inelastic strain is then controlled by the glide, while its rate is determined by the climb. The climb-plus-glide mechanism can be related to the recovery-hardening hypothesis. The hardening results from the resistance to glide due to interaction of moving dislocations with other dislocations, precipitates, etc. The recovery mechanism is the diffusion controlled climb which releases the glide barriers. The climb-plus-glide based creep rate models can be found in [119, 120, 228]. The common result is the power-law creep

$$\dot{\varepsilon}^{cr}_{eq} \propto \left(\frac{\sigma_{eq}}{G}\right)^n \exp\left(-\frac{Q}{RT}\right) \tag{2.2.80}$$

Equation (2.2.80) can be used to fit experimental data for a range of stresses up to $10^{-3}G$. The exponent n varies from 3 to about 10 for metallic materials. At higher stresses above $10^{-3}G$ the power law (2.2.80) breaks down. The measured strain rate is greater than the Eq. (2.2.80) predicts. Within the range of the power-law break down a transition from the climb-plus-glide to the glide mechanism is assumed [120]. The following empirical equation can be applied, e.g. [120, 228],

$$\dot{\varepsilon}^{cr}_{eq} \propto \left[\sinh\left(\alpha\frac{\sigma_{eq}}{G}\right)\right]^n \exp\left(-\frac{Q}{RT}\right), \tag{2.2.81}$$

where α is a material constant. If $\alpha\sigma_{eq}/G < 1$ then (2.2.81) reduces to (2.2.80).

At higher temperatures ($T/T_m > 0.7$) diffusion mechanisms control the creep rate. The deformation occurs at much lower stresses and results from diffusion of vacancies. The mechanism of grain boundary diffusion (Coble creep) assumes diffusive transport of vacancies through and around the surfaces of grains. The deviatoric part of the stress tensor changes the chemical potential of atoms at the grain boundaries. Because of different orientations of grain boundaries a potential gradient occurs. This gradient is the driving force for the grain boundary diffusion. The diffusion through the matrix (bulk diffusion) is the dominant creep mechanism (Nabarro-Herring creep) for temperatures close to the melting point. For details concerning the Coble and the Nabarro-Herring creep models we refer to [119, 228]. These models predict the diffusion controlled creep rate to be a linear function of the stress.

In addition to the dislocation and the diffusion creep, the grain boundary sliding is the important mechanism for poly-crystalline materials. This mechanism occurs because the grain boundaries are weaker than the ordered crystalline structure of the grains [228, 278]. Furthermore, the formation of voids and micro-cracks on grain boundaries contributes to the sliding. The whole deformation rate depends on the grain size and the grain aspect ratio (ratio of the grain dimensions parallel and perpendicular to the tensile stress direction). Samples with a larger grain size usually exhibit a lower strain rate.

2.3 Primary Creep and Creep Transients

In structural analysis applications it is often desirable to consider stress redistributions from the beginning of the creep process up to the creep with constant rate. Let us note, that in a statically undetermined structure stress redistributions take place even if primary creep is ignored. In the case of rapid changes of external loading one must take into account transient effects of the material behavior. Let us discuss some experimental results related to creep under variable multi-axial loading conditions. The majority of multi-axial creep tests have been performed on thin-walled tubes under combined action of tension (compression) force and torque. In this case the uniform stress state $\boldsymbol{\sigma} = \sigma \boldsymbol{n} \otimes \boldsymbol{n} + \tau(\boldsymbol{n} \otimes \boldsymbol{m} + \boldsymbol{m} \otimes \boldsymbol{n})$ is assumed, where σ and τ are calculated from the force and torque as well as the geometry of the cross section (see Sect. 1.1.2). Figure 2.4 presents a sketch of experimental data for type 304 steel ($2\frac{1}{4}$Cr-1Mo) at 600°C [151]. A tube was loaded the first 5 hours by the constant tension force and the constant torque. After that the direction of the force was reversed while the torque kept constant. The normal strain vs. time creep curve under compressive force after the reversal differs substantially from the reference creep curve under tensile force, Fig. 2.4a. The absolute value of the strain rates before and after the reversal differs significantly. Furthermore, the shear strain vs. time creep curve is influenced by the reversal of the axial force, Fig. 2.4b.

Figure 2.5 shows a sketch of experimental results obtained in [255] for INCONEL Alloy 617 (NiCr22Co12Mo) tubes at 900°C under cyclic torsion. Every 100 h the applied torque was reversed leading to the change of the sign of the shear stress. The inelastic shear strain accumulated after each cycle of positive (negative) torque decreases rapidly after few cycles of reversals. Similar behavior is reported

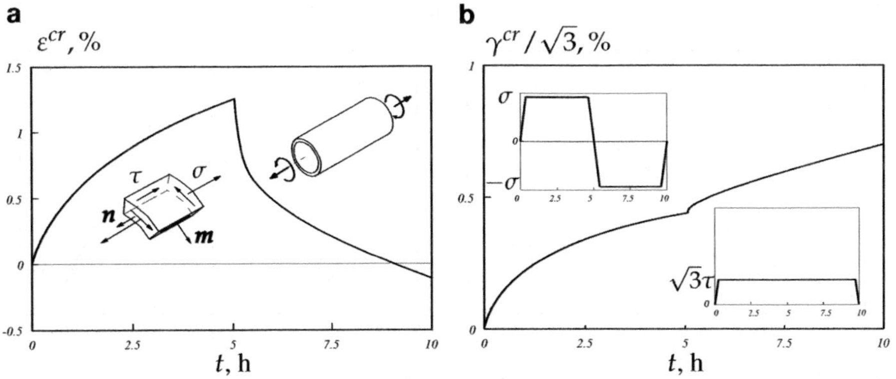

Fig. 2.4 Transient creep at combined tension and torsion. Effect of the normal stress reversal. **a** Normal strain vs. time, **b** shear strain vs. time (after [151])

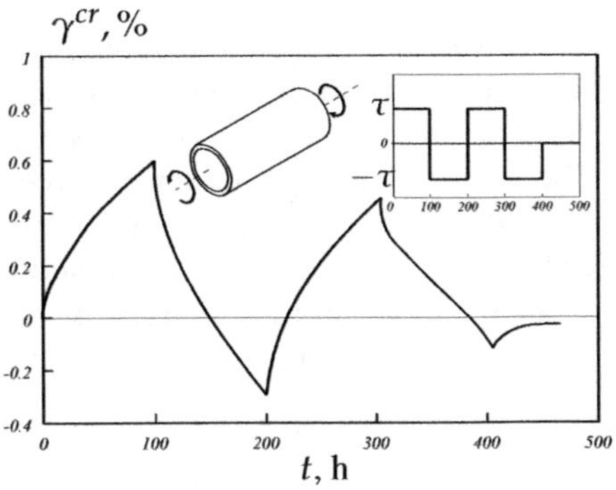

Fig. 2.5 Creep under shear stress reversals (after [255])

in [245] for the type 304 steel, where, in addition, the effect of thermal exposure before and during the loading is discussed. Creep behavior of steels is usually accompanied by the thermally induced evolution of structure of carbide precipitates (coarsening or new precipitation). The effect of ageing has a significant influence on the transient creep of steels as discussed in [245]. For example, the decrease of inelastic shear strain under alternating torsion was not observed if tubular specimens were subjected to the thermal exposure within the time interval of 500 h before the loading.

Additional effects have been observed in the case of reversals of the applied torque combined with the constant tension force, Fig. 2.6. First, the axial strain response is significantly influenced by the cyclic torsion. Second, the rate of the shear strain depends on the sign of the applied torque. Such a response indicates the anisotropic nature of the hardening processes.

Multi-axial creep behavior is significantly influenced by the deformation history. As an example, Fig. 2.7 presents a sketch of results reported in [160] for type 304 stainless steel. Tubular specimens were first loaded up to the stress σ_1 leading to

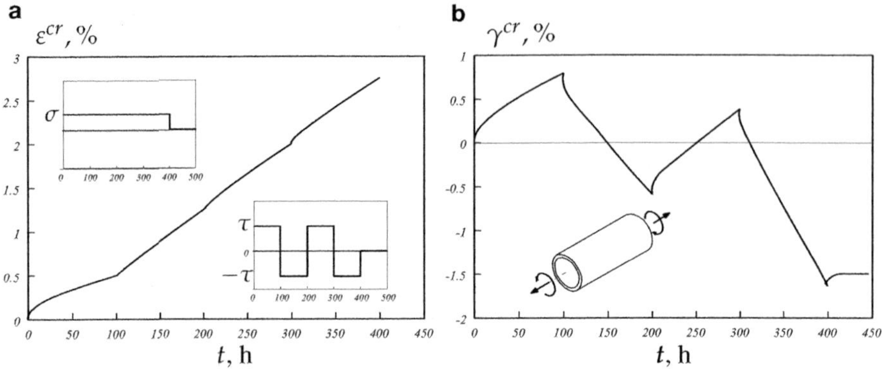

Fig. 2.6 Creep at combined tension and torsion - effect of the shear stress reversals. **a** Normal strain vs. time, **b** shear strain vs. time (after [255])

the plastic strain of 3%. After that the specimens were unloaded to σ_0. Subsequent creep tests have been performed under combined constant normal strain σ and shear strain τ. Different stress states leading to the same value of the von Mises stress $\sigma_{vM} = \sqrt{\sigma^2 + 3\tau^2} = \sigma_0$ were realized. The results show that the tensile creep curve of the material after plastic pre-straining differs significantly from the creep curve of the "virgin material" (curve a). Furthermore, the von Mises creep strain vs. time curves after plastic pre-straining depend significantly on the type of the applied stress state (compare, for example, tension, curve a, torsion, curve b, and compression, curve e).

In this section we discuss phenomenological models to describe primary creep and creep transients under multi-axial stress states. We start with models of time and strain hardening. After that we introduce the concept of kinematic hardening which is widely used for the characterization of transient creep effects under constant and varying loading. Our purpose is to discuss general ideas rather than enter into details of empirical functions of stress and temperature as well as different forms of evolution equations for hardening variables (the so-called hardening rules). Regarding the hardening rules one may consult the comprehensive reviews [90, 244] and monographs [179, 191, 214, 309]. For classification and assessment of different unified models of plasticity-creep interaction we refer to [151, 152].

2.3.1 Time and Strain Hardening

The time hardening model assumes a relationship between the equivalent creep rate, the equivalent stress and the time at fixed temperature, i.e.

$$f_t(\dot{\varepsilon}^{cr}_{eq}, \sigma_{eq}, t) = 0$$

The strain hardening model postulates a relationship between the equivalent creep rate, the equivalent creep strain and the equivalent stress at fixed temperature. In this case

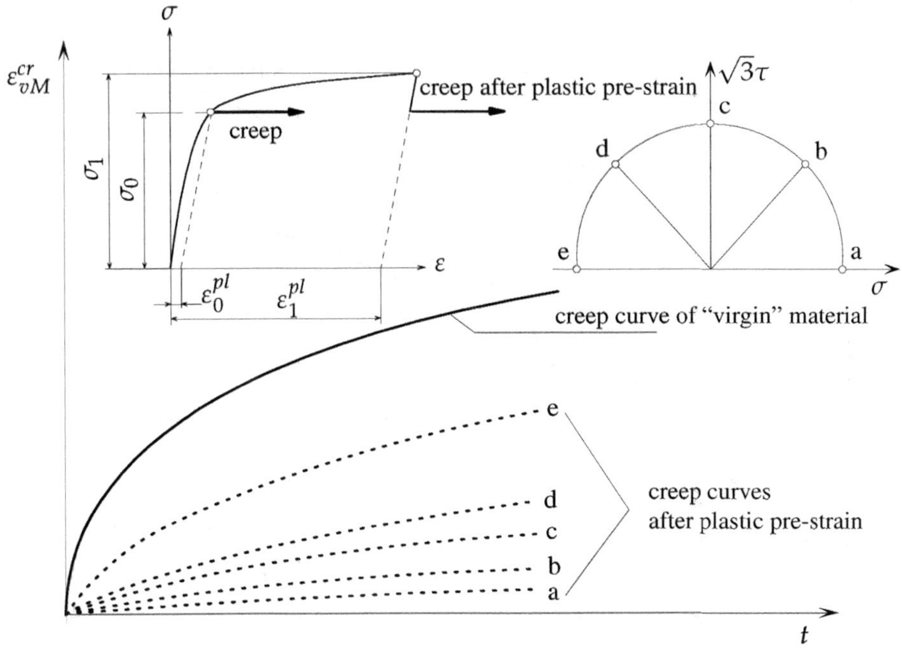

Fig. 2.7 Effect of initial tensile plastic strain on subsequent creep behavior under combined tension and torsion, for details see [160]

$$f_s(\dot{\varepsilon}^{cr}_{eq}, \varepsilon^{cr}_{eq}, \sigma_{eq}) = 0$$

Figure 2.8 illustrates the uni-axial creep response after reloading (stress jump from σ_1 to σ_2 at $t = t_r$). Based on the time hardening model the strain rate at $t \geq t_r$ is determined by the stress σ_2 and the time t_r only. Thus the creep curve for $t \geq t_r$ can be obtained by translation of the curve BC to the point D. Following the strain hardening model the strain rate depends on the stress and the accumulated strain. The creep curve after the stress jump can be determined by translating the curve AC (the creep curve for the stress σ_2 starting from the creep strain ε^{cr}_A accumulated in time t_r) along the time axis. It can be shown that for specific functions of stress, time and strain as well as under the assumption of the constant stress and temperature the strain and the time hardening models lead to the same description. For example, if we set

$$\dot{\varepsilon}^{cr}_{eq} = a\sigma^n_{eq}t^m \tag{2.3.1}$$

according to the time hardening with a, n and m as the material constants the integration with respect to the time variable assuming $\sigma_{eq} = const$ and $\varepsilon^{cr}_{eq} = 0$ at $t = 0$ leads to

$$\varepsilon^{cr}_{eq} = a\sigma^n_{eq}\frac{1}{m+1}t^{m+1} \tag{2.3.2}$$

On the other hand applying the strain hardening model, the creep equation can be formulated as

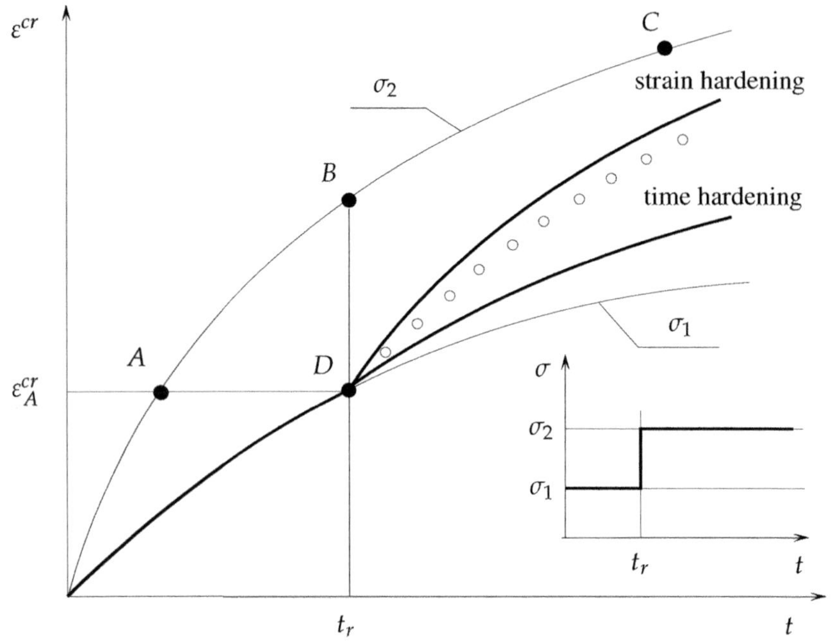

Fig. 2.8 Creep response at variable loading (the open circles denote typical experimental values)

$$\dot{\varepsilon}^{cr}_{eq} = b\sigma^k_{eq}(\varepsilon^{cr}_{eq})^l \tag{2.3.3}$$

Taking into account (2.3.2) the time variable can be eliminated from (2.3.1). As a result the following relations between the material constants can be obtained

$$b = [a(m+1)^m]^{\frac{1}{m+1}}, \quad k = \frac{n}{m+1}, \quad l = \frac{m}{m+1}$$

Vice versa, the strain hardening equation (2.3.2) can be integrated for the special choice of k and l and for $\sigma_{eq} = const$. Again, if $\varepsilon^{cr}_{eq} = 0$ at $t = 0$ we obtain (2.3.2).

Applying the time hardening model the von Mises-Odqvist creep theory (see Sect. 2.2) can be generalized as follows

$$\dot{\boldsymbol{\varepsilon}}^{cr} = \frac{3}{2}a\sigma^{n-1}_{vM}t^m\mathbf{s} \tag{2.3.4}$$

By analogy one can formulate the creep constitutive equation with the strain hardening

$$\dot{\boldsymbol{\varepsilon}}^{cr} = \frac{3}{2}b\sigma^{k-1}_{vM}(\varepsilon^{cr}_{vM})^l\mathbf{s} \tag{2.3.5}$$

The time and the strain hardening models provide simple empirical description of the uni-axial creep curve within the range of primary creep and are still popular in characterizing the material behavior, e.g. [140, 148, 176]. Despite the simplicity, both the models suffer from significant limitations, even if applied stress and temperature are constant. The disadvantage of the time hardening model is that the time variable appears explicitly in equation (2.3.1) for the creep rate. An additional drawback is that the constants m and l take usually the values $-1 < m < 0$, $-1 < l < 0$ as the result of curve fitting. If $\varepsilon^{cr}_{eq} = 0$ at $t = 0$ then Eq. (2.3.3) provides an infinite starting creep rate. One can avoid this problem in a time-step based numerical procedure assuming a small non-zero creep equivalent strain at the starting time step. Finally, both models can be applied only for the case of the constant or slowly varying stresses. Transient creep effects under rapid changes of loading and particularly in the case of stress reversals cannot be described.

Further details of time and strain hardening models can be found in [178, 257]. In [178] a modified strain hardening model is proposed based on the idea of creep strain origins.

2.3.2 Kinematic Hardening

The common approach in describing transient creep effects under complex loading paths is the introduction of internal state variables and appropriate evolution equations (the so-called hardening rules). The scalar-valued internal state variables are applied to characterize isotropic hardening and ageing processes. An example will be discussed in Sect. 2.4.1.3. Several "non-classical" effects observed in tests under non-proportional loading have motivated the use of tensor-valued variables (usually second rank tensors).

The idea of kinematic hardening (translation of the yield surface in the stress space) originates from the theory of plasticity and has been proposed by Prager [264]. In the creep mechanics the kinematic hardening was introduced by Malinin and Khadjinsky [209, 210]. The starting point is the additive decomposition of the stress tensor into two parts: $\boldsymbol{\sigma} = \bar{\boldsymbol{\sigma}} + \boldsymbol{\alpha}$, where $\bar{\boldsymbol{\sigma}}$ is called the active or the effective part of the stress tensor and $\boldsymbol{\alpha}$ denotes the additional or translation part of the stress tensor (back stress tensor). The introduced tensors can be further decomposed into spherical and deviatoric parts

$$\begin{aligned} \bar{\boldsymbol{\sigma}} &= \frac{1}{3}\operatorname{tr}\bar{\boldsymbol{\sigma}}\boldsymbol{I} + \bar{\boldsymbol{s}}, & \operatorname{tr}\bar{\boldsymbol{s}} &= 0, \\ \boldsymbol{\alpha} &= \frac{1}{3}\operatorname{tr}\boldsymbol{\alpha}\boldsymbol{I} + \boldsymbol{\beta}, & \operatorname{tr}\boldsymbol{\beta} &= 0, \\ \boldsymbol{\sigma} &= \frac{1}{3}(\operatorname{tr}\bar{\boldsymbol{\sigma}} + \operatorname{tr}\boldsymbol{\alpha})\boldsymbol{I} + \boldsymbol{s}, & \boldsymbol{s} &= \bar{\boldsymbol{s}} + \boldsymbol{\beta} \end{aligned} \tag{2.3.6}$$

It is assumed that the inelastic strain rate is determined by the active part of the stress tensor. The creep potential is then a function of the active part of the stress tensor, i.e. $W = W(\bar{\boldsymbol{\sigma}}) = W(\boldsymbol{\sigma} - \boldsymbol{\alpha})$, e.g. [252]. As in the case of the classical isotropic creep (Sect. 2.2.1.1) only the second invariant of the deviator $\bar{\boldsymbol{s}}$ is considered. Introducing the von Mises equivalent stress

$$\bar{\sigma}_{vM} \equiv \sqrt{\frac{3}{2}\bar{\boldsymbol{s}} \cdot\cdot\, \bar{\boldsymbol{s}}} = \sqrt{\frac{3}{2}(\boldsymbol{s} - \boldsymbol{\beta}) \cdot\cdot\, (\boldsymbol{s} - \boldsymbol{\beta})} \tag{2.3.7}$$

the flow rule (2.1.6) leads to the following constitutive equation

$$\dot{\boldsymbol{\varepsilon}}^{cr} = \frac{3}{2}\frac{\dot{\varepsilon}^{cr}_{vM}}{\bar{\sigma}_{vM}}\bar{\boldsymbol{s}}, \quad \dot{\varepsilon}_{vM} \equiv \sqrt{\frac{2}{3}\dot{\boldsymbol{\varepsilon}}^{cr} \cdot\cdot\, \dot{\boldsymbol{\varepsilon}}^{cr}} \tag{2.3.8}$$

The equivalent creep rate can be expressed by the use of stress and temperature functions discussed in Sect. 2.2.3. For example, with the power law stress function and the Arrhenius temperature dependence

$$\dot{\varepsilon}^{cr}_{vM} = a\bar{\sigma}^{n}_{vM}, \quad a = a_0 \exp\left(-\frac{Q}{RT}\right) \tag{2.3.9}$$

Equations (2.3.8) contain the deviatoric part of the back stress $\boldsymbol{\beta}$. This internal state variable is defined by the evolution equation and the initial condition. In [207, 208] the following evolution equation is postulated

$$\dot{\boldsymbol{\beta}} = \frac{2}{3}b\dot{\boldsymbol{\varepsilon}}^{cr} - \frac{g(\alpha_{vM})}{\alpha_{vM}}\boldsymbol{\beta} \tag{2.3.10}$$

with

$$\alpha_{vM} \equiv \sqrt{\frac{3}{2}\boldsymbol{\beta} \cdot\cdot\, \boldsymbol{\beta}}$$

For the function g various empirical relations were proposed. One example is [207, 208]

$$g(\alpha_{vM}) = c\alpha^{n}_{vM}, \quad c = c_0 \exp\left(-\frac{Q_r}{RT}\right)$$

Equation (2.3.10) is the multi-axial utilization of the Bailey-Orowan recovery hypothesis, see Sect. 2.2.3, b and c_0 are material constants and Q_r is the activation energy of recovery.

Let us show how the model behaves for the uni-axial homogeneous stress state $\boldsymbol{\sigma}(t) = \sigma(t)\boldsymbol{n} \otimes \boldsymbol{n}$, where $\sigma(t)$ is the magnitude of the applied stress and $\boldsymbol{n}$ is the unit vector. With $\boldsymbol{\alpha}(0) = \boldsymbol{0}$ one can assume that $\boldsymbol{\alpha}(t)$ is coaxial with the stress tensor. Therefore one can write [207, 208]

$$\boldsymbol{\alpha} = \alpha\boldsymbol{n} \otimes \boldsymbol{n}, \quad \boldsymbol{\beta} = \alpha\left(\boldsymbol{n} \otimes \boldsymbol{n} - \frac{1}{3}\boldsymbol{I}\right), \quad \bar{\sigma}_{vM} = |\sigma - \alpha|, \quad \alpha_{vM} = |\alpha|$$

From Eqs. (2.3.9) and (2.3.10) follows

$$\begin{aligned} \dot{\varepsilon}^{cr} &= a\,\mathrm{sign}(\sigma - \alpha)|\sigma - \alpha|^{n}, \quad \dot{\varepsilon}^{cr} \equiv \boldsymbol{n} \cdot \dot{\boldsymbol{\varepsilon}}^{cr} \cdot \boldsymbol{n}, \\ \dot{\alpha} &= b\dot{\varepsilon}^{cr} - c\,\mathrm{sign}\alpha|\alpha|^{n} \end{aligned} \tag{2.3.11}$$

Let us assume that $\sigma(t) = \sigma_0 > 0$, $\alpha(0) = 0$, $\sigma_0 - \alpha > 0$. In addition, let us introduce the variable $H = \alpha/\sigma_0$. From (2.3.11) we obtain

$$\begin{aligned} \dot{\varepsilon}^{cr} &= a\sigma_0^{n}(1 - H)^{n}, \\ \dot{H} &= \sigma_0^{n-1}[ba(1 - H)^{n} - cH^{n}] \end{aligned} \tag{2.3.12}$$

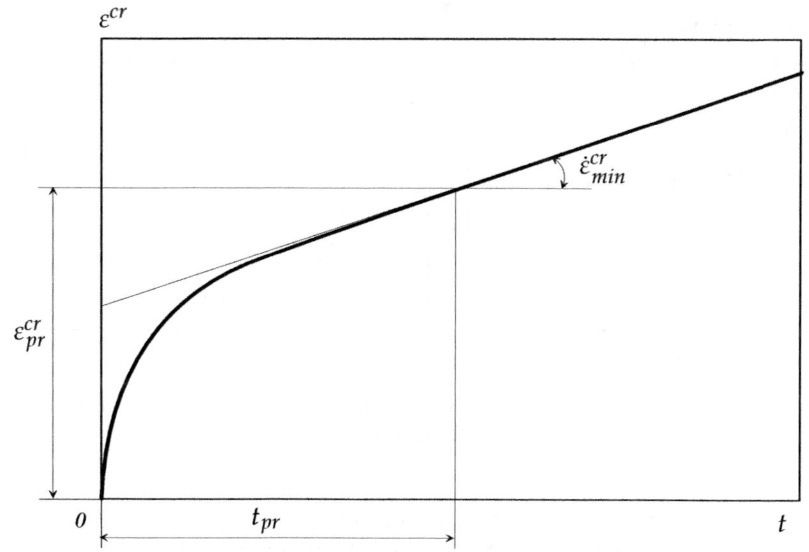

Fig. 2.9 Primary and secondary creep stages of a uni-axial creep curve

The constitutive and evolution Eqs. (2.3.12) describe the primary and the secondary stages of a uni-axial creep curve, Fig. 2.9. In the considered case of the uni-axial tension the parameter $0 \leq H < H_* < 1$ is equal to zero at the beginning of the creep process and increases over time. In the steady state $H = H_*$, where H_* is the saturation value. From the second equation in (2.3.12) we obtain

$$H_* = \frac{1}{1 + \mu^{\frac{1}{n}}}, \quad \mu \equiv \frac{c}{ab} \tag{2.3.13}$$

The minimum creep rate in the steady state is calculated by

$$\dot{\varepsilon}^{cr}_{min} = a\sigma_0^n (1 - H_*)^n = \tilde{a}\sigma_0^n, \quad \tilde{a} \equiv a(1 - H_*)^n \tag{2.3.14}$$

The constants $\tilde{a}$ and n can be obtained from the experimental data of steady-state creep. For the given value of H_* the second equation in (2.3.12) can be integrated providing the duration time of primary creep t_{pr} (see Fig. 2.9)

$$t_{pr} = \frac{\varphi(H_*)}{ba\sigma_0^n}, \quad \varphi(H_*) = \int\limits_0^{H_*} \frac{dH}{(1 - H)^n - \mu H^n}$$

From the first equation in (2.3.12) the creep strain ε^{cr}_{pr} follows at $t = t_{pr}$ (see Fig. 2.9) as

$$\varepsilon^{cr}_{pr} = \frac{\sigma_0}{b} \int\limits_0^{H_*} \frac{(1 - H)^n dH}{(1 - H)^n - \mu H^n}$$

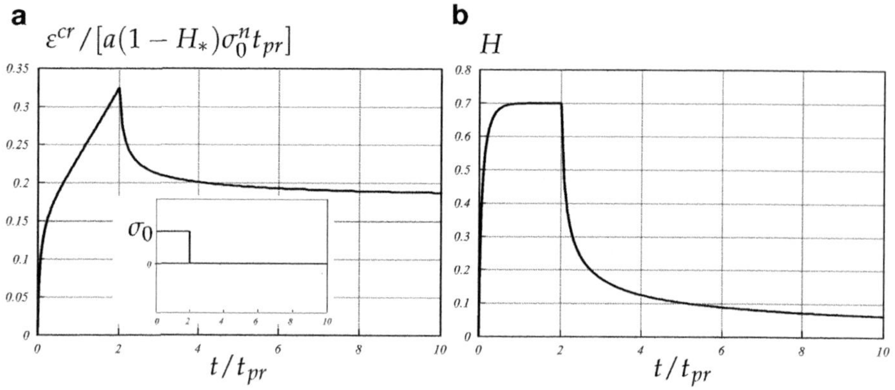

Fig. 2.10 Uni-axial creep after unloading - simulations based on Eqs. (2.3.15) for the case $n = 3$ and $H_* = 0.7$. **a** Creep strain vs. time, **b** hardening variable vs. time.

The above equations can be used for the identification of material constants.

To discuss the model predictions for the case of the uni-axial cyclic loading let us introduce the following dimensionless variables

$$\tilde{\sigma} = \frac{\sigma(t)}{\sigma_0}, \quad \tau = \frac{t}{t_{pr}}, \quad \epsilon = \frac{\varepsilon^{cr}}{a(1-H_*)\sigma_0^n t_{pr}},$$

where σ_0 denotes the constant stress value in the first loading cycle. Equations (2.3.11) take the form

$$\begin{aligned} \frac{d\epsilon}{d\tau} &= a\text{sign}(\tilde{\sigma} - H)\frac{|\tilde{\sigma} - H|^n}{1 - H*}, \\ \frac{dH}{d\tau} &= \varphi(H_*)\left[\text{sign}(\tilde{\sigma} - H)|\tilde{\sigma} - H|^n - \text{sign}(H)\left(\frac{1-H_*}{H_*}\right)^n |H|^n\right] \end{aligned} \qquad (2.3.15)$$

Figures 2.10 and 2.11 illustrate the results of the numerical integration of (2.3.15) with $n = 3$, $H_* = 0.7$ and the initial conditions $\epsilon(0) = 0$ and $H(0) = 0$. In the first case presented in Fig. 2.10 we assume $\sigma = \sigma_0$ within the time interval $[0; 2t_{pr}]$, so that the hardening variable increases up to the saturation value and remains constant. The creep curve exhibits both the primary and the secondary stages, Fig. 2.10a. At $t = 2t_{pr}$ we assume a spontaneous unloading, i.e. $\sigma = 0$. We observe that the model (2.3.15) is able to describe the creep recovery (see Fig. 1.3b). Figure 2.11 presents the numerical results for the case of cyclic loading. Three loading cycles with the constant stresses $\pm\sigma_0$ and the holding time $\Delta t = 2t_{pr}$, Fig. 2.11a, are considered. We observe that the model (2.3.15) predicts identical creep responses for the first and the third loading cycle.

Let us give some comments on the model predictions under multi-axial stress states. To this end we consider the loading case with the constant stress deviator $\boldsymbol{s}$ within a given interval of time $[t_0, t]$. Equations (2.3.8) and (2.3.10) can be rewritten as follows

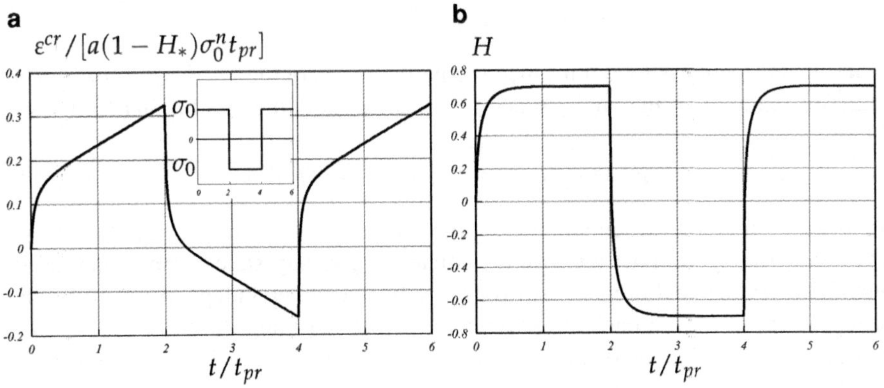

Fig. 2.11 Uni-axial creep under cyclic loading - simulations based on Eqs. (2.3.15) for the case $n = 3$ and $H_* = 0.7$. **a** Creep strain vs. time, **b** hardening variable vs. time.

$$
\begin{aligned}
\dot{\boldsymbol{\varepsilon}}^{cr} &= \frac{3}{2}\frac{f(\bar{\sigma}_{vM})}{\bar{\sigma}_{vM}}(\mathbf{s} - \boldsymbol{\beta}),\\
\dot{\boldsymbol{\beta}} &= b\frac{f(\bar{\sigma}_{vM})}{\bar{\sigma}_{vM}}(\mathbf{s} - \boldsymbol{\beta}) - \frac{g(\alpha_{vM})}{\alpha_{vM}}\boldsymbol{\beta}
\end{aligned}
\qquad (2.3.16)
$$

In the steady creep state $\boldsymbol{\beta} = \boldsymbol{\beta}_*$, where $\boldsymbol{\beta}_*$ is the saturation value of the back stress deviator. From the second equation in (2.3.16) it follows

$$
b\frac{f(\bar{\sigma}_{vM_*})}{\bar{\sigma}_{vM_*}}(\mathbf{s} - \boldsymbol{\beta}_*) = \frac{g(\alpha_{vM_*})}{\alpha_{vM_*}}\boldsymbol{\beta}_*, \qquad (2.3.17)
$$

where

$$
\bar{\sigma}_{vM_*} = \sqrt{\frac{3}{2}(\mathbf{s} - \boldsymbol{\beta}_*) \cdot\cdot (\mathbf{s} - \boldsymbol{\beta}_*)}, \quad \alpha_{vM_*} = \sqrt{\frac{3}{2}\boldsymbol{\beta}_* \cdot\cdot \boldsymbol{\beta}_*}
$$

The double inner product of (2.3.17) with itself results in

$$
[bf(\bar{\sigma}_{vM_*})]^2 = [g(\alpha_{vM_*})]^2
$$

Since $f(\bar{\sigma}_{vM_*}) > 0$ and $g(\alpha_{vM_*}) > 0$ we obtain

$$
bf(\bar{\sigma}_{vM_*}) = g(\alpha_{vM_*}) \qquad (2.3.18)
$$

From (2.3.17) it follows

$$
\boldsymbol{\beta}_* = \frac{\alpha_{vM_*}}{\bar{\sigma}_{vM_*} + \alpha_{vM_*}}\mathbf{s} \quad \Rightarrow \quad \bar{\sigma}_{vM_*} = \sigma_{vM} + \alpha_{vM_*} \qquad (2.3.19)
$$

Now the steady state value of the back stress deviator can be calculated

$$
\boldsymbol{\beta}_* = \alpha_{vM_*}\frac{\mathbf{s}}{\sigma_{vM}} \qquad (2.3.20)
$$

Let us assume power functions for f and g. Then from (2.3.18) we obtain

$$ba(\sigma_{vM} - \alpha_{vM_*})^n = c\alpha_{vM_*}^n$$

As in the uni-axial case we introduce the hardening variable $H = \alpha_{vM}/\sigma_{vM}$. The saturation value H_* is then determined by (2.3.13). From the first Eq. in (2.3.16) we obtain

$$\dot{\boldsymbol{\varepsilon}}_{st}^{cr} = \frac{3}{2}\tilde{a}\sigma_{vm}^{n-1}\boldsymbol{s}, \quad \tilde{a} \equiv a(1 - H_*)^n \tag{2.3.21}$$

We observe that the kinematic hardening model (2.3.16) results in the classical Norton-Bailey-Odqvist constitutive equation of steady-state creep discussed in Sect. 2.2.1. This model predicts isotropic steady-state creep independently from the initial condition for the back stress deviator $\boldsymbol{\beta}$. Furthermore, different stress states leading to the same value of the von Mises equivalent stress will provide the same steady state value of the equivalent creep rate.

As an example let us consider a thin-walled tube subjected to the axial force ant the torque (see Sect. 1.1.2). Let us assume that in the first loading cycle the force and the torque are kept constant over a certain period of time such that the creep rates attain the steady state values. The stress deviator has the following form

$$\boldsymbol{s}_1 = \sigma(\boldsymbol{k} \otimes \boldsymbol{k} - \frac{1}{3}\boldsymbol{I}) + \tau(\boldsymbol{e}_\varphi \otimes \boldsymbol{k} + \boldsymbol{k} \otimes \boldsymbol{e}_\varphi), \tag{2.3.22}$$

where σ is the normal stress, τ is the shear stress and the unit vectors $\boldsymbol{k}$ and $\boldsymbol{e}_\varphi$ designate the axial and the circumferential direction, respectively (cp. Fig. 1.7). In addition, we assume that in the second loading cycle the tube is loaded by the same tensile force but the reversed constant shear stress. In this case the stress deviator is

$$\boldsymbol{s}_2 = \sigma(\boldsymbol{k} \otimes \boldsymbol{k} - \frac{1}{3}\boldsymbol{I}) - \tau(\boldsymbol{e}_\varphi \otimes \boldsymbol{k} + \boldsymbol{k} \otimes \boldsymbol{e}_\varphi), \tag{2.3.23}$$

From Eqs. (2.3.22) and (2.3.23) we find that

$$\boldsymbol{s}_2 = \boldsymbol{Q} \cdot \boldsymbol{s}_1 \cdot \boldsymbol{Q}^T, \quad \boldsymbol{Q} = 2\boldsymbol{k} \otimes \boldsymbol{k} - \boldsymbol{I} \tag{2.3.24}$$

The kinematic hardening model (2.3.16) predicts the following relation between the steady state creep rates in the first and the second loading cycle

$$\dot{\boldsymbol{\varepsilon}}_{st_2}^{cr} = \boldsymbol{Q} \cdot \dot{\boldsymbol{\varepsilon}}_{st_1}^{cr} \cdot \boldsymbol{Q}^T$$

Consequently, the normal strain rates and the magnitudes of the shear strain rates will be the same at the end of the loading cycles.

The model (2.3.16) is applied in [208, 252] for the description of creep for different materials under simple or non-proportional loading conditions. It is demonstrated that the predictions agree with experimental results. However, in many cases deviations from the Norton-Bailey-Odqvist type steady-state creep can be observed in experiments. For example, in the case shown in Fig. 2.6 the steady state shear creep rate changes significantly after the shear stress reversals, although the von Mises equivalent stress remains constant. The results presented in Fig. 2.7 indicate that the initial hardening state due to plastic pre-strain is the reason for the stress state dependence of the subsequent creep behavior. This effect cannot be described by the model (2.3.16).

The models with the back stress of the type (2.3.16) are usually termed to be the models with anisotropic hardening, e.g. [208]. The type of anisotropy is then determined by the symmetry group of the back stress tensor or deviator. The symmetry group of any symmetric second rank tensor includes always eight elements, e.g. [205]. For the tensor $\boldsymbol{\beta}$ the symmetry elements are

$$\boldsymbol{Q}_{\beta} = \pm \boldsymbol{n}_1 \otimes \boldsymbol{n}_1 \pm \boldsymbol{n}_2 \otimes \boldsymbol{n}_2 \pm \boldsymbol{n}_3 \otimes \boldsymbol{n}_3, \qquad (2.3.25)$$

where $\boldsymbol{n}_i$ are the principal axes. The creep potential formulated in terms of the stress deviator and the back stress deviator satisfies the following condition

$$W(\boldsymbol{s}, \boldsymbol{\beta}) = W(\boldsymbol{Q}_{\beta} \cdot \boldsymbol{s} \cdot \boldsymbol{Q}_{\beta}^T, \boldsymbol{Q}_{\beta} \cdot \boldsymbol{\beta} \cdot \boldsymbol{Q}_{\beta}^T) = W(\boldsymbol{Q}_{\beta} \cdot \boldsymbol{s} \cdot \boldsymbol{Q}_{\beta}^T, \boldsymbol{\beta})$$

Consequently the creep potential is an orthotropic function of the stress deviator with the symmetry group defined by (2.3.25). As shown in [75] the kinematic hardening models of the type (2.3.16) leads to a restrictive form of orthotropic inelastic behavior. In order to demonstrate this let us write down the back stress deviator in the following form

$$\begin{aligned} \boldsymbol{\beta} &= \beta_1 \boldsymbol{n}_1 \otimes \boldsymbol{n}_1 + \beta_2 \boldsymbol{n}_2 \otimes \boldsymbol{n}_2 - (\beta_1 + \beta_2) \boldsymbol{n}_3 \otimes \boldsymbol{n}_3 \\ &= \beta_1 (\boldsymbol{n}_1 \otimes \boldsymbol{n}_1 - \boldsymbol{n}_3 \otimes \boldsymbol{n}_3) + \beta_2 (\boldsymbol{n}_2 \otimes \boldsymbol{n}_2 - \boldsymbol{n}_3 \otimes \boldsymbol{n}_3), \end{aligned}$$

where β_1 and β_2 are the principal values and $\boldsymbol{n}_1$, $\boldsymbol{n}_2$ and $\boldsymbol{n}_3$ are the principal directions of $\boldsymbol{\beta}$. For the given back stress deviator $\boldsymbol{\beta}$ the equivalent stress (2.3.7) takes the form

$$\begin{aligned} \bar{\sigma}_{vM}^2 &= 3\tilde{J}_1^2 \left(1 - \frac{\beta_1}{\tilde{J}_1}\right)^2 + 3\tilde{J}_2^2 \left(1 - \frac{\beta_2}{\tilde{J}_2}\right)^2 + \frac{3}{2}\tilde{J}_1 \tilde{J}_2 \left(1 - \frac{\beta_1}{\tilde{J}_1}\right)\left(1 - \frac{\beta_2}{\tilde{J}_2}\right) \\ &+ 3I_{\boldsymbol{n}_1\boldsymbol{n}_2}^2 + 3I_{\boldsymbol{n}_1\boldsymbol{n}_3}^2 + 3I_{\boldsymbol{n}_2\boldsymbol{n}_3}^2, \end{aligned} \qquad (2.3.26)$$

where the invariants $\tilde{J}_i$ are defined by Eqs. (2.2.53) and the invariants $I_{\boldsymbol{n}_i\boldsymbol{n}_j}$ are defined by Eqs. (2.2.48). Steady state creep with initial orthotropic symmetry is discussed in Sect. 2.2.2. In this case the von Mises type equivalent stress includes 6 invariants and 6 independent material constants. The equivalent stress (2.3.26) contains all 6 orthotropic invariants. However, the last three terms (three shear stresses with respect to the three planes of the orthotropic symmetry) are not affected by the hardening. Furthermore, in the steady state range these terms vanish since the back stress deviator $\boldsymbol{\beta}_*$ is coaxial with the stress deviator according to (2.3.20).

The possibilities to improve the predictions of the kinematic hardening model are:

- Introduction of additional state variables like isotropic hardening variable, e.g. [90], ageing variable, e.g. [245], or damage variables, e.g. [104]. Models with damage variables will be discussed in Sect. 2.4.

- Formulation of the creep potential as a general isotropic function of two tensors $\boldsymbol{\sigma}$ and $\boldsymbol{\alpha}$. Such an approach is proposed in [75] for the case of plasticity and includes different special cases of kinematic hardening,
- Consideration of the initial anisotropy of the material behavior, e.g. [151].

Creep models with kinematic hardening of the type (2.3.8) and different specific forms of the hardening evolution equation are discussed in [161, 162, 208, 245, 252, 279] among others. For the description of creep and creep-plasticity interaction at complex loading conditions a variety of unified models is available including the hardening variables as second rank tensors. For details we refer to [179, 181, 191, 214]. Several unified models are reviewed and evaluated in [151, 152]. The historical background of the development of non-linear kinematic hardening rules is presented in [90].

2.4 Tertiary Creep and Creep Damage

Tertiary creep stage is the final part of the creep process. In a uni-axial creep curve tertiary creep is observed as the increase of the creep rate. The shape of the final part of the creep curve and the duration of the tertiary creep stage depend on the material composition, the stress level and the temperature. For some structural steels, the tertiary creep is the major part of the whole creep process, e.g. [108, 249].

The origins of tertiary creep are progressive damage processes including the formation, growth and coalescence of voids on grain boundaries, coarsening of precipitates and environmental effects. The voids may nucleate earlier during the creep process, possibly at primary creep stage or even after spontaneous deformation. The initially existing micro-defects have negligible influence on the creep rate. As their number and size increase with time, they weaken the material providing the decrease in the load-bearing capacity. The coalescence of cavities or propagation of micro-cracks lead to the final fracture. Creep fracture is usually inter-granular [37]. Dyson [102] distinguishes three main categories of creep damage: the strain induced damage, the thermally induced damage and the environmentally induced damage. The strain induced damage may be classified as follows [104]

- excessive straining at constant load,
- grain boundary cavitation and
- progressive multiplication of the dislocation substructure

The first two damage mechanisms occur in all poly-crystalline materials, while the third one is essential for nickel-based super-alloys.

The thermally induced damage mechanisms include material ageing processes. They may lead to the loss of strength and contribute to the nucleation and growth of cavities. An example of the thermally induced ageing for ferritic steels includes the coarsening of carbide precipitates (increase of volume fraction of carbide precipitates or new precipitation), e.g. [258]. The rate of ageing does not depend on the applied stress, but is influenced by the temperature. It can be identified by exposing test-pieces to thermal environment.

The environmentally induced damage (corrosion, oxidation, etc.) appears due to the attack of chemical species contained within the surrounding medium. The environmental damage rate can be inversely related to the test-piece (component) dimensions [102].

The dominance of a creep damage mechanism depends on the alloy composition, on the fabrication route and on the service conditions. For several metals and alloys, fracture mechanism maps are available [37]. By analogy with the deformation mechanism maps, regions with different fracture modes are indicated depending on the stress and the temperature ranges.

Physical modeling of creep damage is complicated by the fact that many different mechanisms may operate and interact in a specific material under given loading conditions. This interaction should be taken into account in the damage rate equations. Models related to the grain boundary cavitation are discussed and reviewed in [158, 278].

The characterization of tertiary creep under multi-axial stress states is the important step in a creep analysis of engineering structures. A life-time prediction of a specific load bearing component designed for creep, or a residual life-time estimation of a structure operating at elevated temperature requires a model which takes into account tertiary creep and damage evolution under multi-axial stress states.

The damage rate and consequently the creep rate are determined by the stress level, the accumulated damage and the temperature. These dependencies can be established based on experimental data from the uni-axial creep tests. If the material is subjected to multi-axial loading, the kind of stress state has a significant influence on the damage growth. Tension and compression lead to different creep rates. Different stress states corresponding to the same von Mises equivalent stress lead, in general, to different equivalent tertiary creep rates while the equivalent strain rate in the secondary stage is approximately the same. These facts are established from the data of creep tests under combined tension and torsion, e.g. [174, 175], as well as from biaxial and triaxial creep tests [289, 290]. Stress state effects must be considered in the damage evolution equation. In Sect. 2.4.1 we discuss various possibilities to characterize the tertiary creep behavior by means of scalar-valued damage parameters. Under non-proportional loading conditions, the additional factor is the influence of the damage induced anisotropy. Examples are creep tests under combined tension and alternating torsion, e.g. [224], and creep tests under biaxial loading with alternating direction of the first principal stress [290]. In both cases the assumption of isotropic creep behavior and the scalar measure of damage lead to disagreement with experimental observations. In Sect. 2.4.2 we review some experimental results illustrating the damage induced anisotropy and discuss creep-damage models with tensor-valued damage variables.

2.4.1 Scalar-Valued Damage Variables

Many microstructural observations show the directional effect of creep damage. For example, during a cyclic torsion test on copper, voids nucleate and grow predominantly on those grain boundaries, which are perpendicular to the first principal direction of the stress tensor, e.g. [137]. Creep damage has therefore an anisotropic nature and should be characterized by a tensor. However, if the initially isotropic material is subjected to constant or monotonic loading the influence of the damage anisotropy on the observed creep behavior, i.e. the strain vs. time curves, is not significant. If the state of damage is characterized by a tensor (see Sect. 2.4.2) then such a tensor can be assumed to be coaxial with the stress tensor under monotonic loading conditions. In such a case only the scalar damage measures will enter the creep constitutive equation. Below we introduce different models of tertiary creep including the phenomenological, the so-called micromechanically consistent and mechanism based models. The effect of damage is described by means of scalar valued damage parameters and corresponding evolution equations. The stress state influences are expressed in the equivalent stress responsible for the damage evolution.

2.4.1.1 Kachanov-Rabotnov Model. The phenomenological creep-damage equations were firstly proposed by L. Kachanov [153] and Rabotnov [269]. A new internal variable has been introduced to characterize the "continuity" or the "damage" of the material. The geometrical interpretation of the continuity variable starts from changes in the cross-section area of a uni-axial specimen. Specifying the initial cross-section area of a specimen by A_0 and the area of voids, cavities, micro-cracks, etc. by A_D, the Kachanov's continuity is defined as follows (see [155])

$$\psi = \frac{A_0 - A_D}{A_0}$$

The value $\psi = 1$ means the virgin, fully undamaged state, the condition $\psi = 0$ corresponds to the fracture (completely damaged cross-section).

Rabotnov [269, 270, 272] introduced the dual damage variable ω. In [270] he pointed out that the damage state variable ω "may be associated with the area fraction of cracks, but such an interpretation is connected with a rough scheme and is therefore not necessary". Rabotnov assumed that the creep rate is additionally dependent on the current damage state. The constitutive equation should have the form

$$\dot{\varepsilon}^{cr} = \dot{\varepsilon}^{cr}(\sigma, \omega)$$

Furthermore, the damage processes can be reflected in the evolution equation

$$\dot{\omega} = \dot{\omega}(\sigma, \omega), \quad \omega|_{t=0} = 0, \quad \omega < \omega_*,$$

where ω_* is the critical value of the damage parameter for which the material fails. With the power functions of stress and damage the constitutive equation may be formulated as follows

$$\dot{\varepsilon}^{cr} = \frac{a\sigma^n}{(1-\omega)^m} \tag{2.4.1}$$

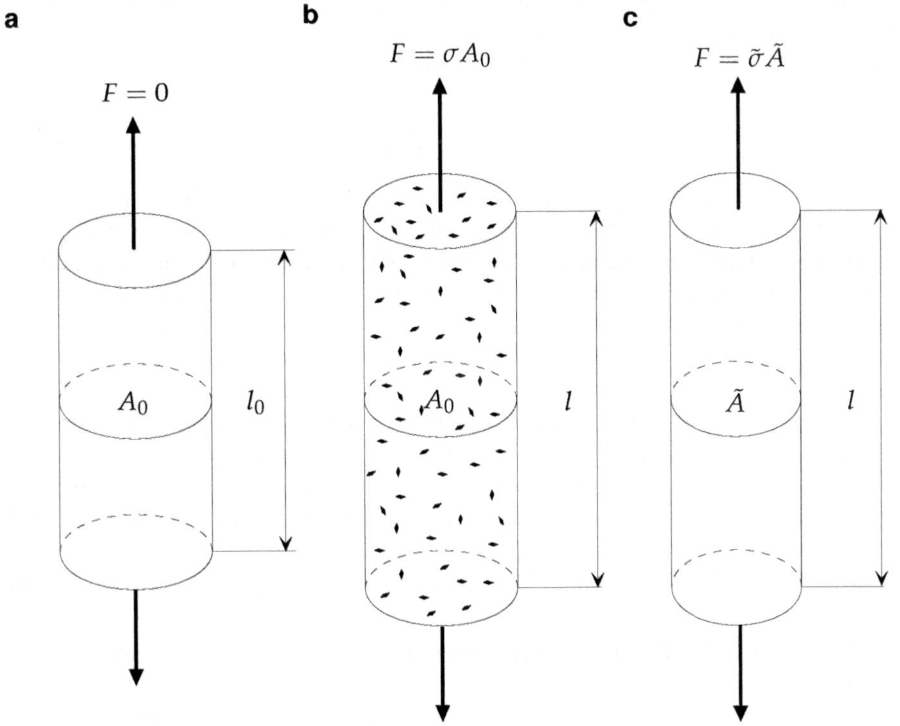

Fig. 2.12 Strain and damage of a bar. **a** Initial state, **b** damaged state, **c** fictitious undamaged state

Similarly, the damage rate can be expressed by

$$\dot{\omega} = \frac{b\sigma^k}{(1-\omega)^l} \tag{2.4.2}$$

These equations contain the material dependent constants: a, b, n, m, l, k. It is easy to prove that for the damage free state ($\omega = 0$), the first equation results in the power law creep constitutive equation.

Setting $m = n$ the first equation can be written as

$$\dot{\varepsilon}^{cr} = a\tilde{\sigma}^n, \tag{2.4.3}$$

where $\tilde{\sigma} = \sigma/(1-\omega)$ is the so-called net-stress or effective stress. In this case (2.4.3) is a generalization of the Norton-Bailey secondary creep law for the description of tertiary creep process. Lemaitre and Chaboche [191] proposed the effective stress concept to formulate constitutive equations for damaged materials based on available constitutive equation for "virgin" materials. An interpretation can be given for a tension bar, Fig. 2.12. Here A_0 denotes the initial cross-section area of the bar, Fig. 2.12a. From the given tensile force F the stress can be computed as $\sigma = F/A_0$.

The axial strain for the loaded bar $\varepsilon = (l - l_0)/l_0$ can be expressed as a function of the stress and the actual damage $\varepsilon = f(\sigma, \omega)$, Fig. 2.12b. For the effective cross-section $\tilde{A} = A_0 - A_D$ the effective stress is

$$\tilde{\sigma} = \frac{F}{\tilde{A}} = \frac{\sigma}{1 - \omega} \tag{2.4.4}$$

Now a fictitious undamaged bar with a cross-section area $\tilde{A}$, Fig. 2.12c, having the same axial strain response as the actual damaged bar $\varepsilon = f(\tilde{\sigma}) = f(\sigma, \omega)$ is introduced. The strain equivalence principle [189] states that any strain constitutive equation for a damaged material may be derived in the same way as for a virgin material except that the usual stress is replaced by the effective stress. Thus the constitutive equation for the creep rate (2.4.3) is the power law generalized for a damaged material.

Equations (2.4.1) and (2.4.2) can be only applied to the case of constant temperature. To generalize them to the non-isothermal conditions the material constants a and b should be replaced by the functions of temperature. Assuming the Arrhenius type temperature dependence (see Sect. 2.2.3) the following relations can be utilized

$$a(T) = a_0 \exp\left(-\frac{Q_a}{RT}\right), \quad b(T) = b_0 \exp\left(-\frac{Q_b}{RT}\right), \tag{2.4.5}$$

where Q_a and Q_b are the activation energies of creep and damage processes, respectively.

To identify the material constants in Eqs. (2.4.1), (2.4.2) and (2.4.5) experimental data of uni-axial creep up to rupture for certain stress and temperature ranges are required. The identification procedure is presented in [168], for example. To discuss the necessary data let us limit to the case of fixed temperature. Furthermore, let us assume $m = n$ in Eq. (2.4.1). Then the uni-axial creep model takes the following form

$$\dot{\varepsilon}^{cr} = a\left(\frac{\sigma}{1 - \omega}\right)^n, \quad \dot{\omega} = \frac{b\sigma^k}{(1 - \omega)^l} \tag{2.4.6}$$

With $\omega = 0$ the first equation describes the steady-state creep. The minimum (steady-state) creep rate is defined by the power law function of the applied stress

$$\dot{\varepsilon}^{cr}_{min} = a\sigma^n \tag{2.4.7}$$

In the steady-state creep range the creep curves are approximated by straight lines, Fig. 2.13a. From the family of creep curves the minimum creep rate vs. stress curve can be obtained. A sketch for such a curve in a double logarithmic scale is presented in Fig. 2.13b. For a certain stress range $\log \dot{\varepsilon}^{cr}_{min}$ can be approximated by a linear function of $\log \sigma$. The material constants a and n can be then determined from the steady-state creep. Let $\dot{\varepsilon}^{cr}_{min1}$ and $\dot{\varepsilon}^{cr}_{min2}$ be minimum creep rates for the constant stresses σ_1 and σ_2, respectively. Then the material constants can be estimated from

$$n = \frac{\log(\dot{\varepsilon}^{cr}_{min1}/\dot{\varepsilon}^{cr}_{min2})}{\log(\sigma_1/\sigma_2)}, \quad a = \frac{\dot{\varepsilon}^{cr}_{min1}}{\sigma_1^n} = \frac{\dot{\varepsilon}^{cr}_{min2}}{\sigma_2^n} \tag{2.4.8}$$

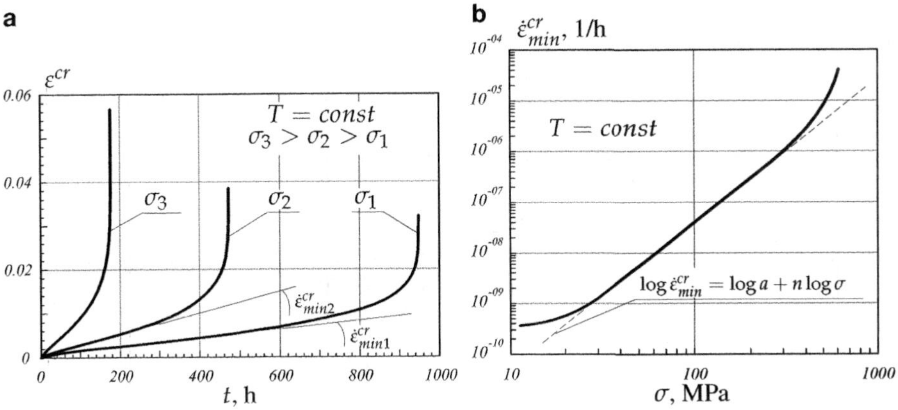

Fig. 2.13 Phenomenological description of uni-axial creep. **a** Creep strain vs. time curves, **b** minimum creep rate vs. stress curve

For a constant stress σ the damage evolution equation in (2.4.6) can be integrated as follows

$$\int_0^{\omega_*} (1-\omega)^l d\omega = \int_0^{t_*} b\sigma^k dt,$$

where t_* is the time to fracture of the specimen. Setting $\omega_* = 1$ and performing the integration one can obtain

$$t_* = \frac{1}{(l+1)b\sigma^k} \tag{2.4.9}$$

This equation describes the failure time vs. applied stress relation. For a number of metals and alloys the experimental data of the long term strength can be approximated by a straight line in a double logarithmic scale. From (2.4.9) follows

$$\log t_* = -(\log C + k \log \sigma), \quad C = b(l+1) \tag{2.4.10}$$

A typical long term strength curve is sketched in Fig. 2.14. Let us note, that the linear approximation (2.4.10) is only valid for a specific stress range, Fig. 2.14. Based on Eq. (2.4.10) and the long term strength curve the following relations can be obtained

$$k = \frac{\log(t_{*2}/t_{*1})}{\log(\sigma_1/\sigma_2)}, \quad b(l+1) = \frac{1}{t_{*1}\sigma_1^k} = \frac{1}{t_{*2}\sigma_2^k},$$

where t_{*1} and t_{*2} are times to fracture corresponding to the applied stresses σ_1 and σ_2, respectively. Integration of the second Eq. (2.4.6) by use of Eq. (2.4.9) provides the damage parameter as a function of time

$$\omega(t) = 1 - \left(1 - \frac{t}{t_*}\right)^{\frac{1}{l+1}} \tag{2.4.11}$$

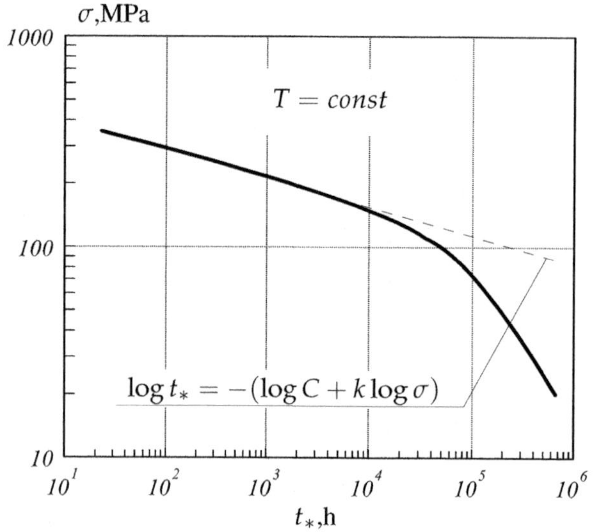

Fig. 2.14 Long-term strength curve

With Eq. (2.4.11) the creep rate equation (2.4.6) can be integrated leading to the creep strain vs. time dependence

$$\varepsilon^{cr}(t) = \frac{a\sigma^{n-k}}{b(l+1-n)}\left[1 - \left(1 - \frac{t}{t_*}\right)^{\frac{l+1-n}{l+1}}\right] \tag{2.4.12}$$

From Eq. (2.4.12) it follows that the constant l must satisfy the condition $l > n - 1$ providing the positive strain for the positive stress values. By setting $t = t_*$ the creep strain before the fracture, i.e. $\varepsilon^{cr}_* = \varepsilon^{cr}(t_*)$, can be calculated as

$$\varepsilon^{cr}_* = \frac{a\sigma^{n-k}}{b(l+1-n)} \tag{2.4.13}$$

For $n > k$ the fracture strain increases with an increase in the stress value. Such a dependence is usually observed for many alloys in the case of moderate stresses. From Eqs. (2.4.13), (2.4.10) and (2.4.7) the following relations can be obtained

$$\varepsilon^{cr}_* = \frac{\dot{\varepsilon}^{cr}_{min} t_*}{1 - \frac{n}{l+1}}, \quad \dot{\varepsilon}^{cr}_{min} t_* = \frac{a}{b(l+1)}\sigma^{n-k} \tag{2.4.14}$$

In the special case $n = k$ the second equation in (2.4.14) reads

$$\dot{\varepsilon}^{cr}_{min} t_* = \frac{a}{b(l+1)} = const$$

This is the Monkman-Grant relationship which states, that for a given material the product of the minimum creep rate and the time to fracture is a material constant.

We observe, that the Monkman-Grant relationship follows from the Kachanov-Rabotnov model if the slopes of the minimum creep rate vs. stress and the stress vs. time to fracture dependencies coincide in the double logarithmic scale. In this case the strain before the creep fracture (creep ductility) should be stress independent, as it follows from the first equation in (2.4.14).

With Eqs. (2.4.14) the creep strain vs. time dependence (2.4.12) takes the form

$$\varepsilon^{cr}(t) = \frac{\dot{\varepsilon}^{cr}_{min} t_*}{1 - \frac{n}{l+1}} \left[1 - \left(1 - \frac{t}{t_*} \right)^{1 - \frac{n}{l+1}} \right] \tag{2.4.15}$$

We observe that the constant l controls the shape of the tertiary creep stage. For $n/(l+1) \ll 1$ Eq. (2.4.15) can be approximated by

$$\varepsilon^{cr}(t) = \dot{\varepsilon}^{cr}_{min} t, \quad 0 \leq t \leq t_*$$

In this case the tertiary creep stage is not observable. Instead of (2.4.6) one may apply the simplified constitutive model, where the influence of creep damage on the creep rate is ignored, i.e.

$$\dot{\varepsilon}^{cr} = a\sigma^n, \quad \dot{\omega} = \frac{b\sigma^k}{(1-\omega)^l} \tag{2.4.16}$$

In a slightly different form Eqs (2.4.16) were originally proposed by L. Kachanov [153] under assumption that for brittle materials the damage processes have negligible influence on the creep rate.

Figure 2.15 provides the plots of Eq. (2.4.15) with respect to the normalized creep strain and normalized time for different values of $r \equiv n/(l+1)$. We observe that even for $r = 0.2$ the creep rate is almost constant. The increase of r leads to the increase of tertiary creep rate, the increase of the "duration" of the tertiary stage and increase of the fracture strain.

The phenomenological model (2.4.6) characterizes the effect of damage evolution and describes the tertiary creep in a uni-axial test. For a number of metals and alloys material constants are available, see e.g. [21, 72, 80, 135, 144, 145, 146, 147, 167, 174, 190, 191, 222]. Instead of the power law functions of stress or damage it is possible to use another kind of functions, e.g. the hyperbolic sine functions in both the creep and damage evolution equations. In addition, by the introduction of suitable hardening functions or internal hardening variables, the model can be extended to consider primary creep.

In applying (2.4.6) to the analysis of structures one should bear in mind that the material constants are estimated from experimental creep curves, usually available for a narrow range of stresses. The linear dependencies between $\log \dot{\varepsilon}^{cr}_{min}$ and $\log \sigma$ or between $\log t_*$ and $\log \sigma$ do not hold for wide stress ranges. For example, it is known from materials science that for higher stresses the damage mode may change from inter-granular to trans-granular, e.g. [37]. Alternatively, tertiary creep can be described by the introduction of several internal variables which are responsible

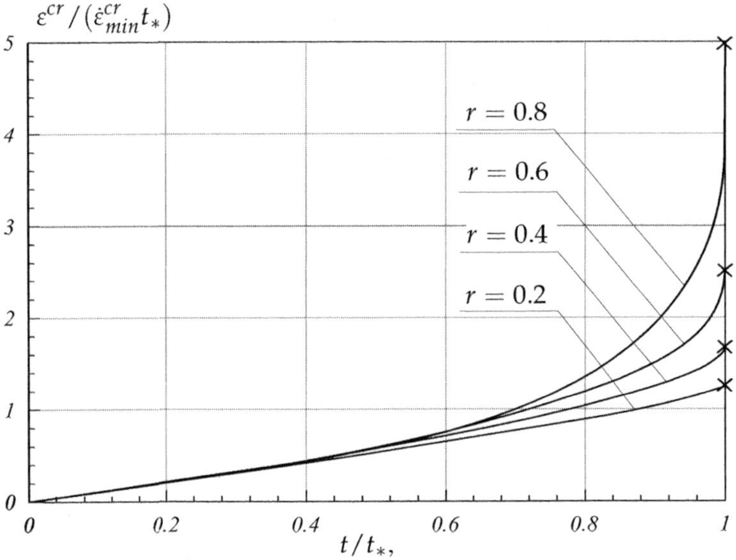

Fig. 2.15 Creep curves for different values of $r = n/(l+1)$

for different interacting damage mechanisms. Examples for such models will be discussed later.

The model (2.4.6) is a system of two ordinary differential equations, which must be integrated over time in order to obtain the current creep strain and damage. For the analysis of statically indeterminate structures the integration must be performed numerically, even in the case of uni-axial stress states. In some cases one may assume that the tertiary creep rate does not lead to significant stress redistribution. Therefore one may neglect the damage variable in the constitutive equation for the creep rate and utilize the model (2.4.16) instead of (2.4.6), e.g. [283]. The damage evolution equation can be integrated separately providing the time to fracture estimation for the given constant stress in the steady-state creep range.

To generalize Eqs (2.4.1) and (2.4.2) to the multi-axial stress states Rabotnov [270] assumed that

- the creep process is determined by the effective stress tensor $\tilde{\boldsymbol{\sigma}} = \boldsymbol{f}(\boldsymbol{\sigma}, \omega)$,
- the creep potential for the damaged material has the same form as for the secondary creep

For example, the Norton-Bailey-Odqvist creep potential (2.1.8) can be generalized as follows

$$W(\tilde{\boldsymbol{\sigma}}) = \frac{\sigma_0}{n+1}\left(\frac{\tilde{\sigma}_{vM}}{\sigma_0}\right)^{n+1}, \quad \tilde{\sigma}_{vM} = \sqrt{\frac{3}{2}\tilde{\boldsymbol{s}} \cdot\cdot\, \tilde{\boldsymbol{s}}}, \quad \tilde{\boldsymbol{s}} = \tilde{\boldsymbol{\sigma}} - \frac{1}{3}\mathrm{tr}\,\tilde{\boldsymbol{\sigma}}\boldsymbol{I} \tag{2.4.17}$$

In [270, 271] Rabotnov proposed the following effective stress tensor

$$\tilde{\boldsymbol{\sigma}} = \frac{\sigma_I}{1-\omega}\boldsymbol{n}_I \otimes \boldsymbol{n}_I + \sigma_{II}\boldsymbol{n}_{II} \otimes \boldsymbol{n}_{II} + \sigma_{III}\boldsymbol{n}_{III} \otimes \boldsymbol{n}_{III} \tag{2.4.18}$$

for the case of distinct principal values of the stress tensor $\sigma_I > \sigma_{II} > \sigma_{III}$ and $\sigma_I > 0$. Equation (2.4.18) implies that the effect of damage is only essential in the planes perpendicular to the first principal direction associated with the maximum tensile stress. In this case the constitutive equation for the creep rate would have the form

$$\dot{\boldsymbol{\varepsilon}}^{cr} = \dot{\lambda}(\tilde{\sigma}_{vM})\left[\boldsymbol{s} + \frac{\omega}{1-\omega}\sigma_I\left(\boldsymbol{n}_I \otimes \boldsymbol{n}_I - \frac{1}{3}\boldsymbol{I}\right)\right], \tag{2.4.19}$$

Equation (2.4.19) suggests that the proportionality between the creep rate tensor and the stress deviator is violated in the tertiary creep range. Leckie and Hayhurst [188] analyzed experimental data of combined tension and torsion for copper and aluminium alloys. They concluded that the strain trajectories (see Fig. 1.9) are not sensibly affected by the damage processes. Therefore, the creep rate can be assumed in the form

$$\dot{\boldsymbol{\varepsilon}}^{cr} = \dot{\lambda}(\tilde{\sigma}_{vM})\boldsymbol{s}$$

Leckie and Hayhurst proposed to generalize the von Mises type secondary creep equation (see Sect. 2.2.1) as follows

$$\dot{\boldsymbol{\varepsilon}}^{cr} = \frac{3}{2}a\left(\frac{\sigma_{vM}}{1-\omega}\right)^n \frac{\boldsymbol{s}}{\sigma_{vM}} \tag{2.4.20}$$

Equation (2.4.20) can be also derived applying the strain equivalence principle [191] and the effective stress tensor in the form $\tilde{\boldsymbol{\sigma}} = \boldsymbol{\sigma}/(1-\omega)$.

The next step is the formulation of the damage evolution equation. By analogy with the uni-axial case, the damage rate should have a form

$$\dot{\omega} = \dot{\omega}(\boldsymbol{\sigma}, \omega)$$

The dependence on the stress tensor can be expressed by means of the "damage equivalent stress" $\sigma^{\omega}_{eq}(\boldsymbol{\sigma})$ which allows to compare tertiary creep and long term strength under different stress states. With the damage equivalent stress, the uni-axial equation (2.4.2) can be generalized as follows

$$\dot{\omega} = \frac{b(\sigma^{\omega}_{eq})^k}{(1-\omega)^l} \tag{2.4.21}$$

The material constants a, b, n, k and l can be identified from uni-axial creep curves. In order to find a suitable expression for the damage equivalent stress, the data from multi-axial creep tests up to rupture are required. In general, σ^{ω}_{eq} can be formulated in terms of three invariants of the stress tensor, for example the basic invariants (see Sect. 2.2.1)

$$\sigma^{\omega}_{eq} = \sigma^{\omega}_{eq}[I_1(\boldsymbol{\sigma}), I_2(\boldsymbol{\sigma}), I_3(\boldsymbol{\sigma})]$$

Similarly to the uni-axial case, see Eq. (2.4.9), the damage evolution equation (2.4.21) can be integrated assuming that the stress tensor is constant during the test. As a result, the relationship between the time to creep fracture and the equivalent stress can be obtained

$$t_* = \frac{1}{(l+1)b}\left(\sigma_{eq}^{\omega}\right)^{-k} \tag{2.4.22}$$

Sdobyrev [295] carried out long term tests on tubular specimens made from alloys EI-237B (Ni-based alloy) and EI-405 (Fe-based alloy) under tension, torsion and combined tension-torsion. The results of the tests are collected for different temperatures with the help of equivalent stress vs. fracture time plots. The following dependence was established

$$\frac{1}{2}(\sigma_I + \sigma_{vM}) = f(\log t_*) \tag{2.4.23}$$

Sdobyrev found that the linear function f provides a satisfactory description of the experimental data. The equivalent stress responsible to the long term strength at elevated temperatures is then $\sigma_{eq}^* = \frac{1}{2}(\sigma_I + \sigma_{vM})$. Based on different mechanisms which control creep failure, the influence of three stress state parameters (the mean stress $\sigma_m = I_1/3$, the first positive principal stress or the maximum tensile stress $\sigma_{maxt} = (\sigma_I + |\sigma_I|)/2$ and the von Mises stress) is discussed by Trunin in [322]. The Sdobyrev criterion was extended as follows

$$\sigma_{eq}^* = \frac{1}{2}(\sigma_{vM} + \sigma_{maxt})\, a^{1-2\eta}, \quad \eta = \frac{3\sigma_m}{\sigma_{vM} + \sigma_{maxt}}, \tag{2.4.24}$$

where a is a material constant. For special loading cases this equivalent stress yields

- uni-axial tension

$$\sigma_{eq}^* = \sigma, \quad \eta = \frac{1}{2}$$

- uni-axial compression

$$\sigma_{eq}^* = \frac{\sigma a^3}{2}, \quad \eta = -1$$

- pure torsion

$$\sigma_{eq}^* = \frac{\sqrt{3}+1}{2}\tau a, \quad \eta = 0$$

The constant a can be calculated from the ultimate stress values leading to the same fracture time for a given temperature. For example, if the ultimate tension and shear stresses are σ_u and τ_u, respectively, then

$$a = \frac{2}{\sqrt{3}+1}\frac{\sigma_u}{\tau_u}$$

Hayhurst [135] proposed the following relationship

$$t_* = A(\alpha\sigma_{maxt} + \beta I_1 + \gamma\sigma_{vM})^{-\chi}, \tag{2.4.25}$$

where A and χ are material constants, $I_1 = 3\sigma_m$ and $\alpha + \beta + \gamma = 1$. Comparing this equation with Eq. (2.4.22) one can obtain

$$A = \frac{1}{(l+1)b}, \quad \chi = k, \quad \sigma_{eq}^{\omega} = \alpha\sigma_{maxt} + \beta I_1 + \gamma\sigma_{vM} \tag{2.4.26}$$

Hayhurst introduced the normalized stress tensor $\bar{\boldsymbol{\sigma}} = \boldsymbol{\sigma}/\sigma_0$ and the normalized time to fracture $\bar{t}_* = t_*/t_{*0}$, where t_{*0} is the time to fracture in a uni-axial test conducted at the stress σ_0. From Eqs (2.4.9) and (2.4.22) it follows

$$\bar{t}_* = \left(\frac{\sigma_{eq}^{\omega}}{\sigma_0}\right)^{-k} = (\bar{\sigma}_{eq}^{\omega})^{-k}$$

By setting the normalized rupture time equal to unity, the equation $\bar{\sigma}_{eq}^{\omega} = 1$ follows, which is connecting the stress states leading to the equal rupture time. In [135] the data of biaxial tests (biaxial tension test, combined tension and torsion of tubular specimens) are collected for different materials. It was found convenient to present the results in terms of the isochronous rupture surface, which is the plot of the equation $\bar{\sigma}_{eq}^{\omega} = 1$ for the specified values of α and β in the normalized stress space. For plane stress states the isochronous rupture loci can be presented in the normalized principal stress axes. Examples for different materials are presented in [135]. The coefficients α and β are specific for each material and, in addition, they may depend on the temperature. Figure 2.16 shows the isochronous rupture loci for three special cases: $\bar{\sigma}_{eq}^{\omega} = \bar{\sigma}_{maxt}$, $\bar{\sigma}_{eq}^{\omega} = \bar{\sigma}_{vM}$ and $\bar{\sigma}_{eq}^{\omega} = 3\bar{\sigma}_m$. The first two represent the limit cases of the material behavior [188].

A more general expression for the damage equivalent stress can be formulated by the use of three invariants of the stress tensor. With the first invariant I_1, the von Mises equivalent stress σ_{vM} and

$$\sin 3\xi = -\frac{27}{2}\frac{(\boldsymbol{s}\cdot\boldsymbol{s})\cdot\cdot\,\boldsymbol{s}}{\sigma_{vM}^3}, \quad -\frac{\pi}{6} \leq \xi \leq \frac{\pi}{6},$$

as a cubic invariant, the following equivalent stress has been proposed in [30]

$$\sigma_{eq}^{\omega} = \lambda_1\sigma_{vM}\sin\xi + \lambda_2\sigma_{vM}\cos\xi + \lambda_3\sigma_{vM} + \lambda_4 I_1 + \lambda_5 I_1\sin\xi + \lambda_6 I_1\cos\xi \tag{2.4.27}$$

The identification of coefficients $\lambda_i, i = 1,\ldots,6$ requires six independent tests. Equation (2.4.27) contains a number of known failure criteria as special cases, see [30]. For example, setting $\lambda_1 = \lambda_2 = \lambda_4 = \lambda_5 = \lambda_6 = 0$ the equation provides the von Mises equivalent stress. Taking into account

$$\sigma_I - \frac{1}{3}\left[2\sigma_{vM}\sin\left(\xi + \frac{2\pi}{3}\right) + I_1\right] = -\frac{1}{3}\sigma_{vM}\sin\xi + \frac{\sqrt{3}}{3}\sigma_{vM}\cos\xi + \frac{1}{3}I_1$$

and with

$$\lambda_1 = -\frac{1}{6}, \quad \lambda_2 = \frac{\sqrt{3}}{6}, \quad \lambda_3 = \frac{1}{2}, \quad \lambda_4 = \frac{1}{6}, \quad \lambda_5 = \lambda_6 = 0$$

one can obtain $\sigma_{eq}^{\omega} = \frac{1}{2}(\sigma_I + \sigma_{vM})$. With

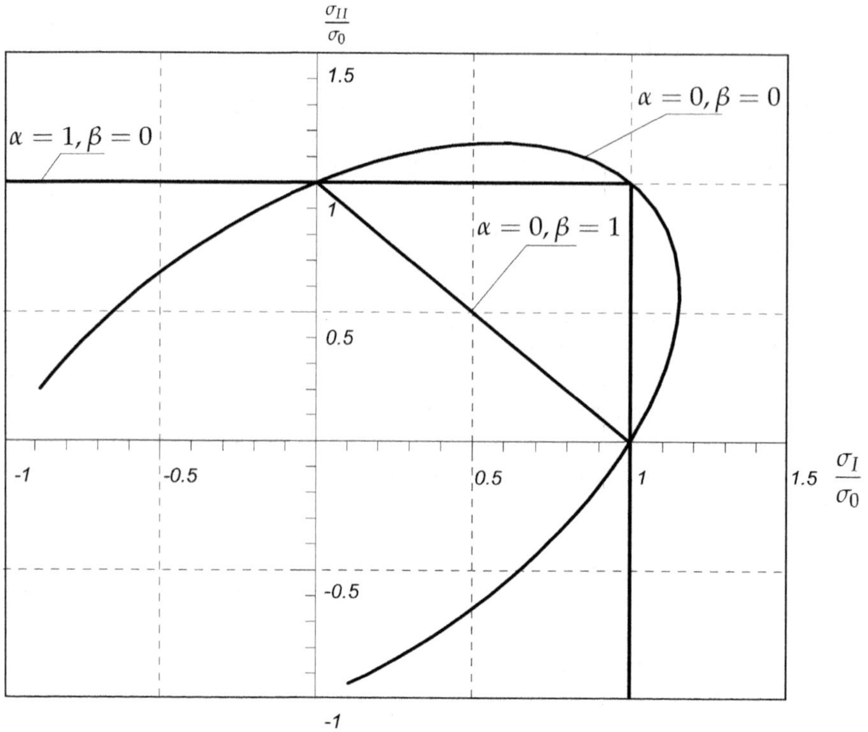

Fig. 2.16 Plane stress isochronous rupture loci, for details see [135]

$$\lambda_1 = -\frac{1}{3}\alpha, \quad \lambda_2 = \frac{\sqrt{3}}{3}\alpha, \quad \lambda_3 = \beta, \quad \lambda_4 = 1 - \frac{2}{3}\alpha - \beta, \quad \lambda_5 = \lambda_6 = 0$$

Eq. (2.4.27) yields $\sigma_{eq}^{\omega} = \alpha\sigma_I + \beta\sigma_{vM} + (1 - \alpha - \beta)I_1$. Other examples are discussed in [6].

In order to identify the material constants, e.g., a in (2.4.24) or α and β in (2.4.25), the values of the ultimate stresses leading to the same failure time for different stress states are necessary. Therefore series of independent creep tests up to rupture are required. For each kind of the test the long term strength curve (stress vs. time to fracture curve), see Fig. 2.14, must be obtained. For example, a series of torsion tests (at least two) under different stress values should be performed. Usually, experimental data from creep tests under complex stress states are limited and the scatter of the experimental results is unavoidable. Therefore, the constitutive and the evolution equation (2.4.20) and (2.4.21) with the two-parametric damage equivalent stress (2.4.26) are widely used in modeling tertiary creep. Examples of material constants as well as structural mechanics applications can be found in [21, 72, 80, 135, 145, 146, 147, 167, 174] among others.

2.4.1.2 Micromechanically-Consistent Models. The creep constitutive equation (2.4.20) includes the effect of damage by means of the equivalent stress concept. An alternative approach to formulate the creep constitutive equation can be based on micromechanics. Rodin and Parks [284] considered an infinite block from incompressible isotropic material containing a given distribution of cracks and subjected to a far field homogeneous stress. As a measure of damage they used $\rho = a^3 N/V$, where N is the number of cracks (voids) in a volume V and a is the averaged radius of a crack. Assuming power law creep, they found that the creep potential for such a material has the following form

$$W(\boldsymbol{\sigma}, \rho, n) = \frac{\dot{\varepsilon}_0 \sigma_0}{n+1} f\big(\zeta(\boldsymbol{\sigma}), \rho, n\big) \left(\frac{\sigma_{vM}}{\sigma_0}\right)^{n+1}, \qquad (2.4.28)$$

where $\dot{\varepsilon}_0$ is the reference creep rate, σ_0 is the reference stress and n is a material constant. $\zeta(\boldsymbol{\sigma})$ is a function representing the influence of the kind of stress state. In [284] the following particular expression is proposed

$$\zeta(\boldsymbol{\sigma}) = \frac{\sigma_I}{\sigma_{vM}},$$

where σ_I is the maximum principal stress. The creep potential (2.4.28) and the flow rule (2.1.6) give

$$\begin{aligned} \dot{\boldsymbol{\varepsilon}}^{cr} &= \frac{\partial W}{\partial \boldsymbol{\sigma}} = \frac{\partial W}{\partial \sigma_{vM}} \frac{\partial \sigma_{vM}}{\partial \boldsymbol{\sigma}} + \frac{\partial W}{\partial \zeta} \frac{\partial \zeta}{\partial \boldsymbol{\sigma}} \\ &= \dot{\varepsilon}_0 \left(\frac{\sigma_{vM}}{\sigma_0}\right)^n \left[\frac{3}{2}\left(f - \frac{\zeta f_{,\zeta}}{n+1}\right) \frac{\boldsymbol{s}}{\sigma_{vM}} + \frac{f_{,\zeta}}{n+1} \boldsymbol{n}_I \otimes \boldsymbol{n}_I\right], \end{aligned} \qquad (2.4.29)$$

where $\boldsymbol{n}_I$ is the first principal direction of the stress tensor. The function f must satisfy the following convexity condition [284]

$$f f_{,\zeta\zeta} - \frac{n}{n+1} f_{,\zeta}^2 > 0,$$

The form of the function f is established for the assumed particular distribution of cracks and by use of a self-consistent approach. In [285] the following expression is proposed

$$\begin{aligned} f(\zeta, \rho, n) &= \left[1 + \alpha(\rho, n)\zeta^2\right]^{\frac{n+1}{2}}, \\ \alpha(\rho, n) &= \frac{2\rho}{n+1} + \frac{(2n+3)\rho^2}{n(n+1)^2} + \frac{(n+3)\rho^3}{9n(n+1)^3} + \frac{(n+3)\rho^4}{108n(n+1)^4} \end{aligned}$$

Models of the type (2.4.29) are popular in materials science related literature, e.g. [124, 217]. They are based on micromechanical considerations and therefore seem to be more preferable for creep-damage analysis. However, only idealized damage states, e.g. dilute non-interacting cracks or voids with a given density and specific distribution can be considered. Furthermore, at present there is no micromechanically-consistent way to establish the form of the evolution equation for the assumed damage variable. Different empirical equations are proposed in the literature. For example, Mohrmann and Sester [217] assume that the cavity nucleation is strain controlled and recommend the following equation

$$\frac{\rho}{\rho_f} = \left(\frac{\varepsilon_{vM}}{\varepsilon_f}\right)^{\gamma},$$

where ρ_f, ε_f and γ are material constants which should be identified from "macroscopic" creep responses.

Bassani and Hawk [40] proposed to use a phenomenological damage parameter ω (see Sect. 2.4.1.1) instead of ρ. The function f is then postulated as follows

$$f(\zeta, \omega, n) = \frac{1}{(1-\omega)^k}\left(1 - \alpha_0\omega + \alpha_0\omega\zeta^2\right)^{\frac{n+1}{2}} \tag{2.4.30}$$

Here

$$\zeta = (1-\alpha_1)\frac{\sigma_I}{\sigma_{vM}} + \alpha_1\frac{\sigma_H}{\sigma_{vM}}$$

and k, n, α_0 and α_1 are material constants. From Eqs (2.4.29) and (2.4.30) follows

$$\begin{aligned} \dot{\boldsymbol{\varepsilon}}^{cr} &= \dot{\varepsilon}_0\left(\frac{\sigma_{vM}}{\sigma_0}\right)^n \frac{1}{(1-\omega)^k}(1-\alpha_0\omega+\alpha_0\omega\zeta^2)^{\frac{n-1}{2}} \times \\ &\times \left\{\frac{3}{2}(1-\alpha_0\omega)\frac{\boldsymbol{s}}{\sigma_{vM}} + \alpha_0\omega\zeta[(1-\alpha_1)\boldsymbol{n}_I\otimes\boldsymbol{n}_I + \alpha_1\boldsymbol{I}]\right\} \end{aligned} \tag{2.4.31}$$

With $\alpha_0 = 0$, $\alpha_1 = 1$ and $k = n$ Eq. (2.4.31) yields the Kachanov-Rabotnov type constitutive equation (2.4.20). By setting $\alpha_0 = 1, k = (n+1)/2$ and $\omega \ll 1$ Eq. (2.4.31) approximates the Rodin and Parks micro-mechanical based model [284]. For the case $k = n$, $\alpha_0 = 1$ and $\alpha_1 = 1$ the constitutive equation for the creep rate can be presented as follows

$$\dot{\boldsymbol{\varepsilon}}^{cr} = \dot{\varepsilon}_0\left[\frac{\sigma_{vM}}{\sigma_0(1-\omega)}\right]^n (1-\omega+\omega\zeta^2)^{\frac{n-1}{2}}\left[\frac{3}{2}(1-\omega)\frac{\boldsymbol{s}}{\sigma_{vM}} + \omega\zeta\boldsymbol{I}\right], \quad \zeta = \frac{\sigma_H}{\sigma_{vM}} \tag{2.4.32}$$

Equation (2.4.32) is applied in [40] to the finite element simulation of creep crack growth.

From Eq. (2.4.31) one can calculate the volumetric creep rate

$$\dot{\varepsilon}_V = \operatorname{tr}\dot{\boldsymbol{\varepsilon}}^{cr} = \dot{\varepsilon}_0\left(\frac{\sigma_{vM}}{\sigma_0}\right)^n \frac{1}{(1-\omega)^k}(1-\alpha_0\omega+\alpha_0\omega\zeta^2)^{\frac{n-1}{2}}\left[\alpha_0\omega\zeta(1+2\alpha_1)\right]$$

We observe that for $\alpha_0 \neq 0$ the damage growth induces dilatation. Creep constitutive equations (2.4.29) or (2.4.31) include the first principal direction of the stress tensor. It should be noted that the dyad $\boldsymbol{n}_I \otimes \boldsymbol{n}_I$ can be only found if $\sigma_I \neq 0$, $\sigma_I \neq \sigma_{II}$ and $\sigma_I \neq \sigma_{III}$. In this case, e.g. [205]

$$\boldsymbol{n}_I \otimes \boldsymbol{n}_I = \frac{1}{(\sigma_I - \sigma_{II})(\sigma_I - \sigma_{III})} \left[\boldsymbol{\sigma}^2 - (\operatorname{tr} \boldsymbol{\sigma} - \sigma_I)\boldsymbol{\sigma} + \frac{\det \boldsymbol{\sigma}}{\sigma_I} \boldsymbol{I} \right] \qquad (2.4.33)$$

Inserting (2.4.33) into (2.4.29) or into (2.4.31) we observe that not only the volumetric strain but also second order effects (see Sect. 2.2.1 for discussion) are "induced" by damage.

2.4.1.3 Mechanism-Based Models. The constitutive and evolution equations (2.4.20) and (2.4.21) are formulated in terms of power law functions of stress. It is known from materials science that the power law stress dependence guarantees the correct description of the creep rate only for a specific stress range (see Sect 2.2.3). In addition, the power law stress and damage functions used in Eqs. (2.4.20) and (2.4.21) may lead to numerical problems in finite element simulations of creep in structures with stress concentrations or in attempts to predict the creep crack growth [198, 288].

The uni-axial creep tests are usually performed under increased stress and temperature levels in order to accelerate the creep process. For the long term analysis of structures the material model should be able to predict creep rates for wide stress ranges including moderate and small stresses. Within the materials science many different damage mechanisms which may operate depending on the stress level and the temperature are discussed, e.g., [102]. Each of the damage mechanisms can be considered by a state variable with an appropriate kinetic equation.

Another way for the formulation of a creep-damage constitutive model is the so-called mechanism-based approach. The internal state variables are introduced according to those creep and damage mechanisms which dominate for a specific material and specific loading conditions. Furthermore, different functions of stress and temperature proposed in materials science can be utilized. The form and the validity frame of such a function depend on many factors including the stress and temperature levels, type of alloying, grain size, etc. The materials science formulations do not provide the values of material constants (only the bounds are given). They must be identified from the data of standard tests, e.g. uni-axial creep test. Examples of mechanism-based models can be found in [136, 137, 176, 250, 258]. Here we discuss the model proposed by Perrin and Hayhurst in [258] for a 0.5Cr-0.5Mo-0.25V ferritic steel in the temperature range 600 – 675°C.

The starting point is the assumption that the rate of the local grain boundary deformation is approximately a constant fraction of the overall deformation rate. From this follows that the constitutive equations for the overall creep rate can be formulated in terms of empirical relationships between the local grain boundary deformation rate and the stress, the temperature, the cavitation rate, etc.

For ferritic steels the nucleation of cavities has been observed at carbide particles on grain boundaries due to the local accumulation of dislocations. The nucleation kinetics can be therefore related to the local deformation. Furthermore, the cavity nucleation depends on the stress state characterized by σ_I / σ_{vM}. Cane [86] observed that the area fraction of intergranular cavities in the plane normal to the applied stress increases uniformly with the accumulated creep strain. He proposed that the nucleation and growth can be combined into an overall measure of cavitation. The cavitated area fraction A_f can be related to the von Mises equivalent creep strain, the von Mises equivalent stress and the maximum principal stress by the equation

$$A_f = D\varepsilon_{vM} \left(\frac{\sigma_I}{\sigma_{vM}} \right)^{\mu}, \tag{2.4.34}$$

where D and μ are constants depending on the material microstructure. Perrin and Hayhurst define the damage state variable ω as the cavitated area fraction. The failure condition in a uni-axial creep test is the complete cavitation of all grain boundaries normal to the applied stress. The cavitated area fraction at failure is approximately $1/3$. Therefore, the critical state at which the material fails, can be characterized by $\omega_* = 1/3$.

The important mechanism of creep damage for the ferritic steel under consideration is the temperature dependent coarsening of carbide precipitates. First, the carbide precipitates restrict the deformation of the grain interior and second, they provide sites for nucleation of cavities. Following Dyson [102], the particle coarsening can be characterized by the state variable $\phi = 1 - l_i/l$ related to the initial (l_i) and current (l) spacing of precipitates. The kinetic equation is derived from the coarsening theory [102, 104]

$$\dot{\phi} = \left(\frac{K_c}{3} \right) (1 - \phi)^4 \tag{2.4.35}$$

with K_c as the material dependent constant for a given temperature. The rate of the coarsening variable is independent from the applied stress and can be integrated with respect to time.

The primary creep is characterized by the work hardening due to the formation of the dislocation substructure. For this purpose a scalar hardening state variable H is introduced. This variable varies from zero to a saturation value H_*, at which no further hardening takes place. The proposed evolution equation is

$$\dot{H} = \frac{h_c \dot{\varepsilon}^{cr}_{vM}}{\sigma_{vM}} \left(1 - \frac{H}{H_*} \right) \tag{2.4.36}$$

with h_c as the material constant.

The creep rate is controlled by the climb plus glide deformation mechanism. For the stress dependence of the creep rate, the hyperbolic sine stress function is used. The materials science arguments for the use of hyperbolic sine function instead of power law function are discussed, for example, by Dyson and McLean [105]. With the assumed mechanisms of hardening, cavitation and ageing as well as the corresponding state variables the following equation for the von Mises creep rate is proposed

$$\dot{\varepsilon}^{cr}_{vM} = A \sinh \frac{B\sigma_{vM}(1 - H)}{(1 - \phi)(1 - \omega)} \tag{2.4.37}$$

Equations (2.4.34) – (2.4.37) are formulated under assumption of constant temperature. The influence of the temperature on the processes of creep deformation, creep cavitation and coarsening can be expressed by Arrhenius functions with appropriate activation energies. Further details of the physical motivation are discussed in [258]. The following set of constitutive and evolution equations has been proposed

$$\begin{aligned}
\dot{\boldsymbol{\varepsilon}}^{cr} &= \frac{3}{2}\frac{\boldsymbol{s}}{\sigma_{vM}} A \sinh \frac{B\sigma_{vM}(1-H)}{(1-\phi)(1-\omega)}, \\
\dot{H} &= \frac{h_c \dot{\varepsilon}^{cr}_{vM}}{\sigma_{vM}}\left(1 - \frac{H}{H_*}\right), \\
\dot{\phi} &= \left(\frac{K_c}{3}\right)(1-\phi)^4, \\
\dot{\omega} &= DN\dot{\varepsilon}^{cr}_{vM}\left(\frac{\sigma_I}{\sigma_{vM}}\right)^{\mu}, \\
A &= A_0 B \exp\left(-\frac{Q_A}{RT}\right), \quad B = B_0 \exp\left(-\frac{Q_B}{RT}\right), \\
K_c &= \frac{K_{c0}}{B^3}\exp\left(-\frac{Q_{K_c}}{RT}\right), \quad D = D_0 \exp\left(-\frac{Q_D}{RT}\right),
\end{aligned} \tag{2.4.38}$$

where $N = 1$ for $\sigma_I > 0$ and $N = 0$ for $\sigma_I \leq 0$. A_0, B_0, D_0, K_{c0}, h_c, H_*, Q_A, Q_B, Q_D and Q_{K_c} are material constants which must be identified from uni-axial creep tests. The material constant μ, the so-called stress state index, can be determined from multi-axial creep rupture data. These constants are identified in [258] based on the experimental data of uni-axial creep over the stress range of 28 – 110 MPa and over the temperature range of 615 – 690°C. In [259] Eqs. (2.4.38) are applied to model creep in different zones of a weldment at 640°C including the weld metal, the heat-affected zone and the parent material.

It should be noted that Eqs. (2.4.38) are specific for the considered material and can only be applied with respect to the dominant mechanisms of the creep deformation and damage evolution. Further examples of mechanism based material models are presented in [251] for a nickel-based super-alloy and in [176] for an aluminium alloy.

2.4.1.4 Models Based on Dissipation. Sosnin [303, 304] proposed to characterize the material damage by the specific dissipation work. The following damage variable has been introduced

$$q = \int_0^t \sigma \dot{\varepsilon}^{cr} d\tau \tag{2.4.39}$$

For the variable q the evolution equation was postulated

$$\dot{q} = f_\sigma(\sigma) f_T(T) f_q(q)$$

For the multi-axial stress and strain states this variable is defined as follows

$$q = \int_0^t \boldsymbol{\sigma} \cdot\cdot\, \dot{\boldsymbol{\varepsilon}}^{cr} d\tau$$

In [304] Sosnin presented experimental data for various titanium and aluminium alloys in a form of q vs. time curves. He found that a critical value q_* exists at which the material fails under creep conditions. The value q_* does not depend on the kind of the applied stress state.

For isotropic materials the creep rate equation can be formulated as follows (see Sect. 2.2.1)

$$\dot{\boldsymbol{\varepsilon}}^{cr} = \frac{3}{2}\frac{P}{\sigma_{vM}}\mathbf{s}, \quad P = \boldsymbol{\sigma} \cdot\cdot\, \dot{\boldsymbol{\varepsilon}}^{cr} = \sigma_{vM}\dot{\varepsilon}^{cr}_{vM} \tag{2.4.40}$$

Sosnin assumed the dissipation power P to be a function of the von Mises equivalent stress, the temperature and the internal state variable q as follows

$$\dot{q} \equiv P = f_\sigma(\sigma_{vM}) f_T(T) f_q(q)$$

In many cases the following empirical equation provides a satisfactory agreement with experimental results

$$\dot{q} = \frac{b\sigma^n_{vM}}{q^k(q_*^{k+1} - q^{k+1})^m}, \tag{2.4.41}$$

where b, n, k, m and q_* are material constants. In [304] experimental data obtained from uni-axial tests and tests on tubular specimens under combined tension and torsion are presented. Particularly the results of combined tension and torsion tests show that the q versus t curves do not depend on the kind of the stress state. The material constants are identified for titanium alloys OT-4, BT-5 and BT-9, for the aluminium alloy D16T and for the steel 45. In [31] the Sosnin's dissipation damage measure is applied to the description of creep-damage of the titanium alloy OT-4 and the aluminium alloy D16T considering stress state effects. In [351] Życzkowski calculated the dissipation power P starting from the Kachanov-Rabotnov constitutive equation (2.4.20). He found that for a class of materials it is possible to express the damage evolution equation (2.4.21) in terms of the dissipation power. He concluded that this approach allows to reduce the number of material constants to be determined from creep tests.

2.4.2 Damage-Induced Anisotropy

For many metals and alloys the dominant damage mechanism is the nucleation and growth of cavities and formation of micro-cracks. Cavities nucleate on grain boundaries having different orientations. At the last stage before creep rupture the coalescence of cavities and the formation of oriented micro-cracks is observed. The preferential direction of micro-cracks depends on the material microstructure and on the kind of the applied stress. For example, micrographs of copper specimens tested under torsion show that the micro-cracks dominantly occur on the grain boundaries whose normals coincide with the direction of the maximum positive principal stress [137, 139, 218]. The strongly oriented micro-cracks may induce anisotropic creep responses particularly at the last stage of the creep process. Creep behavior of the

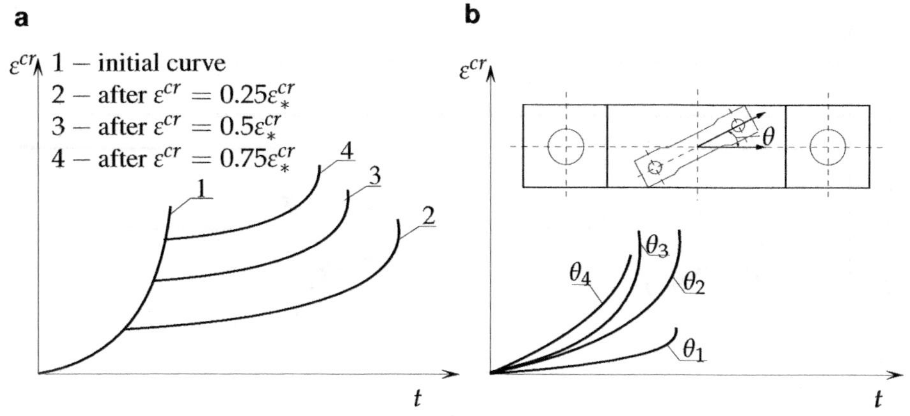

Fig. 2.17 Uni-axial creep tests with different orientations of the loading direction. **a** Creep curve for a flat specimen and creep curves for small specimens after different prestraining, **b** creep curves for different loading directions after pre-straining of $0.75\varepsilon^{cr}_*$ (after [66, 108])

austenitic steel X8 CrNiMoNb 1616 and the ferritic steel 13 CrMo 4 4 is experimentally studied in [66, 108] with respect to different loading orientations. Figure 2.17 schematically presents the results of testing. Uni-axial creep tests were carried out on flat specimens at different stress and temperature levels. In order to establish the influence of the creep history (pre-loading and pre-damage), series of flat specimens were tested up to different values of the creep strain. The values of the creep pre-straining were $\varepsilon^{cr} = 0.25\varepsilon^{cr}_*; 0.5\varepsilon^{cr}_*; 0.75\varepsilon^{cr}_*$, where ε^{cr}_* is the creep strain at fracture. After unloading, small specimens were manufactured from the pre-strained flat specimens with different orientation to the loading axis, Fig. 2.17b. The uni-axial tests performed on the small specimens show that the creep responses depend on the angle of the orientation θ. In [108] it is demonstrated that for small specimens pre-strained up to $0.25\varepsilon^{cr}_*$ the creep response is not sensitive to the angle θ. The significant dependence of the creep curves and the fracture times on the angle θ has been observed for specimens pre-strained up to $0.75\varepsilon^{cr}_*$.

In [224] creep tests were carried out on thin-walled copper tubes under combined tension and torsion. The loading history and the creep responses are schematically presented in Fig. 2.18. During the first cycle the specimens were preloaded by constant normal and shear stresses within the time interval $[0, t_1]$. In the second cycle from t_1 up to creep rupture the specimens were loaded under the same constant normal stress but the reversed constant shear stress. The stress state after the reversal is characterized by the change of the principal directions. The angle between the first principal direction in the reference loading cycle and after the reversal can be controlled by the values of the normal and the shear stresses. Creep curves for different angles are presented in [224]. It is demonstrated that the creep-damage model with a scalar damage parameter, see Sect. 2.4.1, is not able to predict the creep behavior after the shear stress reversal. Particularly, it significantly underestimates the

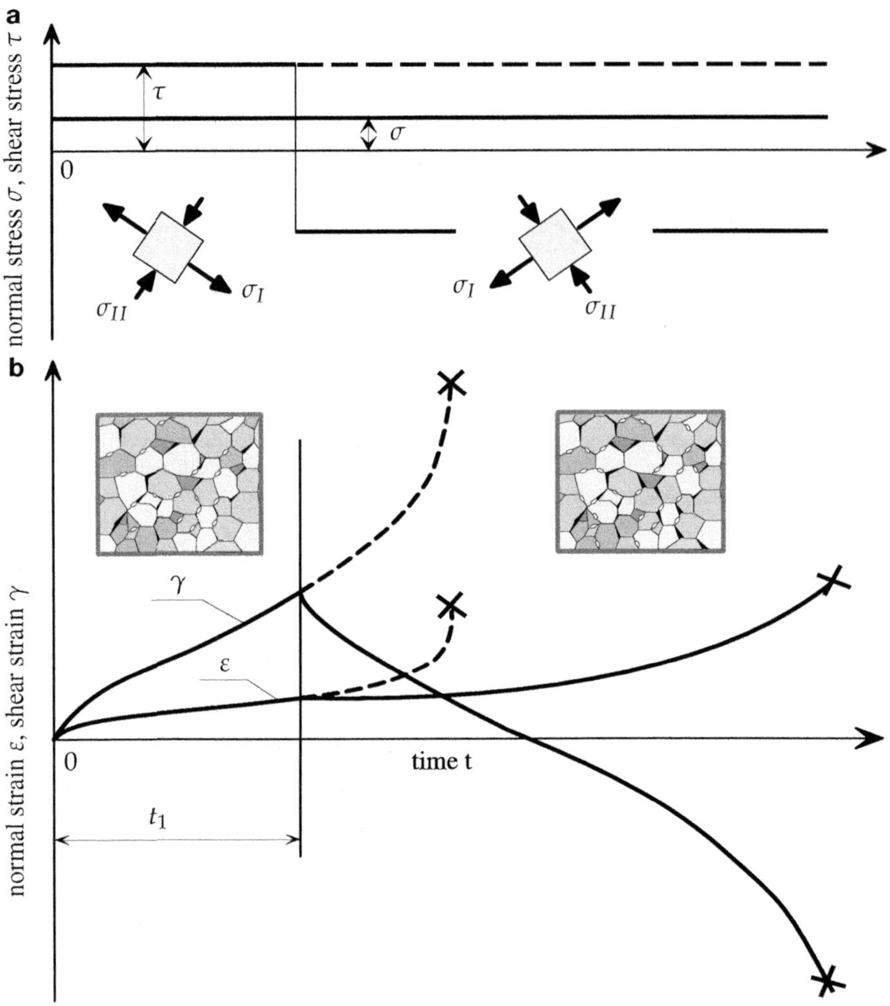

Fig. 2.18 Creep tests at combined tension and torsion. **a** Loading history, **b** creep responses (after [224])

fracture time in all loading cases. Similar results are discussed in [225] based on tests on Nimonic 80A.

The introduced examples of experimental observations indicate that the creep rate and the lifetime of a specimen additionally depend on the orientation of micro-defects with respect to the principal axes of the stress tensor. One way to consider such a dependence is the use of a tensor-valued damage parameter. A second rank damage tensor was firstly introduced by Vakulenko and M. Kachanov [324] for the description of elastic-brittle damage. The first attempt to use a tensor-valued damage parameter in creep mechanics is due to Murakami and Ohno [221, 223].

They considered a characteristic volume V in the material having N wedge cracks and specified the area of the grain boundary occupied by the kth crack by dA_g^k. They assumed that the state of damage can be characterized by the following second rank symmetric tensor

$$\boldsymbol{\Omega} = \frac{3}{A_g(V)} \sum_{k=1}^{N} \int_V [\boldsymbol{m}^k \otimes \boldsymbol{m}^k + w^k(\boldsymbol{I} - \boldsymbol{m}^k \otimes \boldsymbol{m}^k)] dA_g^k, \qquad (2.4.42)$$

where $\boldsymbol{m}^k$ is the unit normal vector to the kth crack and $A_g(V)$ is the total area of all grain boundaries in V. w^k characterizes the effect of the kth crack on the area reduction in the planes whose normals are perpendicular to $\boldsymbol{m}^k$. Specifying the three principal values of $\boldsymbol{\Omega}$ by $\Omega_j, j = 1, 2, 3$, and the corresponding principal directions by the unit vectors $\boldsymbol{n}_j$ the damage tensor can be formulated in the spectral form

$$\boldsymbol{\Omega} = \sum_{j=1}^{3} \Omega_j \boldsymbol{n}_j \otimes \boldsymbol{n}_j \qquad (2.4.43)$$

The principal values of the damage tensor Ω_j are related to the cavity area fractions in three orthogonal planes with the unit normals $\pm\boldsymbol{n}_j$. The cases $\Omega_j = 0$ and $\Omega_j = 1$ correspond to the undamaged state and the creep-rupture in the jth plane, respectively. By analogy with the uni-axial bar (see Fig. 2.12) Murakami and Ohno introduced a fictitious undamaged configuration in a solid by means of effective infinitesimal area elements. From three orthogonal planes having the unit normals $-\boldsymbol{n}_j$ an infinitesimal tetrahedron is constructed with area elements $-\tilde{\boldsymbol{n}}_i d\tilde{A}_i$ and $\tilde{\boldsymbol{n}} d\tilde{A}$ so that

$$\tilde{\boldsymbol{n}} d\tilde{A} = \sum_{j=1}^{3} \boldsymbol{n}_j d\tilde{A}_j = \sum_{j=1}^{3} (1 - \Omega_j) \boldsymbol{n}_j dA_j \qquad (2.4.44)$$

With $\Omega_j \boldsymbol{n}_j = \boldsymbol{n}_j \cdot \boldsymbol{\Omega} = \boldsymbol{\Omega} \cdot \boldsymbol{n}_j$

$$\tilde{\boldsymbol{n}} d\tilde{A} = (\boldsymbol{I} - \boldsymbol{\Omega}) \cdot \boldsymbol{n} dA \qquad (2.4.45)$$

The stress vector acting in the plane with the unit normal $\boldsymbol{n}$ can be specified by $\boldsymbol{\sigma}_{(\boldsymbol{n})}$. The resultant force vector acting in the plane dA is

$$dA\boldsymbol{\sigma}_{(\boldsymbol{n})} = dA\boldsymbol{n} \cdot \boldsymbol{\sigma} = d\tilde{A}\tilde{\boldsymbol{n}} \cdot (\boldsymbol{I} - \boldsymbol{\Omega})^{-1} \cdot \boldsymbol{\sigma} = d\tilde{A}\tilde{\boldsymbol{n}} \cdot \tilde{\boldsymbol{\sigma}}, \quad \tilde{\boldsymbol{\sigma}} \equiv (\boldsymbol{I} - \boldsymbol{\Omega})^{-1} \cdot \boldsymbol{\sigma}, \qquad (2.4.46)$$

where $\tilde{\boldsymbol{\sigma}}$ is the effective stress tensor. Introducing the so-called damage effect tensor $\boldsymbol{\Phi} \equiv (\boldsymbol{I} - \boldsymbol{\Omega})^{-1}$ one can write

$$\tilde{\boldsymbol{\sigma}} = \boldsymbol{\Phi} \cdot \boldsymbol{\sigma} \qquad (2.4.47)$$

According to the strain equivalence principle [191], the constitutive equation for the virgin material, for example the constitutive equation for the secondary creep, can be generalized to the damaged material replacing the Cauchy stress tensor $\boldsymbol{\sigma}$ by the net-stress tensor $\tilde{\boldsymbol{\sigma}}$. The net stress tensor (2.4.47) is non-symmetric. Introducing the symmetric part

$$\tilde{\boldsymbol{\sigma}}^S = \frac{1}{2}(\boldsymbol{\sigma} \cdot \boldsymbol{\Phi} + \boldsymbol{\Phi} \cdot \boldsymbol{\sigma}) \tag{2.4.48}$$

the secondary creep equation is generalized as follows [225]

$$\dot{\boldsymbol{\varepsilon}}^{cr} = \frac{3}{2} a \tilde{\sigma}_{vM}^{n-1} \tilde{\boldsymbol{s}}^S, \quad \tilde{\boldsymbol{s}}^S = \tilde{\boldsymbol{\sigma}}^S - \frac{1}{3} \mathrm{tr} \tilde{\boldsymbol{\sigma}}^S \boldsymbol{I}, \quad \tilde{\sigma}_{vM} = \sqrt{\frac{3}{2} \tilde{\boldsymbol{s}}^S \cdot\cdot\, \tilde{\boldsymbol{s}}^S} \tag{2.4.49}$$

The rate of the damage tensor is postulated as a function of the stress tensor and the current damage state. The following evolution equation is proposed in [224] for the description of creep damage of copper

$$\dot{\boldsymbol{\Omega}} = b[\alpha \tilde{\sigma}_I^S + (1-\alpha) \tilde{\sigma}_{vM}^S]^k (\boldsymbol{n}_I^{\tilde{\sigma}} \cdot \boldsymbol{\Phi} \cdot \boldsymbol{n}_I^{\tilde{\sigma}})^l \boldsymbol{n}_I^{\tilde{\sigma}} \otimes \boldsymbol{n}_I^{\tilde{\sigma}}, \tag{2.4.50}$$

where b, α, k and l are material constants and the unit vector $\boldsymbol{n}_I^{\tilde{\sigma}}$ denotes the direction corresponding to the first positive principal stress $\tilde{\sigma}_I$. The constitutive and evolution equations (2.4.49) and (2.4.50) have been applied in [225] for the description of creep-damage behavior of Nimonic 80A. The second rank damage tensor (2.4.43) and the net stress (2.4.48) have been used in [224] with McVetty-type creep equations for the prediction of creep-damage of copper. The results suggest that the model with the damage tensor provides better agreement with experimental data if compared to the model with a scalar-valued damage parameter. In [223] the following damage evolution equation is utilized

$$\dot{\boldsymbol{\Omega}} = b[\alpha \tilde{\sigma}_I^S + \beta \tilde{\sigma}_m + (1-\alpha-\beta) \tilde{\sigma}_{vM}^S]^k (\mathrm{tr}\, \boldsymbol{\Phi}^2)^{l/2} \left[\eta \boldsymbol{I} + (1-\eta) \boldsymbol{n}_I^{\tilde{\sigma}} \otimes \boldsymbol{n}_I^{\tilde{\sigma}}\right], \tag{2.4.51}$$

where β and η are material constants. This equation takes into account the influence of the mean stress on the damage rate. Furthermore, the isotropic part of the damage tensor associated with the growth of voids is included.

To discuss the damage tensor (2.4.43) let us consider a uni-axial homogeneous stress state $\boldsymbol{\sigma} = \sigma_0 \boldsymbol{m} \otimes \boldsymbol{m}$ with $\sigma_0 > 0$ and $\boldsymbol{m} = const$. Let us specify $\boldsymbol{\Omega} = \boldsymbol{0}$ as the initial condition. The evolution equation (2.4.50) takes the form

$$\dot{\boldsymbol{\Omega}}(t) = \dot{\omega}(t) \boldsymbol{m} \otimes \boldsymbol{m}, \quad \dot{\omega} = \frac{b \sigma_0^k}{(1-\omega)^{k+l}}, \quad \omega(0) = 0 \tag{2.4.52}$$

The equation for the scalar ω can be integrated as shown in Sect. 2.4.1.1. As a result one can find the relation between the time to fracture and the stress σ_0. Based on this relation and experimental data one can estimate the values of material constants b, k and l (Sect. 2.4.1.1). According to the introduced damage measure (2.4.43) the damage state $\boldsymbol{\Omega} = \omega \boldsymbol{m} \otimes \boldsymbol{m}$ corresponds to the case of uniformly distributed penny-shaped cracks (circular planes) with the unit normals $\boldsymbol{m}$.

Now let us assume that the damage state $\boldsymbol{\Omega} = \omega_0 \boldsymbol{m} \otimes \boldsymbol{m}$, $0 < \omega_0 < 1$ is induced as a result of the constant stress $\boldsymbol{\sigma} = \sigma_0 \boldsymbol{m} \otimes \boldsymbol{m}$ exerted over a period of time $[0, t_1]$ and in the next loading cycle the stress $\boldsymbol{\sigma} = \sigma_0 \boldsymbol{p} \otimes \boldsymbol{p}$, $\boldsymbol{p} \cdot \boldsymbol{m} = 0$ is applied. In this case the solution of (2.4.50) can be written down as follows

$$\boldsymbol{\Omega}(t) = \omega_0 \boldsymbol{m} \otimes \boldsymbol{m} + \omega_1(t) \boldsymbol{p} \otimes \boldsymbol{p}, \quad \dot{\omega}_1 = \frac{b\sigma_0^k}{(1-\omega_1)^{k+l}}, \quad \omega_1(0) = 0 \quad (2.4.53)$$

Equation (2.4.53) predicts that in the second cycle the material behaves like a virgin material. Then the total time to fracture can be calculated as follows

$$t_* = t_1 + \frac{1}{(k+l+1)b\sigma_0^k}$$

We observe that t_* does not depend on the damage value ω_0 induced during the first loading cycle. The rate of nucleation and growth of new voids (cracks) on the planes orthogonal to $\boldsymbol{p}$ will not be affected by cracks formed in the first loading cycle. Furthermore, if a compressive stress, i.e., $\boldsymbol{\sigma} = -\sigma_0 \boldsymbol{p} \otimes \boldsymbol{p}$ is applied in the second cycle the model predicts no damage accumulation.

Let us note that the evolution equations (2.4.50) and (2.4.51) can only be applied if $\tilde{\sigma}_I \neq 0$, $\tilde{\sigma}_I \neq \tilde{\sigma}_{II}$ and $\tilde{\sigma}_I \neq \tilde{\sigma}_{III}$. In this case the dyad $\boldsymbol{n}_I^{\tilde{\sigma}} \otimes \boldsymbol{n}_I^{\tilde{\sigma}}$ can be found from the identity (2.4.33). For the stress states $\boldsymbol{\sigma} = a_0 \boldsymbol{I}$ or $\boldsymbol{\sigma} = a\boldsymbol{p} \otimes \boldsymbol{p} + b(\boldsymbol{I} - \boldsymbol{p} \otimes \boldsymbol{p})$, $a < b$, there is an infinite number of first principal directions. Such stress states are typical for engineering structural components. For example, the stress state of the type $\boldsymbol{\sigma} = a\boldsymbol{p} \otimes \boldsymbol{p} + b(\boldsymbol{I} - \boldsymbol{p} \otimes \boldsymbol{p})$ arises in the midpoint of a transversely loaded square plate with all for edges to be fixed (e.g. supported or clamped edges), [16]. In the loaded (top) surface of such a plate $b < a < 0$ while in the bottom surface $b > a$, $a < 0, b > 0$. Stress states of the same type arise in different rotationally symmetric problems of structural mechanics. For analysis of such problems a modified form of the evolution equation (2.4.51) is required [122].

Various forms of creep-damage constitutive equations with second rank damage tensors have been utilized. In [15] the effective stress tensor

$$\tilde{\boldsymbol{\sigma}} = \boldsymbol{\Phi}^{1/2} \cdot \boldsymbol{\sigma} \cdot \boldsymbol{\Phi}^{1/2} \quad (2.4.54)$$

originally proposed in [94], is applied to formulate the creep-damage constitutive equation. Mechanisms of damage activation and deactivation are taken into account. The model predictions are compared with experimental data of creep in copper. In [265, 266, 267, 268] a second rank damage tensor is applied for the modeling of creep of nickel-based single crystal super-alloys SRR 99 and CMSX-6 at 760° C. The proposed constitutive equations take into account both the initial anisotropy and the damage induced anisotropy.

The symmetry group of a symmetric second rank tensor includes at least eight elements (see Sect. 2.3.2). With the second rank damage tensor and the effective stress tensors (2.4.48) or (2.4.54) only restrictive forms of orthotropic tertiary creep can be considered (a similar situation is discussed in Sect. 2.3.2). Therefore in many publications it is suggested to introduce higher order damage tensors. For different definitions of damage tensors one may consult [11, 13, 59, 177, 189, 298]. A critical review is given in [291]. At present, the available experimental data on creep responses do not allow to verify whether the orthotropic symmetry is an appropriate symmetry assumption for the modeling of anisotropic creep-damage processes.

From the micro-structural point of view one may imagine rather complex three-dimensional patterns of voids and cracks which nucleate and propagate as the result of multi-axial non-proportional loadings. An attempt to predict these patterns would result in a complex mathematical model with a large (or even infinite) number of internal variables including tensors of different rank. A model to characterize different patterns of cracks may be based on the orientation distribution function, orientation averaging and the so-called orientation tensors. This approach is widely used in different branches of physics and materials science for the statistical modeling of oriented micro-structures. Examples include fiber suspensions [187], mixtures [115], polymers and polymer composites [24, 315]. The application of orientational averaging to characterize damage states under creep conditions is discussed in [218, 247, 308].

Finally let us note, that the material behavior at elevated temperature and non-proportional loading is a complex interaction of different deformation and damage mechanisms such as hardening, recovery, creep-damage, fatigue-damage, etc. Several unified models utilize constitutive equations of creep with kinematic and/or isotropic hardening and include damage effects by means of the effective stress concept and the strain equivalence principle. In [161] the Malinin-Khadjinsky kinematic hardening rule, see Sect. 2.3.2 and isotropic Kachanov-Rabotnov type damage variable are discussed. The damage rate is additionally governed by the magnitude of the hardening variable, so that the coupling effect of damage and strain hardening/softening can be taken into account. It is shown that the kinematic hardening coupled with isotropic damage predicts well the effect of longer life-time after the stress reversal. In [101] the Chaboche-Rousselier visco-plasticity model is modified to predict the coupled creep-plasticity-damage behavior. The scalar damage variable is introduced as a sum of the accumulated time-dependent and cycle-dependent components. Various approaches to formulate a unified material model within the framework of continuum damage mechanics and thermodynamics of dissipative processes are discussed in [88, 89, 91, 191].

The verification of a unified model with non-linear anisotropic hardening and damage coupling requires a large number of independent tests under non-proportional loading. As a rule, accurate experimental data are rarely available. Furthermore, non-uniform stress and strain fields may be generated in a standard uni-axial specimen under non-proportional cyclic loading conditions [195]. They may be the reason for the large scatter of experimental data and misleading interpretations.

3 Examples of Constitutive Equations for Various Materials

In Chapter 2 we discussed theoretical approaches to the modeling of creep behavior. Chapter 3 deals with applications to various engineering materials. The models include specific forms of the constitutive equation for the creep rate tensor and evolution equations for internal state variables. In addition, constitutive functions of stress and temperature are specified. In order to find a set of material constants, creep tests under constant load and temperature leading to a homogeneous stress state are required. The majority of available experimental data is presented as creep strain versus time curves from standard uni-axial tests. Based on these curves the material constants are identified. It should be taken into account that experimental data may show a large scatter generated by testing a series of specimens removed from the same material. The origins of scatter in creep tests are discussed, for example, in [103]. Furthermore, unlike small strain elasticity, the creep behavior may significantly depend on the kind of processing of specimens, e.g. the heat treatment. As a result, different data sets for the material with the same chemical composition may be found in the literature. For example, one may compare experimental data for 9Cr1Mo (P91) ferritic steel obtained in different laboratories [3, 93, 106, 165, 249, 335].

Section 3.1 provides an overview of constitutive equations recently applied to characterize isotropic creep and long term strength of several alloys. The objective of Sect. 3.2 is to develop a model for anisotropic creep behavior in a weld metal produced by multi-pass welding. To explain the origins of anisotropic creep, a mechanical model for a binary structure composed of "fine-grained" and "coarse-grained" constituents with different creep properties is introduced. The results illustrate the basic features of the stress redistribution and damage growth in the constituents of the weld metal and agree qualitatively with experimental observations. The structural analysis of a welded joint requires a constitutive equation of creep for the weld metal under multi-axial stress states. For this purpose we apply the approaches developed in Sect 2.2.2 to model creep for initially anisotropic materials. The outcome is the a coordinate-free equation for secondary creep formulated in terms of the Norton-Bailey-Odqvist creep potential and three invariants of the stress tensor. The material constants are identified according to the experimental data presented in the literature.

3.1 Models of Isotropic Creep for Several Alloys

Models of isotropic creep are discussed in Sects. 2.2.1, 2.3 and 2.4.1. The creep rate tensor is assumed to be coaxial with the stress tensor (deviator) and the internal state variables characterizing the hardening and damage processes are scalars. The assumption of isotropic creep is usually a good approximation for many metals and alloys in the case of proportional loading. In this section we summarize the phenomenological and mechanism-based material models and present the specific forms of response functions and material constants.

3.1.1 Type 316 Steel

The first example is type 316 stainless steel at 650°C. In [199] the following creep equations are applied

$$\begin{aligned} &\dot{\boldsymbol{\varepsilon}}^{cr} = \frac{3}{2} f_1(\sigma_{vM}) g_1(\omega) \frac{\boldsymbol{s}}{\sigma_{vM}}, \quad \dot{\omega} = f_2\left[\sigma_{eq}^{\omega}(\boldsymbol{\sigma})\right] g_2(\omega), \\ &\boldsymbol{\varepsilon}^{cr}|_{t=0} = \boldsymbol{0}, \quad \omega|_{t=0} = 0, \quad 0 \leq \omega \leq \omega_*, \\ &\boldsymbol{s} = \boldsymbol{\sigma} - \frac{1}{3}\mathrm{tr}\,\boldsymbol{\sigma}\boldsymbol{I}, \quad \sigma_{vM} = \sqrt{\frac{3}{2}\boldsymbol{s} \cdot\cdot\, \boldsymbol{s}} \end{aligned} \tag{3.1.1}$$

Here $\boldsymbol{\varepsilon}^{cr}$ is the creep strain tensor, $\boldsymbol{\sigma}$ is the stress tensor, ω is the scalar-valued damage parameter and σ_{eq}^{ω} is the damage equivalent stress (see Sects. 2.2.1 and 2.4.1). The response functions f_1, f_2, g_1, and g_2 are

$$\begin{aligned} f_1(\sigma) &= a\sigma^n, \quad g_1(\omega) = (1-\omega)^{-n}, \\ f_2(\sigma) &= b\sigma^k, \quad g_2(\omega) = (1-\omega)^{-k} \end{aligned} \tag{3.1.2}$$

The material constants are presented in [199] as follows

$$\begin{aligned} &a = 2.13 \cdot 10^{-13}\,\mathrm{MPa}^{-n}/\mathrm{h}, \quad b = 9.1 \cdot 10^{-10}\,\mathrm{MPa}^{-k}/\mathrm{h}, \\ &n = 3.5, \quad k = 2.8 \end{aligned} \tag{3.1.3}$$

Note, that the constants a and b in (3.1.2) are found for the constant temperature. In the general case they must be replaced by functions of temperature. It is assumed that the damage evolution is controlled by the maximum tensile stress. In this case the damage equivalent stress takes the form

$$\sigma_{eq}^{\omega}(\boldsymbol{\sigma}) = \frac{\sigma_I + |\sigma_I|}{2},$$

where σ_I is the first principal stress. The elastic material behavior is characterized by the following values of the Young's modulus E and the Poisson's ratio ν

$$E = 1.44 \cdot 10^5\,\mathrm{MPa}, \quad \nu = 0.314 \tag{3.1.4}$$

Let us note that the response functions and material constants in Eqs. (3.1.1) can be found in the literature for numerous metals and alloys. Examples are presented in the monographs [80, 191, 208, 257, 262, 272, 298].

3.1.2 Steel 13CrMo4-5

In [296] the creep behavior of steel 13CrMo4-5 at 550°C is described by (3.1.1) with the following response functions

$$f_1(\sigma) = a\sigma^n, \quad g_1(\omega) = 1 - \rho + \rho(1-\omega)^{-n}, \\ f_2(\sigma) = b\sigma^k, \quad g_2(\omega) = (1-\omega)^{-l} \tag{3.1.5}$$

The material constants are

$$a = 1.94 \cdot 10^{-15}\,\text{MPa}^{-n}/\text{h}, \quad b = 3.302 \cdot 10^{-13}\,\text{MPa}^{-k}/\text{h}, \\ n = 4.354, \quad k = 3.955, \quad l = 1.423, \quad \rho = 0.393 \tag{3.1.6}$$

The damage equivalent stress is assumed in the form

$$\sigma^{\omega}_{eq}(\boldsymbol{\sigma}) = \alpha \frac{\sigma_I + |\sigma_I|}{2} + (1-\alpha)\sigma_{vM}$$

with $\alpha = 0.43$. The elastic material constants are $E = 1.6 \cdot 10^5$ MPa and $\nu = 0.3$.

3.1.3 Aluminium Alloy D16AT

Figure 3.1 shows the experimental data of uni-axial creep for the aluminium alloy AlCuMg2[1] at 300°C [167]. The creep behavior is described by (3.1.1) with the following response functions

$$f_1(\sigma) = a\sigma^n, \quad g_1(\omega) = (1-\omega^r)^{-n}, \\ f_2(\sigma) = b\sigma^k, \quad g_2(\omega) = (1-\omega^r)^{-k} \tag{3.1.7}$$

The material constants are estimated as follows [167]

$$a = 0.335 \cdot 10^{-7}\,\text{MPa}^{-n}/\text{h}, \quad b = 1.9 \cdot 10^{-7}\,\text{MPa}^{-k}/\text{h}, \\ n = k = 3, \quad r = 1.4, \quad \omega_* = 0.8 \tag{3.1.8}$$

The multi-axial tertiary creep is assumed to be controlled by the von Mises equivalent stress, i.e. $\sigma^{\omega}_{eq}(\boldsymbol{\sigma}) = \sigma_{vM}$. The elastic material constants are $E = 0.65 \cdot 10^5$ MPa and $\nu = 0.3$.

3.1.4 Aluminium Alloy BS 1472

The experimental data for aluminium alloy BS 1472 at 150± 0.5°C (Al, Cu, Fe, Ni, Mg and Si alloy) are published in [176]. The authors proposed to describe the uni-axial creep curves (loading conditions 227.53, 241.3 and 262 MPa) by use of two approaches. The first approach is based on (3.1.1) and the time hardening function. The proposed model is

[1] The given abbreviation (DIN 1745) correspond to Russian D16AT. The alloy is similar to the American alloy 24ST4.

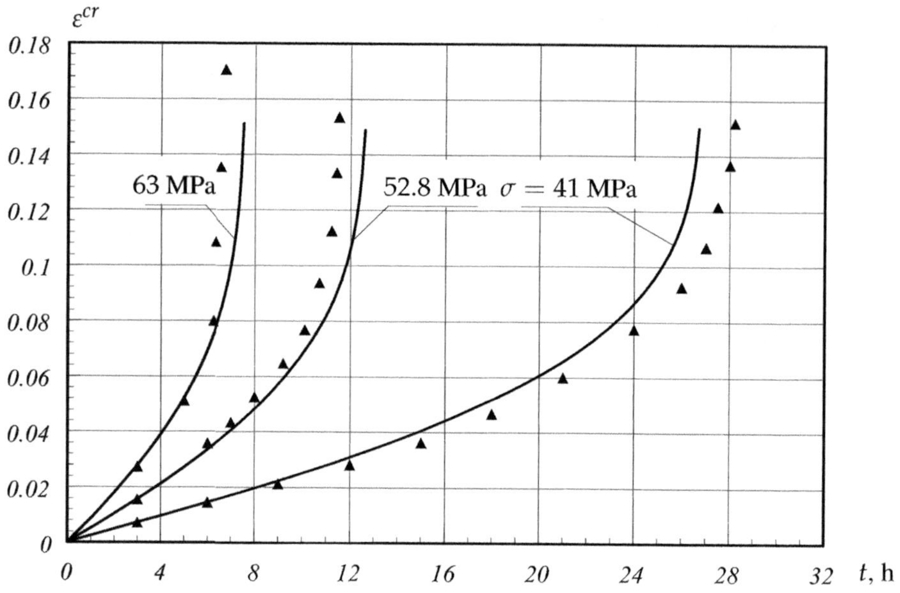

Fig. 3.1 Experimental data and model predictions for the aluminium alloy AlCuMg2 at 300°C (after [167])

$$\dot{\boldsymbol{\varepsilon}}^{cr} = \frac{3}{2}\frac{a\sigma_{vM}^{n-1}}{(1-\omega)^n}\mathbf{s}t^m, \quad \dot{\omega} = \frac{b(\sigma_{eq}^{\omega})^k}{(1-\omega)^l}t^m \tag{3.1.9}$$

with $\sigma_{eq}^{\omega} = \sigma_{vM}$. The material constants in (3.1.9) are identified as follows [176]

$$\begin{aligned} &a = 3.511 \cdot 10^{-31}\ \mathrm{MPa}^{-n}/\mathrm{h}^{m+1}, \quad b = 1.960 \cdot 10^{-23}\ \mathrm{MPa}^{-k}/\mathrm{h}^{m+1}, \\ &n = 11.034, \quad k = 8.220, \quad l = 12.107, \quad m = -0.3099 \end{aligned} \tag{3.1.10}$$

The elastic material constants are $E = 0.71 \cdot 10^5$ MPa and $\nu = 0.3$. Equations (3.1.9) include the time hardening function. One shortcoming of the time hardening model is that the creep behavior depends on the choice of the time scale (see Sect. 2.3.1).

Alternatively the experimental data presented in [176] can be described by the equations

$$\dot{\boldsymbol{\varepsilon}}^{cr} = \frac{3}{2}\frac{a\sigma_{vM}^{n-1}}{(1-\omega)^m}\mathbf{s}, \quad \dot{\omega} = \frac{b(\sigma_{eq}^{\omega})^k}{(1-\omega)^l} \tag{3.1.11}$$

with the following set of material constants

$$\begin{aligned} &a = 1.35 \cdot 10^{-39}\ \mathrm{MPa}^{-n}/\mathrm{h}, \quad b = 3.029 \cdot 10^{-35}\ \mathrm{MPa}^{-k}/\mathrm{h}, \\ &n = 14.37, \quad k = 12.895, \quad l = 12.5, \quad m = 10 \end{aligned} \tag{3.1.12}$$

In the above equations the primary creep effect is neglected. Figure 3.2 presents the experimental results and the predictions by Eqs. (3.1.9) and (3.1.11).

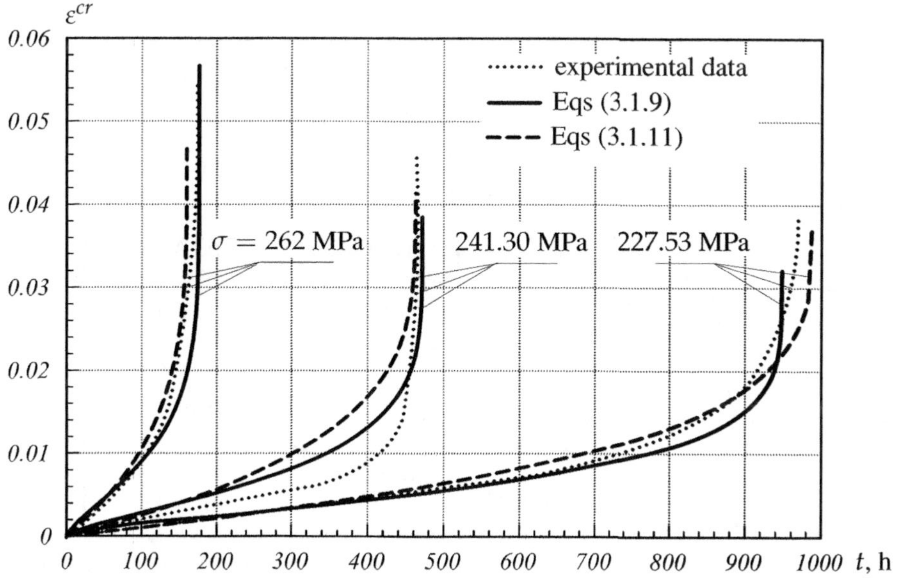

Fig. 3.2 Experimental data and model predictions for the aluminium alloy BS 1472 at $150 \pm 0.5°$C (after [176])

The second approach applied in [176] is based on the mechanism-based model (see Sect. 2.4.1). The model equations can be summarized as follows

$$
\begin{aligned}
\dot{\boldsymbol{\varepsilon}}^{cr} &= \frac{3}{2} \frac{A}{(1-\omega)^n} \frac{\boldsymbol{s}}{\sigma_{vM}} \sinh\left[\frac{B\sigma_{vM}(1-H)}{1-\Phi}\right], \\
\dot{H} &= \frac{h_c}{\sigma_{vM}} \frac{A}{(1-\omega)^n} \sinh\left[\frac{B\sigma_{vM}(1-H)}{1-\Phi}\right]\left(1-\frac{H}{H_*}\right), \\
\dot{\Phi} &= \frac{K_c}{3}(1-\Phi)^4, \\
\dot{\omega} &= \frac{DA}{(1-\omega)^n}\left(\frac{\sigma_I}{\sigma_{vM}}\right)^{\mu} N \sinh\left[\frac{B\sigma_{vM}(1-H)}{1-\Phi}\right], \\
n &= \frac{B\sigma_{vM}(1-H)}{1-\Phi} \coth\left[\frac{B\sigma_{vM}(1-H)}{1-\Phi}\right], \\
N &= 1 \quad \text{for} \quad \sigma_I > 0, \qquad N = 0 \quad \text{for} \quad \sigma_I \leq 0, \\
0 &\leq \omega < 0.3, \qquad 0 \leq \Phi < 1, \qquad 0 \leq H \leq H_*
\end{aligned}
\qquad (3.1.13)
$$

The set of equations (3.1.13) includes the creep constitutive equation and evolution equations with respect to three internal state variables. The hardening variable H is introduced to describe primary creep. The variable Φ characterizes the ageing process. The variable ω is responsible for the grain boundary creep constrained cavitation.

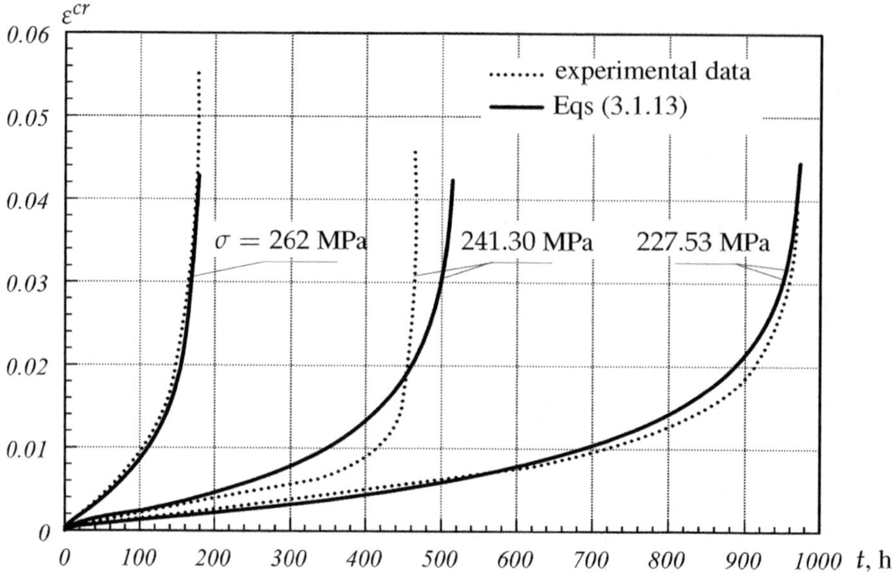

Fig. 3.3 Experimental data and simulations for the aluminium alloy BS 1472 at $150 \pm 0.5°$C (after [176])

The material constants in (3.1.13) may be divided into three groups: the constants h_c and H_* must be obtained from the primary creep stage; A and B characterize the secondary creep (minimum creep rate vs. stress); and K_c and D must be found from the tertiary creep stage. The constant μ is the so-called stress state index, which characterizes the stress state dependence of the damage evolution. The material constants are identified in [176] as follows

$$\begin{aligned} A &= 2.960 \cdot 10^{-11}\,\mathrm{h}^{-1}, & B &= 7.167 \cdot 10^{-2}\,\mathrm{MPa}^{-1}, \\ h_c &= 1.370 \cdot 10^{5}\,\mathrm{MPa}, & H_* &= 0.2032, \\ K_c &= 19.310 \cdot 10^{-5}\,\mathrm{h}^{-1}, & D &= 6.630 \end{aligned} \qquad (3.1.14)$$

Figure 3.3 presents the experimental creep curves and predictions based on (3.1.13). To identify the stress state index experimental data from multi-axial creep tests up to creep rupture are required. The isochronous rupture loci obtained according to Eqs. (3.1.13) for different values of μ in the range $0 \leq \mu \leq 12$ are presented in [176].

This example illustrates that the same experimental data can be described by quite different relations (3.1.9), (3.1.11) and (3.1.13). The model (3.1.13) seems to be more preferable since it is based on material science arguments. One feature of (3.1.13) is the use of a hyperbolic function for the dependence of the minimum creep rate on the stress instead of the power function in (3.1.11). Let us compare how the models (3.1.11) and (3.1.13) describe the secondary creep rate for a wide stress range. For this purpose we assume $\omega \ll 1$ in (3.1.11) leading to the Norton-Bailey creep equation $\dot{\varepsilon}^{cr}_{min} = a\sigma^n$. In (3.1.13) we set $H = H_*$, $\omega \ll 1$ and $\Phi \ll 1$ resulting in $\dot{\varepsilon}^{cr}_{min} = A \sinh[B\sigma(1 - H_*)]$. Figure 3.4 shows the minimum creep

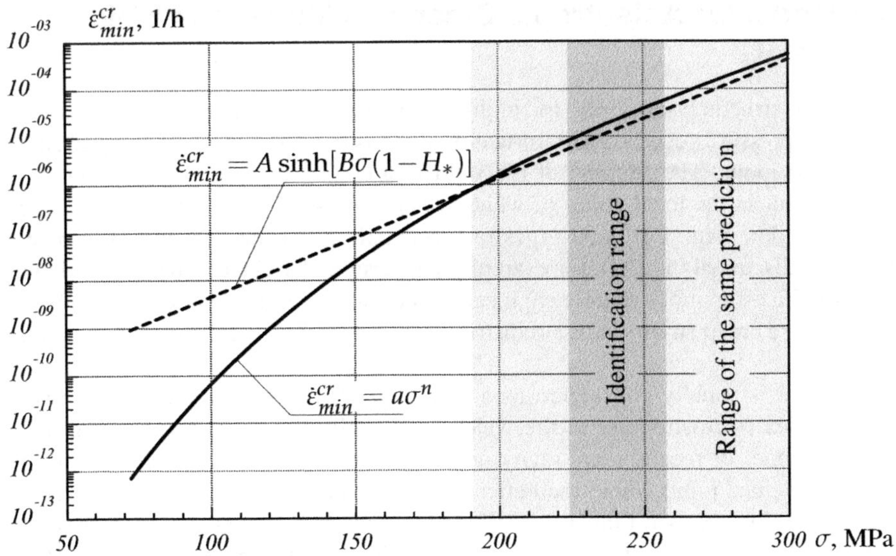

Fig. 3.4 Minimum creep rate vs. stress by use of the power law and sinh functions

rate as a function of stress calculated by the use of material constants (3.1.14) and (3.1.12). We observe that within the stress range 227 – 262 MPa the minimum creep rate vs. stress curves coincide. The coincidence of curves is not surprising since the material constants in both models were identified from creep tests carried out within the stress range 227 – 262 MPa. This stress range is marked in Fig. 3.4 as the identification range. Furthermore, a wider stress range exist, for which the power law and the hyperbolic sine functions provide nearby the same prediction, Fig. 3.4. If a structure is loaded in such a way that the von Mises equivalent stress lies within this range, than both the models would lead to similar results of structural analysis, e.g. time dependent deformations.

However, in most applications one has to analyze statically indeterminate structures. In this case, if the external loads are constant, the stresses may rapidly relax down at the beginning of the creep process. Therefore, the range of moderate and small stress values is important in the structural analysis. For this range the two applied models lead to quite different predictions, Fig. 3.4. In [8, 33] we utilized the models (3.1.11) and (3.1.13) for the structural analysis of pressurized cylindrical shells and transversely loaded rectangular plates. The maximum values of the von Mises equivalent stress in the reference elastic state of structures were within the identification range. The results of creep analysis based on the models (3.1.11) and (3.1.13) qualitatively agree only at the beginning of the creep process as long as the maximum values of the von Mises equivalent stress lay within the range of the same prediction. With the relaxation and redistribution of stresses, the discrepancy between the results increases leading to quite different long term predictions. The differences in estimated life times were of up to a factor 5.

3.2 Model for Anisotropic Creep in a Multi-Pass Weld Metal

For many structures designed for high-temperature applications, e.g., piping systems and pressure vessels, an important problem is the assessment of creep strength of welded joints. The lifetime of the welded structure is primarily determined by the behavior in the local zones of welds, where time-dependent creep and damage processes dominate. Different types of creep failure that have occurred in recent years are discussed in [297], for example. The design of welded structures and their residual life estimations require engineering mechanics models that would be able to characterize creep strains, stress redistributions, and damage evolution in the zones of welds.

A weld is usually considered as a metallurgical notch. The reason for this is the complex microstructure in the weld metal itself and in the neighboring heat-affected zone. In recent years many research activities have been directed to the study of welded joints. First, theoretical and experimental analyses have addressed the welding process with the aim of predicting the formation of the microstructure of the welds and analyzing residual stresses [38]. Second, the behavior of welded joints under the mechanical and thermal loadings was investigated [148]. Here one must consider that the stress-strain response at room temperature is quite different for the weld metal, the heat-affected zone, and the base metal (parent material), particularly if they are loaded beyond the yield limit. At elevated temperatures quite different inelastic strain vs. time curves can be obtained in different zones even in the case of a constant moderate load. Figure 3.5 illustrates zones with different microstructures and the variation in material behavior within the weld.

The results of creep tests of cross-weld specimens [146, 147], and specimens with a simulated microstructure [203, 211, 333, 335] show significant variation in creep properties in different material zones within the weld. Furthermore, they illustrate that the intercritical region of the heat-affected zone is the weakest part of the weld with respect to the creep properties. The material with the heat-affected zone microstructure usually exhibits the highest creep rate and the shortest time to failure if compared to other material zones within the weld for the same load and temperature.

For thick and moderately thick cross sections, multi-pass welding is usually preferred, where many stringer beads are deposited in a defined sequence. As a result of heating and cooling cycles during the welding process, the complex bead-type microstructure of the weld metal is formed, where every single bead consists of columnar, coarse-grained, and fine-grained regions, e.g., [148]. The results of uniaxial creep tests for the weld metal 9CrMoNbV are reported in [144]. They show that the creep strain vs. time curves significantly differ for specimens removed from the weld metal in the longitudinal (welding) direction and the transverse direction. Furthermore, different types of damage were observed for the longitudinal and the transverse specimens.

One possibility for studying the creep behavior in structures is the use of continuum damage mechanics, e.g., [23, 19, 136]. The application of this approach to

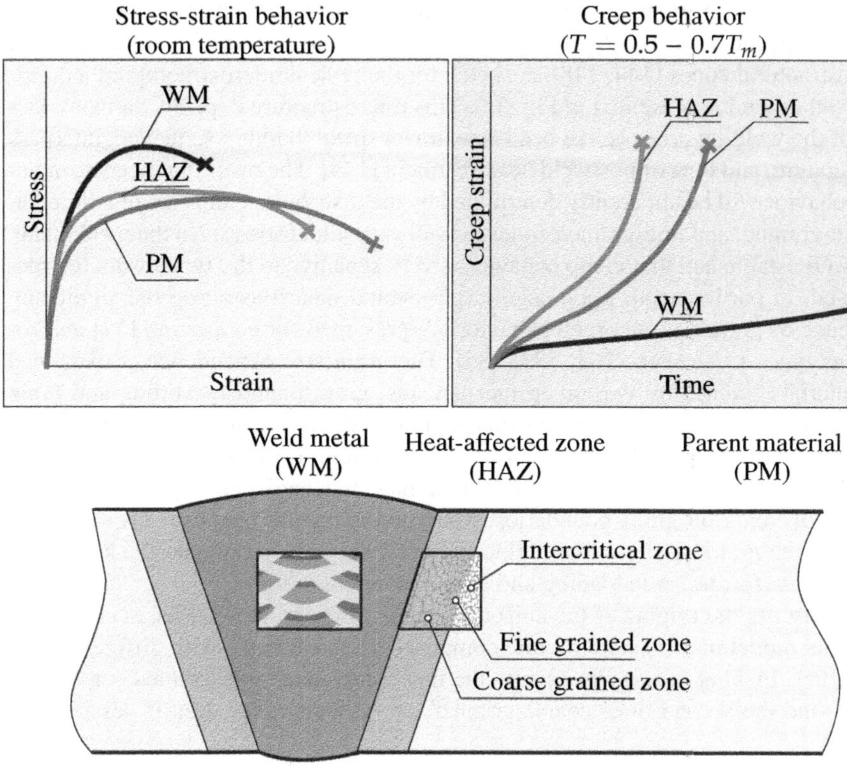

Fig. 3.5 Typical microstructure of welded joint and material behavior

welded joints is discussed in [132, 140, 148], for example. Here the weld is considered as a heterogeneous structure composed of at least three constituents - the weld metal, the heat-affected zone, and the parent material with different creep properties. Constitutive and evolution equations that are able to reflect experimental data of primary, secondary, and tertiary creep in different zones of the welded joint are presented in [106, 132, 140, 148, 333], among others. The results of finite element simulations illustrate stress redistributions, creep strains, and damage evolution in different zones of the weld [106, 132, 140, 148]. Furthermore, they allow to analyze the influence of numerous factors like weld dimensions, types of external loading, and material properties on the creep behavior of welded structures, e.g., [148]. However, as far as we know, the anisotropic creep of multi-pass weld metals has not been considered.

3.2.1 Origins of Anisotropic Creep

A weld bead produced by a single pass welding has a columnar solidification microstructure. During the multi-pass welding many weld beads are deposited in the groove by a defined sequence. As a subsequent weld bead is laid, the part of the metal produced in previous cycles is subjected to the local reheating and cooling. As a result, the weld beads consist of columnar, coarse-grained and fine-grained microstructural zones [144, 148]. A sketch for the typical microstructure of a multi-pass weld metal is presented in Fig. 3.6. This microstructure depends on many factors of the welding process like bead size, travel speed, buildup sequence, interpass temperature, and type of postweld heat treatment [144]. The resulting inelastic material behavior will be apparently determined by the distribution and size of columnar, coarse-grained, and fine-grained zones as well as residual stresses in the weld metal. It is well established that creep behavior is very sensitive to the type of microstructure and, in particular, to grain size. Experimental data illustrating the significant influence of grain size on creep behavior are presented for copper in [172] and for various types of steel in [203, 333, 335]. The grain size dependence is explained in materials science by two creep mechanisms: grain boundary sliding and grain boundary diffusion. These mechanisms operate under moderate loading and within a temperature range of $0.5 < T/T_m < 0.7$, where T_m is the melting temperature [228] (see Sect. 2.2.3). The principal damage mechanism is the nucleation and growth of voids on grain boundaries. Many experimental observations show that the finer the grain structure, the higher the secondary creep rate and the higher the damage rate for the same loading and temperature conditions.

To discuss the origins of the anisotropic creep in a weld metal let us consider a uni-axial model of a binary structure composed of constituents with different creep properties. In what follows let us term the first constituent "fine-grained" or "creep-weak" and the second one "coarse-grained" or "creep-strong." Let us describe the

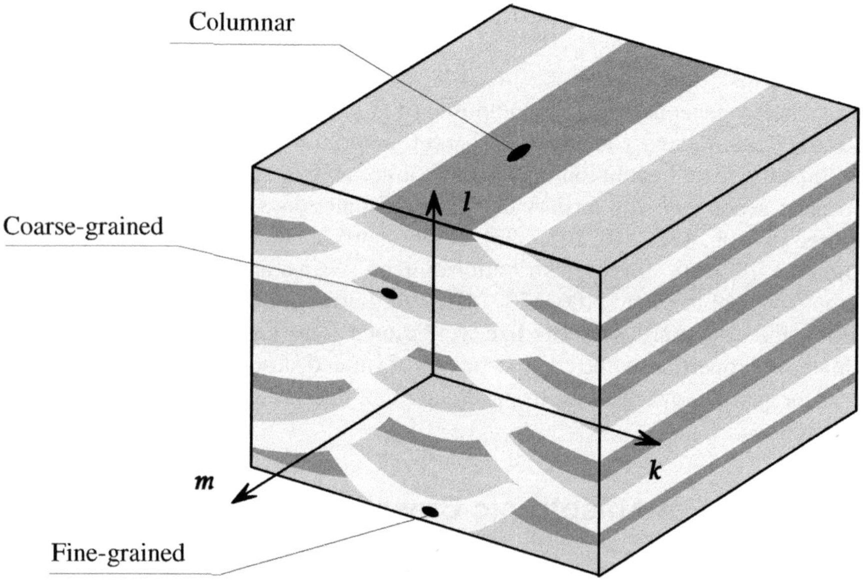

Fig. 3.6 Microstructure of the weld metal (after [148])

creep behavior of the constituents by use of the Kachanov-Rabotnov model (see Sect. 2.4.1.1)

$$\dot{\varepsilon}^{cr} = \frac{a\sigma^n}{(1-\omega)^n}, \qquad \dot{\omega} = \frac{b\sigma^k}{(1-\omega)^l} \tag{3.2.1}$$

In what follows we use the subscripts f and c for the fine-grained and coarse-grained constituents, respectively. For the sake of simplicity we assume that the constituents have the same value of Young's modulus E and the same values of constants n, k and l in (3.2.1). Let us introduce the dimensionless quantities

$$s = \frac{\sigma}{\sigma_0}, \quad \epsilon = \frac{\varepsilon}{\varepsilon_0}, \quad \epsilon^{cr} = \frac{\varepsilon^{cr}}{\varepsilon_0}, \quad \tau = \frac{t}{t_{*_f}}, \tag{3.2.2}$$

where t_{*_f} is the time to fracture of the fine-grained constituent, σ_0 is the reference stress and ε_0 is the elastic strain at σ_0, i.e. $\varepsilon_0 = \sigma_0/E$. Equations (3.2.1) can be formulated for two constituents as follows

$$\left\{ \begin{aligned} \frac{d\epsilon_f^{cr}}{d\tau} &= \tilde{a}\frac{s^n}{(1-\omega_f)^n} \\ \frac{d\omega_f}{d\tau} &= \tilde{b}\frac{s^k}{(1-\omega_f)^l} \end{aligned} \right. , \qquad \left\{ \begin{aligned} \frac{d\epsilon_c^{cr}}{d\tau} &= \alpha\tilde{a}\frac{s^n}{(1-\omega_c)^n} \\ \frac{d\omega_c}{d\tau} &= \beta\tilde{b}\frac{s^k}{(1-\omega_c)^l} \end{aligned} \right. , \tag{3.2.3}$$

where

$$\tilde{a} = \epsilon_{*_f}\left(1 - \frac{n}{l+1}\right), \quad \tilde{b} = \frac{1}{l+1}, \quad \alpha = \frac{\dot{\varepsilon}^{cr}_{min_c}}{\dot{\varepsilon}^{cr}_{min_f}}, \quad \beta = \frac{t_{*_f}}{t_{*_c}}$$

Figure 3.7 illustrates creep curves obtained after integration of (3.2.3) for the cases $n = 3, k = n + 1, l = n + 2, \epsilon_{*_f} = 5, \alpha = 0.15, \beta = 0.25, s = 1$.

Let us consider a connection of constituents in parallel, as is usually the case for composite materials, e.g. [7, 92]. The strains and the strain rates can be assumed to be the same (iso-strain concept)

$$\varepsilon = \varepsilon_f = \varepsilon_c, \quad \dot{\varepsilon} = \dot{\varepsilon}_f = \dot{\varepsilon}_c \tag{3.2.4}$$

We assume that a constant load $F = \sigma_0 A$, Fig. 3.8, is applied to the composite, where A is the cross section area. Specifying by N_f and N_c the internal forces in the constituents so that $N_f + N_c = F$ we can write

$$\sigma_f A_f + \sigma_c A_c = \sigma_0 A, \quad \eta_f \sigma_f + (1-\eta_f)\sigma_c = \sigma_0, \quad \eta_f s_f + (1-\eta_f)s_c = 1 \tag{3.2.5}$$

where $\eta_f = A_f/A$ is the volume fraction of the fine-grained constituent. For the stresses we apply the following constitutive equations

$$\sigma_f = E(\varepsilon - \varepsilon_f^{cr}), \quad \sigma_c = E(\varepsilon - \varepsilon_c^{cr}) \tag{3.2.6}$$

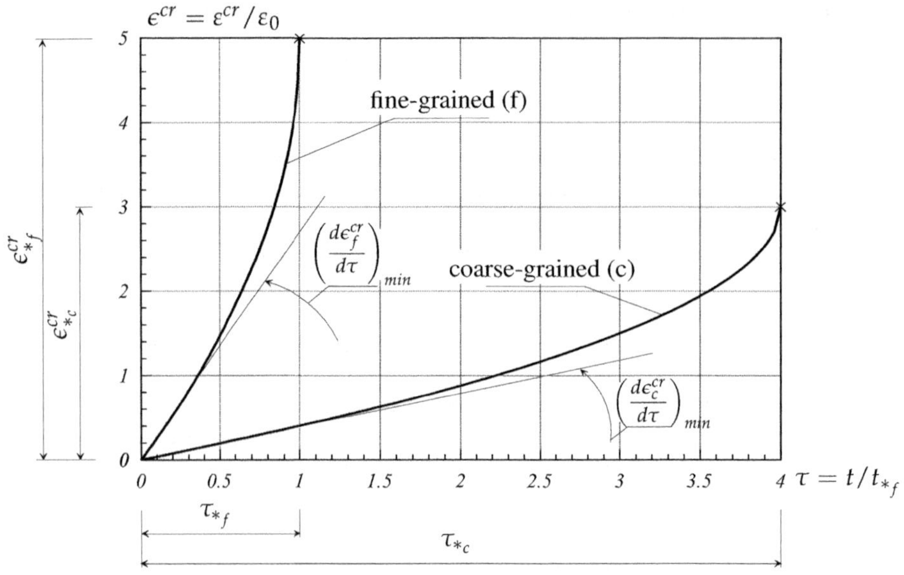

Fig. 3.7 Creep curves for constituents

Based on Eqs. (3.2.3) - (3.2.6) one can formulate a system of ordinary differential equations describing the stress redistribution between constituents. With respect to the stress in the fine-grained constituent the following equation can be obtained

$$\frac{ds_f}{d\tau} = \bar{a}(1 - \eta_f)\left[\frac{\alpha}{(1-\eta_f)^n}\frac{(1-\eta_f s_f)^n}{(1-\omega_c)^n} - \frac{s_f^n}{(1-\omega_f)^n}\right] \qquad (3.2.7)$$

Equation (3.2.7) is numerically solved together with the evolution equations for the damage parameters (3.2.3) and initial conditions $s_f = 1, \omega_f = \omega_c = 0$ providing time variation of the stress s_f. The stress s_c can be then computed from (3.2.5). The results are shown in Fig. 3.8 for the case $\eta_f = 0.3$. In addition, Fig. 3.9 presents creep strains and the damage parameters in the constituents as well as the creep strain of the "composite" $\epsilon^{cr} \equiv \epsilon - 1$. At the beginning of the creep process the creep rate is higher in the fine-grained constituent, Fig. 3.9a. Therefore, the stress in the fine-grained constituent relaxes down while the stress in the coarse-grained constituent increases, Fig. 3.8. If we neglect the influence of damage on the creep process, i.e. set $\omega_f = \omega_c = 0$ in (3.2.7), we obtain the steady-state creep solution. The corresponding results are plotted in Fig. 3.8 by dotted lines. We observe that the maximum value of s_c and the minimum value of s_f in the case of creep-damage almost coincide with the corresponding steady-state values. The steady-state solution for s_f follows from (3.2.7) by setting $\omega_f = \omega_c = 0$ and $\frac{ds_f}{d\tau} = 0$. The corresponding value for s_c is obtained from (3.2.5). The results are

$$s_{f_{min}} = \frac{\alpha^{\frac{1}{n}}}{1 - \eta_f(1 - \alpha^{\frac{1}{n}})}, \qquad s_{c_{max}} = \frac{1}{1 - \eta_f(1 - \alpha^{\frac{1}{n}})}$$

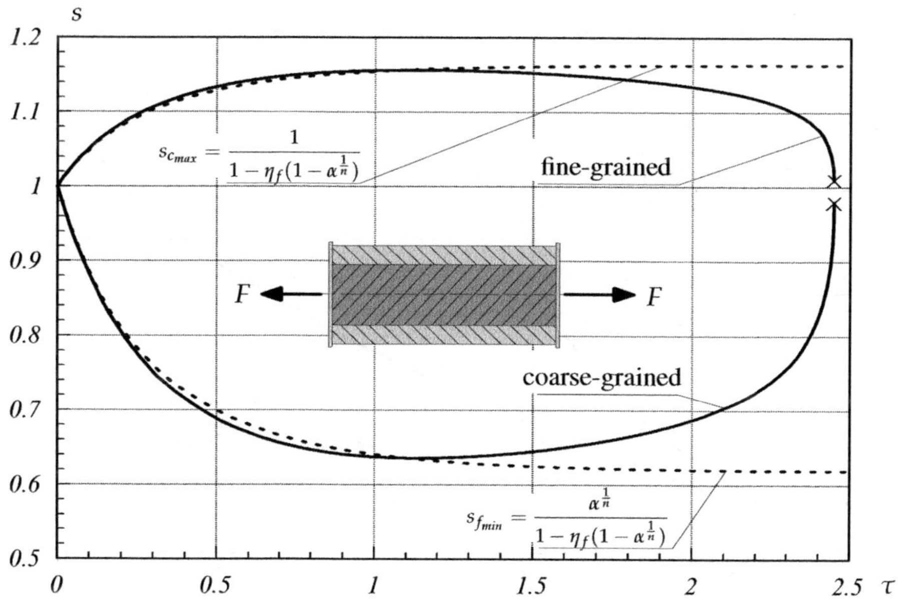

Fig. 3.8 Normalized stresses vs. normalized time for connection of constituents in parallel

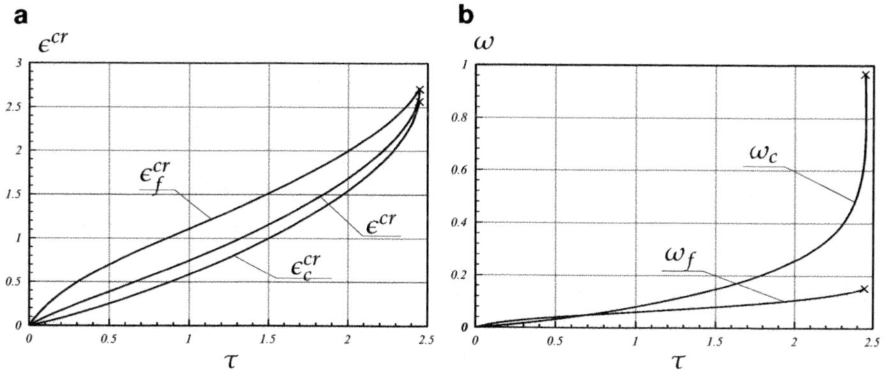

Fig. 3.9 Connection of constituents in parallel. **a** Normalized creep strains vs. normalized time, **b** damage parameters vs. normalized time

We observe that these stress values are determined by the volume fraction of the "fine-grained" constituent η_f and the ratio of minimum creep rates α. The stress value s_c is higher than s_f after the initial stress redistribution. Therefore, the coarse-grained constituent exhibits the higher creep rate and the higher damage rate in the final stage of the creep process, Fig. 3.9. The calculation predicts the failure initiation in the coarse-grained constituent.

In the case of a connection of constituents in series (iso-stress concept) we assume

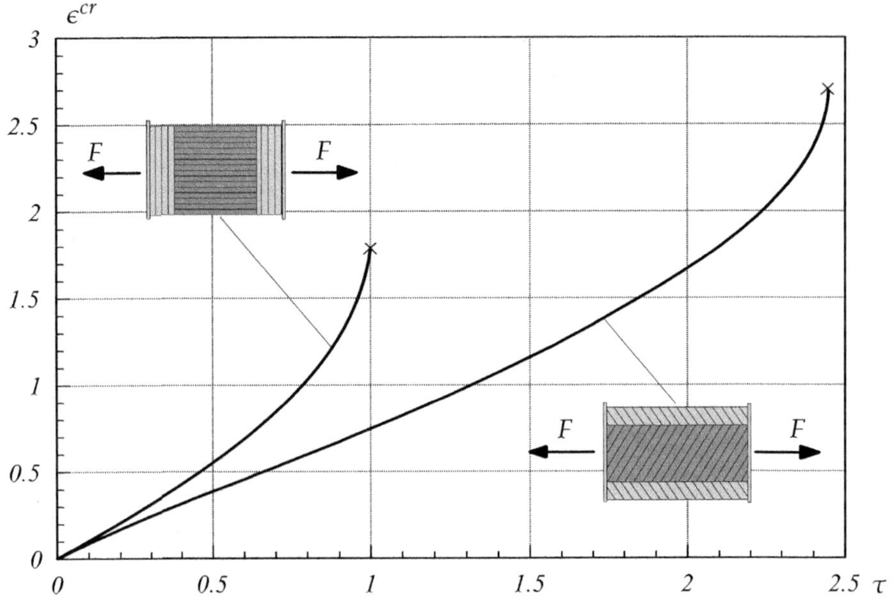

Fig. 3.10 Creep curves for the binary structure in the cases of parallel and series connections of constituents

$$\sigma_0 = \sigma_f = \sigma_c, \quad \varepsilon^{cr} = \eta_f \varepsilon_f^{cr} + (1 - \eta_f)\varepsilon_c^{cr}$$

The results can be obtained by integration (3.2.3) for $s_f = s_c = 1$. The corresponding plots of normalized creep strains are presented in Fig. 3.7. The maximum creep and damage rates are now in the fine-grained constituent. The lifetime of the binary structure is determined by the lifetime of the fine-grained constituent for the given constant stress.

Figure 3.10 shows the creep curves obtained for the two considered cases of the binary structure under the same constant load. The results of the presented model provide an analogy to the creep behavior of a weld metal loaded in the longitudinal (welding) and the transverse directions. The experimental creep curves for the specimen removed from the weld metal in two directions are presented in [144]. They show, that the transverse specimens exhibit higher minimum creep rate. Furthermore, the creep curves for transverse specimens have a much shorter tertiary stage and lower values of fracture strain if compared to curves for specimens removed in the welding direction. The times to fracture for the transverse specimens are much shorter than those for the longitudinal specimens. From the results in Fig. 3.10 we observe that these effects are predicted by the mechanical model of the binary structure. Furthermore, our results for the damage evolution qualitatively agree with the results of microstructural damage observations presented in [144]. For the longitudinal specimens extensive voids and cracks were observed in columnar and coarse-grained regions along the entire specimen length. For the transverse specimens voids and cracks are localized near the fracture surface. The fracture surface has fine-grained structure and the failure propagated through the fine-grained regions of the specimen.

Based on the presented results we may conclude that among many different creep and damage mechanisms which may operate and interact during the creep process an essential role plays the stress redistribution between the creep-weak and creep-strong constituents. For longitudinal specimens this mechanism leads to a prolonged tertiary creep stage. The material behaves like a "more ductile" material, although the damage and failure occur in the "more brittle" creep-strong constituent.

3.2.2 Modeling of Secondary Creep

For the analysis of welded structures a model which is able to reflect anisotropic creep in a weld metal under multi-axial stress states has to be developed. Three-dimensional models for binary or multi-component media are discussed within the framework of continuum mechanics (e.g. [25]). A generalization of the composite model developed in the previous section to the multi-axial stress states would however require the knowledge of creep properties of constituents under multi-axial stress states. Furthermore, creep mechanisms of interaction between constituents, like frictional sliding should be taken into account.

In what follows we assume the weld metal to be a quasi-homogeneous anisotropic material. For a description of creep we prefer the engineering creep mechanics approach, where the creep potential hypothesis, the representation of tensor functions and internal state variables are incorporated (see Chapt. 2). The resulting constitutive equations are compatible with the finite element method and can be utilized in standard finite element codes for structural analysis purposes.

Examples for anisotropic creep behavior and related constitutive equations are presented for single-crystal alloys in [52] and for fiber-reinforced materials in [280]. One problem of anisotropic creep modeling is that the assumed material symmetries (microstructure symmetries) are difficult to verify in creep tests due to the relatively large scatter of experimental data. Furthermore, the material may lose some or even all symmetries during the creep as a consequence of hardening and damage processes.

In our case the material symmetries can be established according to the arrangement of the weld beads in the weld metal. For the structure presented in Fig. 3.6 one can assume the reflection $\boldsymbol{Q}_1 = \boldsymbol{I} - 2\boldsymbol{m} \otimes \boldsymbol{m}$, the rotation $\boldsymbol{Q}_2 = 2\boldsymbol{l} \otimes \boldsymbol{l} - \boldsymbol{I}$ and the reflection $\boldsymbol{Q}_3 = \boldsymbol{Q}_1 \cdot \boldsymbol{Q}_2 = \boldsymbol{I} - 2\boldsymbol{k} \otimes \boldsymbol{k}$ to be the elements of the material symmetry group, where $\boldsymbol{I}$ is the second rank unit tensor and $\boldsymbol{k}$, $\boldsymbol{l}$ and $\boldsymbol{m}$ are orthogonal unit vectors.

However, this material symmetry group is poor for the modeling of creep. Indeed, based on the model discussed in the previous section we can assume that the same creep mechanisms will operate by loading the weld metal in $\boldsymbol{k}$- or $\boldsymbol{l}$-directions. Although the experimental data presented in [144] are available only for specimen removed in $\boldsymbol{m}$- and $\boldsymbol{k}$-directions, one may assume that that the difference between the experimental creep curves by loading in $\boldsymbol{k}$- and $\boldsymbol{l}$-directions will be not essential with respect to the usual scatter of experimental data. Here we assume transversely isotropic creep, where the plane spanned on the vectors $\boldsymbol{k}$ and $\boldsymbol{l}$ is the quasi-isotropy plane.

The models of steady-state creep under the assumption of transverse isotropy are derived in Sec 2.2.2.1 and 2.2.2.2. Here we apply the creep constitutive equation (2.2.43).

3.2.3 Identification of Material Constants

In the equivalent stress expression (2.2.41) the α_i's play the role of dimensionless factors. Three independent uniform stress states should be realized in order to determine α_i. The relevant stress states are

- Uni-axial tension in the direction $\boldsymbol{m}$ (longitudinal tension test). In this case the stress tensor is $\boldsymbol{\sigma} = \sigma_0 \boldsymbol{m} \otimes \boldsymbol{m}$, where $\sigma_0 > 0$ is the magnitude of the applied stress. From (2.2.41) and (2.2.43) follows

$$J_m = \sigma_0, \quad I_{3m} = I_{4m} = 0, \quad \sigma_{eq} = \sigma_0\sqrt{\alpha_1},$$
$$\dot{\boldsymbol{\varepsilon}}^{cr} = \sqrt{\alpha_1}\dot{\varepsilon}_{eq}\left[\boldsymbol{m}\otimes\boldsymbol{m} - \frac{1}{2}(\boldsymbol{I} - \boldsymbol{m}\otimes\boldsymbol{m})\right] \qquad (3.2.8)$$

- Uni-axial tension in the direction $\boldsymbol{k}$ (transverse tension test), i.e. $\boldsymbol{\sigma} = \sigma_0 \boldsymbol{k}\otimes\boldsymbol{k}$, $\sigma_0 > 0$. From (2.2.41) and (2.2.43) we obtain

$$\boldsymbol{s}_p = \frac{1}{2}\sigma_0(\boldsymbol{k}\otimes\boldsymbol{k} - \boldsymbol{l}\otimes\boldsymbol{l}), \quad J_m = -\frac{1}{2}\sigma_0,$$
$$I_{3m} = \frac{1}{4}\sigma_0^2, \quad I_{4m} = 0, \quad \sigma_{eq} = \frac{1}{2}\sigma_0\sqrt{\alpha_1 + 3\alpha_2},$$
$$\dot{\boldsymbol{\varepsilon}}^{cr} = \frac{\dot{\varepsilon}_{eq}}{2\sqrt{\alpha_1+3\alpha_2}}\left[(\alpha_1+3\alpha_2)\boldsymbol{k}\otimes\boldsymbol{k} + (\alpha_1-3\alpha_2)\boldsymbol{l}\otimes\boldsymbol{l} - 2\alpha_1\boldsymbol{m}\otimes\boldsymbol{m}\right] \qquad (3.2.9)$$

- Uniform shear in the plane spanned on $\boldsymbol{m}$ and $\boldsymbol{k}$, i.e. $\boldsymbol{\sigma} = \tau_0(\boldsymbol{m}\otimes\boldsymbol{k} + \boldsymbol{k}\otimes\boldsymbol{m})$, $\tau_0 > 0$. From (2.2.41) and (2.2.43)

$$J_m = I_{3m} = 0, \quad I_{4m} = \tau_0^2, \quad \dot{\boldsymbol{\varepsilon}}^{cr} = \frac{\sqrt{3\alpha_3}}{2}\dot{\varepsilon}_{eq}(\boldsymbol{m}\otimes\boldsymbol{k} + \boldsymbol{k}\otimes\boldsymbol{m}) \qquad (3.2.10)$$

The next step is the form of the creep potential $W(\sigma_{eq})$ or the form of the creep rate vs. stress dependence in the steady-state range. The criteria for the choice of a suitable function are the type of the deformation mechanisms operating for the given stress and temperature range as well as the best fitting of the experimentally obtained strain vs. time curves. Experimental data for the weld metal 9CrMoNbV are presented in [144] for the stress range 87-100 MPa and the constant temperature 650°C. The authors used a power law in order to fit the experimental data for secondary creep of longitudinal and transverse specimens. In this case the Norton-Bailey-Odqvist creep potential can be applied [243]

$$W(\sigma_{eq}) = \frac{a}{n+1}\sigma_{eq}^{n+1}, \quad \dot{\varepsilon}_{eq} = a\sigma_{eq}^{n}, \tag{3.2.11}$$

where a and n are material constants. For the longitudinal direction from (3.2.8) and (3.2.11) it follows

$$\dot{\varepsilon}_L^{cr} \equiv \boldsymbol{m} \cdot \dot{\boldsymbol{\varepsilon}}^{cr}\boldsymbol{m} = a_L\sigma_0^n, \quad a_L \equiv a\alpha_1^{\frac{n+1}{2}} \tag{3.2.12}$$

Taking the longitudinal direction to be the "reference" direction we set in (3.2.12) $\alpha_1 = 1$. From (3.2.9) and (3.2.11) we obtain for the transverse direction

$$\dot{\varepsilon}_T^{cr} \equiv \boldsymbol{k} \cdot \dot{\boldsymbol{\varepsilon}}^{cr} \cdot \boldsymbol{k} = a_T\sigma_0^n, \quad a_T \equiv a\left(\frac{1+3\alpha_2}{4}\right)^{\frac{n+1}{2}} \tag{3.2.13}$$

In [144] the values for the material constants are presented. However, the exponent n is found to be different for the longitudinal and the transverse directions. Different values for n contradict to the creep potential hypothesis employed in the previous section. Here we compute the values for a_L, a_T and n based on the following functional

$$\begin{gathered} F(\tilde{a}_L, \tilde{a}_T, n) = \sum_{i=1}^{k}(\tilde{a}_L + n\tilde{\sigma}_i - \tilde{\dot{\varepsilon}}_{L_i})^2 + \sum_{i=1}^{k}(\tilde{a}_T + n\tilde{\sigma}_i - \tilde{\dot{\varepsilon}}_{T_i})^2, \\ \tilde{a}_L \equiv \log a_L, \quad \tilde{a}_T \equiv \log a_T, \quad \tilde{\sigma} \equiv \log \sigma_0, \\ \tilde{\dot{\varepsilon}}_L \equiv \log \dot{\varepsilon}_L, \quad \tilde{\dot{\varepsilon}}_T \equiv \log \dot{\varepsilon}_T, \end{gathered} \tag{3.2.14}$$

where k is the number of experimental data points. Setting the first variation of F to zero leads to the system of three algebraic equations with respect to $\tilde{a}_L$, $\tilde{a}_T$ and n. As the result we obtain the following set of material constants

$$\begin{aligned} a_L &= 1.377 \cdot 10^{-21}\mathrm{MPa}^{-n}/\mathrm{h}, \\ a_T &= 2.023 \cdot 10^{-21}\mathrm{MPa}^{-n}/\mathrm{h}, \quad n = 8.12 \end{aligned} \tag{3.2.15}$$

Figure 3.11 shows the experimental data presented in [144] and the numerical predictions by use of (3.2.12), (3.2.13) and (3.2.15).

Finally let us summarize the constitutive equation for secondary creep and the set of identified material constants as follows

$$\begin{gathered} \dot{\boldsymbol{\varepsilon}}^{cr} = \frac{3}{2}a\sigma_{eq}^{n-1}\left[J_m\left(\boldsymbol{m}\otimes\boldsymbol{m} - \frac{1}{3}\boldsymbol{I}\right) + \alpha_2\boldsymbol{s}_p + \alpha_3(\boldsymbol{\tau}_m \otimes \boldsymbol{m} + \boldsymbol{m}\otimes\boldsymbol{\tau}_m)\right], \\ \sigma_{eq}^2 - \left(\boldsymbol{m}\cdot\boldsymbol{\sigma}\cdot\boldsymbol{m} - \frac{1}{2}\mathrm{tr}\,\boldsymbol{\sigma}_p\right)^2 + \frac{3}{2}\alpha_2\mathrm{tr}\,\boldsymbol{s}_p^2 + 3\alpha_3\tau_{\boldsymbol{m}}^2, \\ a = 1.377\cdot 10^{-21}\mathrm{MPa}^{-n}/\mathrm{h}, \quad n = 8.12, \quad \alpha_2 = 1.117 \end{gathered} \tag{3.2.16}$$

The weighting factor α_3, which stands for the influence of the transverse shear stress, remains undetermined in (3.2.16). Future work should be directed toward the understanding of creep and damage mechanisms in weld metals and related testing under stress states with nonzero vector $\boldsymbol{\tau}_m$.

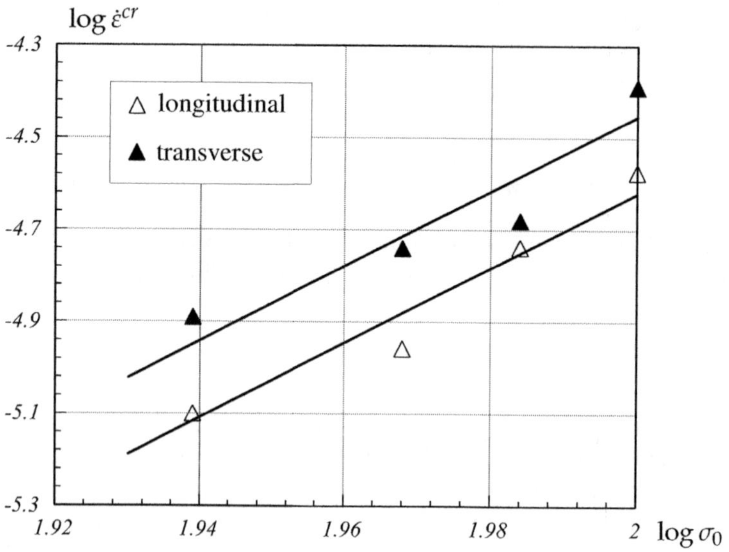

Fig. 3.11 Minimum creep rates vs. stress (experimental data after [144])

Model (3.2.16) is limited only to secondary creep behavior and allows to reproduce only the secondary part of the creep curves presented in [144]. For the description of the whole creep process including the primary and tertiary creep stages, model (3.2.16) can be modified by use of hardening and damage variables.

4 Modeling of Creep in Structures

In Chapters 2 and 3 we introduced constitutive and evolution equations for the modeling of creep in engineering materials. The objective of Chapt. 4 is the application of creep constitutive models to structural analysis. In Sect. 4.1 we start with the discussion of aims and basic steps in modeling of creep in structures. In Sect. 4.2 we formulate initial-boundary value problems describing creep behavior in three-dimensional solids and give an overview on numerical solution procedures. Sections 4.3 - 4.4 are devoted to the review and evaluation of structural mechanics models of beams, plates and shells in the context of their applicability to the analysis of creep and long term strength. For several problems we develop closed-form solutions and special numerical solutions based on the Ritz method. The results are applied to verify finite element solutions obtained by a general purpose finite element code and a user-defined material subroutine. Special numerical examples are selected to illustrate the influence of various discretisztion parameters (mesh size, number of Gauss points, etc.) on the solution accuracy. Furthermore, they allow to compare creep lifetime predictions based on different structural mechanics models and related types of finite elements. To discuss the applicability of the developed techniques to real engineering problems an example of a spatial steam pipeline is presented. Long term behavior of the pipeline under constant internal pressure and constant temperature is simulated by the finite element method. Numerical results are compared with the data from engineering practice.

4.1 General Remarks

The aim of creep modeling is to reflect basic features of creep in structures including the development of inelastic deformations, relaxation and redistribution of stresses as well as the local reduction of material strength (see Sect. 1.2). A model should be able to account for material deterioration processes in order to predict long term structural behavior and to analyze critical zones of creep failure. Structural analysis under creep conditions usually requires the following steps:

1. Assumptions must be made with regard to the geometry of the structure, types of loading and heating as well as kinematical constraints.
2. A suitable structural mechanics model must be applied based on the assumptions concerning kinematics of deformations, types of internal forces (moments) and related balance equations.

3. A reliable constitutive model must be formulated to reflect time dependent creep deformations and processes accompanying creep like hardening/recovery and damage.
4. A mathematical model of the structural behavior (initial-boundary value problem) must be formulated including the material independent equations, constitutive (evolution) equations as well as initial and boundary conditions.
5. Numerical solution procedures to solve non-linear initial-boundary value problems must be developed.
6. The verification of the applied models must be performed including the structural mechanics model, the constitutive model, the mathematical model as well as the numerical methods and algorithms.

The first two steps are common within continuum mechanics and engineering mechanics. Here, mathematical models of idealized solids and structures are developed and investigated. Examples include the models of three-dimensional solids, beams, rods, plates and shells. The idealizations are related to the continuum hypothesis, cross section assumptions, etc. The above models were originally developed within the theory of linear elasticity, e.g. [129, 316]. In creep mechanics they are applied together with constitutive and evolution equations describing idealized creep behavior (e.g. steady-state creep) [80, 142, 178, 208, 241]. As mentioned in Sect. 1.1.2 and Chapt. 2, many structural materials exhibit non-classical creep phenomena such as different creep rates under tension and compression, stress state dependence of tertiary creep, damage induced anisotropy, etc. Consideration of such effects may require various extensions of available structural mechanics models. For example, the concept of the stress free (neutral) plane widely used in the theory of beams and plates becomes invalid in creep mechanics if the material shows different creep rates under tension and compression (see Sect. 1.2). Below we discuss the applicability of classical and refined models of beams, plates and shells to the creep analysis. Bases on several examples we examine the accuracy of cross section assumptions for displacement and stress fields.

The mathematical model of creep in structure is the initial-boundary value problem (IBVP) which usually includes partial differential equations describing kinematics of deformation and balance of forces, ordinary differential equations describing creep processes as well as initial and boundary conditions. The numerical solution can be organized as follows, e.g. [80, 257]. For known values of the creep strain tensor and internal state variables at a fixed time the boundary value problem (BVP) is solved. Here direct variational methods, e.g. the Ritz method, the Galerkin method, the finite element method are usually applied. In addition, a time step procedure is required to integrate constitutive and evolution equations of creep. Below various methods are reviewed and discussed with respect to their efficiency and numerical accuracy.

In recent years the finite element method has become the widely accepted tool for structural analysis. The advantage of the finite element method is the possibility to model and analyze engineering structures with complex geometries, various types of loadings and boundary conditions. General purpose finite element codes

ABAQUS, ANSYS, COSMOS, MARC etc. were developed to solve various problems in solid mechanics. In application to the creep analysis one should take into account that a general purpose constitutive equation which allows to reflect the whole set of creep and damage processes in structural materials over a wide range of loading and temperature conditions is not available at present. Therefore, a specific constitutive model with selected internal state variables, special types of stress and temperature functions as well as material constants identified from available experimental data should be incorporated into the commercial finite element code by writing a user-defined material subroutine. Below the ABAQUS and ANSYS finite element codes are applied to the numerical analysis of creep in structures. In order to consider damage processes the user-defined subroutines are developed and implemented. The subroutines serve to utilize constitutive and evolution equations with damage state variables. In addition, they allow the postprocessing of damage, i.e. the creation of contour plots visualizing damage distributions.

An important question in the creep analysis is that on reliability of the applied models, numerical methods and obtained results. The reliability assessment may require the following verification steps:

- *Verification of developed finite element subroutines.* To assess that the subroutines are correctly coded and implemented, results of finite element computations must be compared with reference solutions of benchmark problems. Several benchmark problems have been proposed in [42] based on an in-house finite element code. Below we recall closed form solutions of steady-state creep in elementary structures, well-known in the creep mechanics literature. To extend these solutions to the primary and tertiary creep ranges we apply the Ritz and the time step methods. The advantage of these problems is the possibility to obtain reference solutions without a finite element discretization. Furthermore, they allow to verify finite element subroutines over a wide range of finite element types including beam, shell and solid type elements.
- *Verification of applied numerical methods.* Here the problems of the suitable finite element type, the mesh density, the time step size and the time step control must be analyzed. They are of particular importance in creep damage related simulations. Below these problems are discussed based on numerical tests and by comparison with reference solutions.
- *Verification of constitutive and structural mechanics models.* This step requires creep tests of model structural components and the corresponding numerical analysis by the use of the developed techniques. Examples of recent experimental studies of creep in structures include beams [80], transversely loaded plates [167, 230], thin-walled tubes under internal pressure [169, 182], pressure vessels [106, 117], circumferentially notched bars [136]. Let us note that the experimental data for model structures are usually limited to short-term creep tests. The finite element codes and subroutines are designed to analyze real engineering structures. Therefore long term analysis of several typical structures should be performed and the results should be compared with data collected from engineering

practice of power and petrochemical plants. Below an example of the creep finite element analysis for a spatial steam pipeline is discussed.

4.2 Initial-Boundary Value Problems and General Solution Procedures

The objective of this section is to discuss the governing mechanical equations describing creep in three-dimensional solids and related numerical solution procedures. The set of equations includes material independent equations, constitutive and evolution equations, see Chapt. 2, as well as the initial and boundary conditions. The formulated IBVP must be solved by numerical methods. Explicit and implicit time integration methods are reviewed with respect to their accuracy and efficiency. Within the time-step procedures, different possibilities are discussed to solve linearized boundary value problems. The attention will be given to the variational formulations and the use of direct variational methods.

4.2.1 Governing Equations

Let us consider a solid occupying the volume V with the surface A. We assume that the solid is fixed on the surface part A_u and loaded by surface forces on the part A_p. The position of a material point within the solid in the reference state is described by the position vector $\boldsymbol{r}(q^i) = \boldsymbol{e}_i q^i$, $i = 1, 2, 3$, where $\boldsymbol{e}_i$ are basis vectors and q^i are coordinates (see Sect. B.1). The corresponding position in the actual state can be characterized by the position vector $\boldsymbol{R}(q^i, t)$ or by the displacement vector $\boldsymbol{u}(q^i, t) = \boldsymbol{R}(q^i, t) - \boldsymbol{r}(q^i)$. The problem is to find the time sequence of the actual configurations $\boldsymbol{R}(q^i, t)$ as a result of external actions for a given time interval and $\forall q^i \in V$. The governing equations are discussed in continuum mechanics e.g. [32, 39, 48, 61, 111, 134, 184, 205]. Constitutive equations describing creep processes have been introduced in Chapt. 2. Besides the kinematical quantities, additional unknowns are the creep strain tensor $\boldsymbol{\varepsilon}^{cr}(q^i, t)$ and the set of internal state variables $H_k(q^i, t), k = 1, \ldots, n$ and $\omega_l(q^i, t), l = 1, \ldots, m$. They are introduced to characterize the current state of the material microstructure and to reflect the entire previous history of the creep process (see Sect. 2.1). To present basic ideas of IBVP we limit our considerations to linearized kinematical equations in the sense of infinitesimal strains and displacements. Furthermore, we assume a classical non-polar continuum, quasi-static processes and isothermal conditions. The related comments were made in Sect. 2.1.

The governing equations can be summarized as follows

- kinematical equations
  - strain-displacement relation

$$\boldsymbol{\varepsilon} = \frac{1}{2}\left(\nabla\boldsymbol{u} + (\nabla\boldsymbol{u})^T\right), \quad q^i \in V, \tag{4.2.1}$$

  where $\boldsymbol{\varepsilon}$ is the tensor of infinitesimal strains.

• compatibility condition

$$\nabla \times (\nabla \times \boldsymbol{\varepsilon})^T = \mathbf{0}, \quad q^i \in V, \tag{4.2.2}$$

– equilibrium conditions

$$\nabla \cdot \boldsymbol{\sigma} + \rho \bar{\boldsymbol{f}} = \mathbf{0}, \quad \boldsymbol{\sigma} = \boldsymbol{\sigma}^T, \quad q^i \in V, \tag{4.2.3}$$

where ρ is the material density and $\bar{\boldsymbol{f}}$ is the density of volumetric forces

– boundary conditions

$$\begin{aligned} \boldsymbol{u} &= \bar{\boldsymbol{u}}, & q^i &\in A_u, \\ \boldsymbol{\sigma} \cdot \boldsymbol{\nu} &= \bar{\boldsymbol{p}}, & q^i &\in A_p, \end{aligned} \tag{4.2.4}$$

where $\bar{\boldsymbol{u}}$ is the given displacement vector and $\bar{\boldsymbol{p}}$ is the vector of given surface forces and $\boldsymbol{\nu}$ is the outward unit normal to A_p. The vectors $\bar{\boldsymbol{f}}$, $\bar{\boldsymbol{p}}$ and $\bar{\boldsymbol{u}}$ can, in general, be functions of coordinates and time.

With the assumption of infinitesimal strains the additive decomposition of the total strain into elastic, thermal and creep parts is usually postulated

$$\boldsymbol{\varepsilon} = \boldsymbol{\varepsilon}^{el} + \boldsymbol{\varepsilon}^{th} + \boldsymbol{\varepsilon}^{cr} \tag{4.2.5}$$

The constitutive equation for the stress tensor can be assumed in the form of the generalized Hooke's law as follows

$$\boldsymbol{\sigma} = {}^{(4)}\mathbf{C} \cdot\cdot (\boldsymbol{\varepsilon} - \boldsymbol{\varepsilon}^{th} - \boldsymbol{\varepsilon}^{cr}) \tag{4.2.6}$$

In the case of isotropic elasticity the tensor ${}^{(4)}\mathbf{C}$ takes the form

$${}^{(4)}\mathbf{C} = \lambda \boldsymbol{I} \otimes \boldsymbol{I} + \mu(\boldsymbol{e}_k \otimes \boldsymbol{I} \otimes \boldsymbol{e}^k + \boldsymbol{e}_i \otimes \boldsymbol{e}_k \otimes \boldsymbol{e}^i \otimes \boldsymbol{e}^k), \tag{4.2.7}$$

where λ and μ are the Lamé's constants

$$\mu = G = \frac{E}{2(1+\nu)}, \quad \lambda = \frac{\nu E}{(1+\nu)(1-2\nu)}$$

E is the Young's modulus, G is the shear modulus and ν is the Poisson's ratio.

If an isotropic solid is heated from the reference temperature T_0 up to T, the thermal part of the strain tensor is

$$\boldsymbol{\varepsilon}^{th} = \alpha_T \Delta T \boldsymbol{I}, \quad \Delta T \equiv T - T_0, \tag{4.2.8}$$

where α_T is the coefficient of the thermal expansion. ΔT can be a function of coordinates and time too.

The constitutive equation for the creep rate and evolution equations for internal state variables are discussed in Chapt. 2. Here we will use the rate equations in the following form

$$
\begin{aligned}
\dot{\boldsymbol{\varepsilon}}^{cr} &= \frac{\partial \Phi\left(\sigma_{eq}(\boldsymbol{\sigma}), H_k, \omega_l; T\right)}{\partial \boldsymbol{\sigma}}, \quad k = 1, \ldots, n, \quad l = 1, \ldots, m, \\
\dot{H}_k &= \dot{H}_k\left(\sigma_{eq}^H(\boldsymbol{\sigma}), H_k, \omega_l; T\right), \quad \dot{\omega}_l = \dot{\omega}_l\left(\sigma_{eq}^{\omega}(\boldsymbol{\sigma}), H_k, \omega_l; T\right)
\end{aligned}
\tag{4.2.9}
$$

The hardening variables H_k and the damage variables ω_l can be scalars or tensors. For the creep strain as well as for the set of hardening and damage variables the initial conditions must be specified. Let us assume

$$
\boldsymbol{\varepsilon}^{cr}\Big|_{t=0} = \mathbf{0}, \quad H_k\Big|_{t=0} = H_k^0, \quad \omega_l\Big|_{t=0} = \omega_l^0,
\tag{4.2.10}
$$

where H_k^0 and ω_l^0 are the initial values of the hardening and the damage parameters. Equations (4.2.1) - (4.2.10) describe the quasi-static creep process in a solid.

4.2.2 Vector-Matrix Representation

To formulate initial-boundary value problems and numerical solution procedures let us rewrite Eqs. (4.2.1) – (4.2.10) in the vector-matrix notation. For the sake of brevity we introduce the Cartesian coordinates x_1, x_2, x_3. The Cartesian components of vectors and tensors can be collected into the following "numerical" vectors and matrices:

Stress vector $\boldsymbol{\sigma}$	$\boldsymbol{\sigma}^T$	=	$[\sigma_{11}\ \sigma_{22}\ \sigma_{33}\ \sigma_{12}\ \sigma_{23}\ \sigma_{31}]$
Strain vector $\boldsymbol{\varepsilon}$	$\boldsymbol{\varepsilon}^T$	=	$[\varepsilon_{11}\ \varepsilon_{22}\ \varepsilon_{33}\ \gamma_{12}\ \gamma_{23}\ \gamma_{31}]$
Displacement vector $\boldsymbol{u}$	$\boldsymbol{u}^T$	=	$[u_1\ u_2\ u_3]$

Vector of creep strains $\boldsymbol{\varepsilon}^{cr}$	$\boldsymbol{\varepsilon}^{cr\,T}$	=	$[\varepsilon_{11}^{cr}\ \varepsilon_{22}^{cr}\ \varepsilon_{33}^{cr}\ \gamma_{12}^{cr}\ \gamma_{23}^{cr}\ \gamma_{31}^{cr}]$
Vector of internal variables $\boldsymbol{\xi}$	$\boldsymbol{\xi}^T$	=	$[H_1\ H_2\ \ldots H_n\ \omega_1\ \omega_2\ \ldots \omega_m]$
Vector of thermal strains $\boldsymbol{\varepsilon}^{th}$	$\boldsymbol{\varepsilon}^{th\,T}$	=	$[\alpha_T \Delta T\ \alpha_T \Delta T\ \alpha_T \Delta T\ 0\ 0\ 0]$

Vector of body forces $\bar{\boldsymbol{f}}$	$\bar{\boldsymbol{f}}^T$	=	$[\bar{f}_1\ \bar{f}_2\ \bar{f}_3]$
Vector of surface forces $\bar{\boldsymbol{p}}$	$\bar{\boldsymbol{p}}^T$	=	$[\bar{p}_1\ \bar{p}_2\ \bar{p}_3]$

Stress vector $\boldsymbol{\sigma}_{\boldsymbol{\nu}}$ on $\boldsymbol{\nu} dA$	$\boldsymbol{\sigma}_{\boldsymbol{\nu}}^T$	$=$	$[\sigma_{\nu_1}\ \sigma_{\nu_2}\ \sigma_{\nu_3}]$
Normal vector $\boldsymbol{\nu}$	$\boldsymbol{\nu}^T$	$=$	$[\nu_1\ \nu_2\ \nu_3]$,
	ν_i	$=$	$\cos(\boldsymbol{\nu}, x_i)$

Transformation matrix $\overset{u}{\boldsymbol{T}}$

$$\overset{u}{\boldsymbol{T}} = \begin{bmatrix} 1 & 0 & 0 \\ 0 & 1 & 0 \\ 0 & 0 & 1 \end{bmatrix}$$

Transformation matrix $\overset{\sigma}{\boldsymbol{T}}$

$$\overset{\sigma}{\boldsymbol{T}} = \begin{bmatrix} \nu_1 & 0 & 0 & \nu_2 & 0 & \nu_3 \\ 0 & \nu_2 & 0 & \nu_1 & \nu_3 & 0 \\ 0 & 0 & \nu_3 & 0 & \nu_2 & \nu_1 \end{bmatrix}$$

Differential matrix $\boldsymbol{D}$

$$\boldsymbol{D} = \begin{bmatrix} \partial_1 & 0 & 0 & \partial_2 & 0 & \partial_3 \\ 0 & \partial_2 & 0 & \partial_1 & \partial_3 & 0 \\ 0 & 0 & \partial_3 & 0 & \partial_2 & \partial_1 \end{bmatrix}$$

Differential matrix $\boldsymbol{D}_1$

$$\boldsymbol{D}_1 = \begin{bmatrix} 0 & \partial_3^2 & \partial_2^2 & 0 & -\partial_2\partial_3 & 0 \\ \partial_3^2 & 0 & \partial_1^2 & 0 & 0 & -\partial_1\partial_3 \\ \partial_2^2 & \partial_1^2 & 0 & -\partial_1\partial_2 & 0 & 0 \\ 0 & 0 & -\partial_1\partial_2 & -\frac{1}{2}\partial_3^2 & \frac{1}{2}\partial_1\partial_3 & \frac{1}{2}\partial_2\partial_3 \\ -\partial_2\partial_3 & 0 & 0 & \frac{1}{2}\partial_1\partial_3 & -\frac{1}{2}\partial_1^2 & \frac{1}{2}\partial_1\partial_2 \\ 0 & -\partial_1\partial_3 & 0 & \frac{1}{2}\partial_2\partial_3 & \frac{1}{2}\partial_1\partial_2 & -\frac{1}{2}\partial_2^2 \end{bmatrix}$$

with

$$\partial_i = \frac{\partial(\ldots)}{\partial x_i}, \quad \partial_i^2 = \frac{\partial^2(\ldots)}{\partial x_i^2}$$

Elasticity matrix (stiffness matrix) $\boldsymbol{E}$

$$\boldsymbol{E} = \begin{bmatrix} (2\mu+\lambda) & \lambda & \lambda & 0 & 0 & 0 \\ & (2\mu+\lambda) & \lambda & 0 & 0 & 0 \\ & & (2\mu+\lambda) & 0 & 0 & 0 \\ & & & \mu & 0 & 0 \\ & & & & \mu & 0 \\ \text{SYM} & & & & & \mu \end{bmatrix}$$

Reciprocal elasticity matrix (compliance matrix) $\boldsymbol{E}^{-1}$

$$\boldsymbol{E}^{-1} = \frac{1}{E}\begin{bmatrix} 1 & -\nu & -\nu & 0 & 0 & 0 \\ & 1 & -\nu & 0 & 0 & 0 \\ & & 1 & 0 & 0 & 0 \\ & & & 2(1+\nu) & 0 & 0 \\ & & & & 2(1+\nu) & 0 \\ \text{SYM} & & & & & 2(1+\nu) \end{bmatrix}$$

With the introduced notations and $\boldsymbol{x}^T = [x_1 \; x_2 \; x_3]$ we can rewrite the governing equations (4.2.1) – (4.2.10) as follows

Kinematical equations:
Strain-displacement relation

$$\boldsymbol{\varepsilon} = \boldsymbol{D}^T\boldsymbol{u}, \quad \boldsymbol{x} \in V \tag{4.2.11}$$

Compatibility condition

$$\boldsymbol{D}_1\boldsymbol{\varepsilon} = \boldsymbol{0}, \quad \boldsymbol{x} \in V \tag{4.2.12}$$

Prescribed boundary displacements $\bar{\boldsymbol{u}}$ on A_u

$$\overset{u}{\boldsymbol{T}}\,\boldsymbol{u} = \bar{\boldsymbol{u}}, \quad \boldsymbol{x} \in A_u \tag{4.2.13}$$

Equilibrium conditions:

$$\boldsymbol{D}\boldsymbol{\sigma} + \bar{\boldsymbol{f}} = \boldsymbol{0}, \quad \boldsymbol{x} \in V \tag{4.2.14}$$

Prescribed surface forces $\bar{\boldsymbol{p}}$ on A_p

$$\overset{\sigma}{\boldsymbol{T}}\,\boldsymbol{\sigma} = \boldsymbol{\sigma}_{\boldsymbol{\nu}} = \bar{\boldsymbol{p}}, \quad \boldsymbol{x} \in A_p \tag{4.2.15}$$

Constitutive and evolution equations:

$$\boldsymbol{\sigma} = \mathbf{E}(\boldsymbol{\varepsilon} - \boldsymbol{\varepsilon}^{th} - \boldsymbol{\varepsilon}^{cr}), \quad \boldsymbol{x} \in V \tag{4.2.16}$$

$$\begin{aligned} \dot{\boldsymbol{\varepsilon}}^{cr} &= \boldsymbol{g}(\boldsymbol{\sigma}, \boldsymbol{\xi}; T) \\ \dot{\boldsymbol{\xi}} &= \boldsymbol{h}(\boldsymbol{\sigma}, \boldsymbol{\xi}; T) \end{aligned} \tag{4.2.17}$$

Initial conditions

$$\boldsymbol{\varepsilon}^{cr}(\boldsymbol{x}, 0) = \mathbf{0}, \quad \boldsymbol{\xi}(\boldsymbol{x}, 0) = \boldsymbol{\xi}_0 \tag{4.2.18}$$

The function $\boldsymbol{g}$ can be formulated if the creep potential Φ is specified, see Sect. 2.1. The vector $\boldsymbol{\xi}$ and and the function $\boldsymbol{h}$ can be defined for the selected internal state variables and the corresponding evolution equations. Examples of hardening variables are presented in Sect. 2.3. Damage variables are discussed in Sect. 2.4.

4.2.3 Numerical Solution Techniques

Let us assume that the creep strain vector and the vector of internal state variables are known functions of the coordinates for a fixed time. With the strain-displacement relations (4.2.11), the constitutive equations (4.2.16) can be written as follows

$$\boldsymbol{\sigma} = \mathbf{E}(\boldsymbol{D}^T \boldsymbol{u} - \boldsymbol{\varepsilon}^{th} - \boldsymbol{\varepsilon}^{cr}) \tag{4.2.19}$$

Taking into account the equilibrium conditions (4.2.14) and the static boundary conditions (4.2.15) we obtain

$$\begin{aligned} \boldsymbol{D}\boldsymbol{E}\boldsymbol{D}^T\boldsymbol{u} &= -\bar{\boldsymbol{f}} + \boldsymbol{D}\boldsymbol{E}\boldsymbol{\varepsilon}^{th} + \boldsymbol{D}\boldsymbol{E}\boldsymbol{\varepsilon}^{cr}, \quad \boldsymbol{x} \in V, \\ \overset{\sigma}{\boldsymbol{T}}\,\boldsymbol{E}\boldsymbol{D}^T\boldsymbol{u} &= \bar{\boldsymbol{p}} + \overset{\sigma}{\boldsymbol{T}}\,\boldsymbol{E}\boldsymbol{\varepsilon}^{th} + \overset{\sigma}{\boldsymbol{T}}\,\boldsymbol{E}\boldsymbol{\varepsilon}^{cr}, \quad \boldsymbol{x} \in A_p \end{aligned} \tag{4.2.20}$$

With the kinematic boundary conditions (4.2.13), the partial differential equations and the boundary conditions (4.2.20) represent the BVP with the displacement vector $\boldsymbol{u}$ as an unknown vector. Introducing the fictitious force vectors corresponding to the given thermal strains and the creep strains at fixed time we can write Eqs. (4.2.20) as follows

$$\begin{aligned} \boldsymbol{D}\boldsymbol{E}\boldsymbol{D}^T\boldsymbol{u} &= -\bar{\boldsymbol{f}} + \boldsymbol{f}^{th} + \boldsymbol{f}^{cr}, \quad \boldsymbol{f}^{th} = \boldsymbol{D}\boldsymbol{E}\boldsymbol{\varepsilon}^{th}, \quad \boldsymbol{f}^{cr} = \boldsymbol{D}\boldsymbol{E}\boldsymbol{\varepsilon}^{cr}, \\ \overset{\sigma}{\boldsymbol{T}}\,\boldsymbol{E}\boldsymbol{D}^T\boldsymbol{u} &= \bar{\boldsymbol{p}} + \boldsymbol{p}^{th} + \boldsymbol{p}^{cr}, \quad \boldsymbol{p}^{th} = \overset{\sigma}{\boldsymbol{T}}\,\boldsymbol{E}\boldsymbol{\varepsilon}^{th}, \quad \boldsymbol{p}^{cr} = \overset{\sigma}{\boldsymbol{T}}\,\boldsymbol{E}\boldsymbol{\varepsilon}^{cr} \end{aligned} \tag{4.2.21}$$

These equations are the equilibrium conditions expressed in terms of three unknown components of the displacement vector. After the solution of Eqs. (4.2.21) one can obtain the six components of the stress vector from Eq. (4.2.19). Inserting the stress vector into the creep constitutive equations (4.2.17) one can calculate the rates of the creep strains and those of the internal variables. Based on the equations introduced, the IBVP of the type $\dot{\boldsymbol{Y}} = \boldsymbol{G}(\boldsymbol{Y})$ can be formulated, where $\boldsymbol{Y}$ includes the vectors of creep strains and internal variables. The operator $\boldsymbol{G}$ involves the solution of the linearized boundary value problem for the fixed creep strains and internal variables. The initial conditions are Eqs. (4.2.18).

An alternative formulation can be based on the compatibility condition (4.2.12). First the constitutive equations (4.2.16) after differentiation with respect to time can be written as

$$\dot{\boldsymbol{\sigma}} = \boldsymbol{E}(\dot{\boldsymbol{\varepsilon}} - \dot{\boldsymbol{\varepsilon}}^{th} - \dot{\boldsymbol{\varepsilon}}^{cr}) = \boldsymbol{E}\left[\dot{\boldsymbol{\varepsilon}} - \dot{\boldsymbol{\varepsilon}}^{th} - \boldsymbol{g}(\boldsymbol{\sigma}, \boldsymbol{\xi}; T)\right]$$

Reordering this equation the total strain vector takes the form

$$\dot{\boldsymbol{\varepsilon}} = \boldsymbol{E}^{-1}\dot{\boldsymbol{\sigma}} + \dot{\boldsymbol{\varepsilon}}^{th} + \boldsymbol{g}(\boldsymbol{\sigma}, \boldsymbol{\xi}; T) \tag{4.2.22}$$

For isothermal processes $\dot{\boldsymbol{\varepsilon}}^{th} = \boldsymbol{0}$. The compatibility condition (4.2.12) can be rewritten in terms of the strain rate vector

$$\boldsymbol{D}_1\dot{\boldsymbol{\varepsilon}} = \boldsymbol{0} \tag{4.2.23}$$

After inserting (4.2.22) into (4.2.23) we obtain

$$\boldsymbol{D}_1\boldsymbol{E}^{-1}\dot{\boldsymbol{\sigma}} + \boldsymbol{D}_1\boldsymbol{g}(\boldsymbol{\sigma}, \boldsymbol{\xi}; T) = \boldsymbol{0} \tag{4.2.24}$$

The six equations (4.2.24) describe the stress redistribution during the creep process. The initial conditions are the solutions of the linear elastic problem for the stresses

$$\boldsymbol{D}_1\boldsymbol{E}^{-1}\boldsymbol{\sigma}(\boldsymbol{x}, 0) = \boldsymbol{0},$$

as well as $\boldsymbol{\xi}(\boldsymbol{x}, 0) = \boldsymbol{\xi}_0$. The IBVP can be formulated again as $\dot{\boldsymbol{Y}} = \boldsymbol{G}(\boldsymbol{Y})$, where $\boldsymbol{Y}$ includes now the stress vector and the vector of internal state variables. The stress redistribution equation (4.2.24) can be also formulated in terms of stress functions. A variety of stress functions can be found in such a way that the equilibrium conditions (4.2.14) are identically satisfied. As an example we can introduce the vector of stress functions $\boldsymbol{\psi}$, so that $\boldsymbol{\sigma} = \boldsymbol{D}_1\boldsymbol{\psi}$. It is easy to verify that in the absence of body forces the equilibrium conditions $\boldsymbol{D}\boldsymbol{\sigma} = \boldsymbol{D}\boldsymbol{D}_1\boldsymbol{\psi} = \boldsymbol{0}$ are identically satisfied. With the stress functions $\boldsymbol{\psi}$ we can write (4.2.24) as follows

$$\boldsymbol{D}_1\boldsymbol{E}^{-1}\boldsymbol{D}_1\dot{\boldsymbol{\psi}} + \boldsymbol{D}_1\boldsymbol{g}(\boldsymbol{D}_1\boldsymbol{\psi}, \boldsymbol{\xi}; T) = \boldsymbol{0}$$

Because there exist identities between the six compatibility conditions (only three of them are independent), see e.g. [129], it is possible to transform the six equations (4.2.24) into three independent equations. For example, one can express six components of the stress vector by three Maxwell's stress functions [129, 260]. After inserting into (4.2.24) one can obtain three equations for three unknown stress functions.

In addition to the displacement formulation (4.2.21) and the stress formulation (4.2.24), it is possible to express the governing equations in terms of displacements and stresses. Such mixed formulations can be useful for solving creep problems of beams, plates and shells.

4.2.3.1 Time Integration Methods. The governing equations include first order time derivatives and the prescribed initial conditions. The unknown displacements in Eqs. (4.2.21) or the unknown stresses in (4.2.24) are functions of coordinates and time. The exact integration of these equations with respect to the time variable is feasible only for one-dimensional problems, e.g. for rods. In the general case of the structural analysis, numerical time integration methods must be applied for solving non-linear IBVP. The commonly used solution technique in mechanics and thermodynamics is the finite difference method. The time derivatives are replaced by finite differences. Starting with the initial conditions (in our case the elastic displacement or stress fields), the finite difference method leads to a step-by-step solution. A variety of time integration algorithms can be found in textbooks on numerical methods, e.g. [96, 110, 130, 294].

Here we discuss some typical examples of time integration procedures mostly used in creep analysis. Let us start with the displacement formulation of the governing equations and neglect the thermal strains for the sake of brevity. The initial condition is the solution of the elasticity problem

$$\boldsymbol{D}\boldsymbol{E}\boldsymbol{D}^T\boldsymbol{u}_0 = -\bar{\boldsymbol{f}}, \quad \boldsymbol{\sigma}_0 = \boldsymbol{E}\boldsymbol{D}^T\boldsymbol{u}_0 \tag{4.2.25}$$

with $\boldsymbol{u}_0 = \boldsymbol{u}(\boldsymbol{x},0)$ and $\boldsymbol{\sigma}_0 = \boldsymbol{\sigma}(\boldsymbol{x},0)$. One way to obtain the displacements and stresses at time $t_1 = t_0 + \Delta t_1$ is to assume that the rates of the creep strains and the internal state variables are approximately constant within the time interval $[t_0, t_1]$. Then for any time interval $[t_n, t_{n+1}]$ and with $\Delta t_n = t_{n+1} - t_n$ we can write

$$\begin{aligned} \boldsymbol{\varepsilon}^{cr}_{n+1} &= \boldsymbol{\varepsilon}^{cr}_n + \Delta\boldsymbol{\varepsilon}^{cr}_n, & \boldsymbol{\xi}_{n+1} &= \boldsymbol{\xi}_n + \Delta\boldsymbol{\xi}_n, \\ \Delta\boldsymbol{\varepsilon}^{cr}_n &= \Delta t_n \boldsymbol{g}(\boldsymbol{\sigma}_n, \boldsymbol{\xi}_n; T_n), & \Delta\boldsymbol{\xi}_n &= \Delta t_n \boldsymbol{h}(\boldsymbol{\sigma}_n, \boldsymbol{\xi}_n; T_n) \end{aligned} \tag{4.2.26}$$

The displacements and stresses at t_{n+1} can be updated using Eqs. (4.2.19) and (4.2.21). The following time integration scheme can be formulated:

```
set n = 0, ε₀^cr = 0, ξ₀ = 0
```

set $n = 0$, $\boldsymbol{\varepsilon}^{cr}_0 = \boldsymbol{0}$, $\boldsymbol{\xi}_0 = \boldsymbol{0}$

solve BVP $\boldsymbol{D}\boldsymbol{E}\boldsymbol{D}^T\boldsymbol{u}_0 = -\bar{\boldsymbol{f}}$, calculate $\boldsymbol{\sigma}_0 = \boldsymbol{E}\boldsymbol{D}^T\boldsymbol{u}_0$

1: calculate

$\Delta\boldsymbol{\varepsilon}^{cr}_n = \Delta t_n \boldsymbol{g}(\boldsymbol{\sigma}_n, \boldsymbol{\xi}_n, T_n)$, $\quad \Delta\boldsymbol{\xi}_n = \Delta t_n \boldsymbol{h}(\boldsymbol{\sigma}_n, \boldsymbol{\xi}_n, T_n)$

$\boldsymbol{\varepsilon}^{cr}_{n+1} = \boldsymbol{\varepsilon}^{cr}_n + \Delta\boldsymbol{\varepsilon}^{cr}_n$, $\quad \boldsymbol{\xi}_{n+1} = \boldsymbol{\xi}_n + \Delta\boldsymbol{\xi}_n$,

solve BVP

$$\begin{aligned} \boldsymbol{D}\boldsymbol{E}\boldsymbol{D}^T\boldsymbol{u}_{n+1} &= -\bar{\boldsymbol{f}} + \boldsymbol{D}\boldsymbol{E}\boldsymbol{\varepsilon}^{cr}_{n+1}, \\ \boldsymbol{\sigma}_{n+1} &= \boldsymbol{E}(\boldsymbol{D}^T\boldsymbol{u}_{n+1} - \boldsymbol{\varepsilon}^{cr}_{n+1}) \end{aligned} \tag{4.2.27}$$

if $t_{n+1} < t_N$ and $\omega_l < \omega_{l*}, l = 1, \ldots, m$ then set $n := n + 1$ go to 1

else finish

The calculations can be repeated within the whole given interval of time $[t_0, t_N]$ by setting $n := n + 1$ in Eqs. (4.2.26). For the creep-damage related analysis it is necessary to prove of whether the critical damage state is achieved. If the damage variable $\omega_l, l = 1, \ldots, m$ attains the critical value ω_{l*} the calculations must be terminated.

The forward difference equations (4.2.26) correspond to the one-step explicit Euler method. This method is widely used in the creep analysis because of simplicity. The accuracy of the method depends on the time step size. Furthermore, this method is conditionally stable that means that the stability is restricted to small time steps. Therefore stable results can be obtained only for $\Delta t \leq \Delta t_{crit}$. There is no general recipe how to control the time step size by the use the one step explicit method. For example, in [348] it is recommended to compute the time step size from the condition that the increment of the creep strain does not exceed one half of the elastic strain, i.e.

$$\Delta t_n \boldsymbol{g}(\boldsymbol{\sigma}_n, \boldsymbol{\xi}_n; T_n) \leq \frac{1}{2} \boldsymbol{E}^{-1} \boldsymbol{\sigma}_n$$

A further restriction is connected with the assumption that the stresses have to be constant within the time interval $[t_n, t_{n+1}]$. Therefore this method can be recommended for structural analysis under constant or monotonic loading and temperature conditions only. In the case of loading jumps or cyclic loading changes very small time steps are necessary in order to provide a stable solution.

One way to improve the accuracy of time-dependent solutions is the use of multi-step methods of the Runge-Kutta type, see e.g. [80, 110, 130]. These explicit methods are conditionally stable as well. However, they provide higher order accuracy if compared with the one-step forward difference method. Furthermore, for the creep-damage related analysis the so-called embedded methods [130], which allow to control the time step size, can be recommended. In [20, 21] the embedded fourth order Kutta-Merson method has been applied to creep problems of shells of revolution.

The next possibility to improve the one-step method is the use of the generalized trapezoidal rule [96]

$$\begin{aligned} \boldsymbol{\varepsilon}^{cr}_{n+1} &= \boldsymbol{\varepsilon}^{cr}_n + \Delta t \left[(1-\theta)\dot{\boldsymbol{\varepsilon}}^{cr}_n + \theta \dot{\boldsymbol{\varepsilon}}^{cr}_{n+1} \right], \\ \boldsymbol{\xi}_{n+1} &= \boldsymbol{\xi}_n + \Delta t \left[(1-\theta)\dot{\boldsymbol{\xi}}_n + \theta \dot{\boldsymbol{\xi}}_{n+1} \right], \end{aligned} \qquad (4.2.28)$$

where θ ($0 \leq \theta \leq 1$) is the parameter controlling the stability. The rule (4.2.28) includes different well-known methods as special cases. Setting $\theta = 0$ the forward difference explicit Euler method (4.2.26) follows. For $\theta > 0$ we obtain a variety of implicit methods: for $\theta = 1/2$ - the trapezoidal rule (Crank-Nicolson method), and for $\theta = 1$ the backward difference method (implicit Euler method). The advantage of the implicit methods is their unconditional stability that means that the solution will be stable independently on the time step size. The price for the unconditional stability is the necessity to solve non-linear equations at each time step. Equations (4.2.28) can be rewritten as follows

$$
\begin{aligned}
\boldsymbol{\varepsilon}^{cr}_{n+1} &= \boldsymbol{\varepsilon}^{cr}_{n} + \Delta\boldsymbol{\varepsilon}^{cr}_{n}, \\
\boldsymbol{\zeta}_{n+1} &= \boldsymbol{\zeta}_{n} + \Delta\boldsymbol{\zeta}_{n}, \\
\Delta\boldsymbol{\varepsilon}^{cr}_{n} &= \Delta t_n \left[(1-\theta)\boldsymbol{g}(\boldsymbol{\sigma}_n, \boldsymbol{\zeta}_n; T_n) + \theta\boldsymbol{g}(\boldsymbol{\sigma}_{n+1}, \boldsymbol{\zeta}_{n+1}, T_{n+1})\right], \\
\Delta\boldsymbol{\zeta}_{n} &= \Delta t_n \left[(1-\theta)\boldsymbol{h}(\boldsymbol{\sigma}_n, \boldsymbol{\zeta}_n; T_n) + \theta\boldsymbol{h}(\boldsymbol{\sigma}_{n+1}, \boldsymbol{\zeta}_{n+1}, T_{n+1})\right]
\end{aligned}
\qquad (4.2.29)
$$

Equations (4.2.29) are non-linear with respect to $\boldsymbol{\zeta}_{n+1}$ for $\theta > 0$. Note that for a material model with strain hardening, the vector $\boldsymbol{\zeta}_n$ includes the equivalent creep strain. In this case Eqs. (4.2.29) are non-linear with respect to $\boldsymbol{\varepsilon}^{cr}_{n+1}$. These equations can be solved using known iteration methods. The simplest possibility is the fixed point iteration method leading to the following scheme at the time step $[t_n, t_{n+1}]$:

set $i = 0$, $\boldsymbol{\varepsilon}^{cr^0}_{n+1} = \boldsymbol{\varepsilon}^{cr}_{n}$, $\boldsymbol{\zeta}^{i}_{n} = \boldsymbol{\zeta}_{n}$, $\boldsymbol{\sigma}^{0}_{n+1} = \boldsymbol{\sigma}_{n}$

1: calculate

$$\Delta\boldsymbol{\varepsilon}^{cr^i}_{n} = \Delta t_n \left[(1-\theta)\boldsymbol{g}(\boldsymbol{\sigma}_n, \boldsymbol{\zeta}_n; T_n) + \theta\boldsymbol{g}(\boldsymbol{\sigma}^{i}_{n+1}, \boldsymbol{\zeta}^{i}_{n+1}; T_{n+1})\right],$$

$$\Delta\boldsymbol{\zeta}^{i}_{n} = \Delta t_n \left[(1-\theta)\boldsymbol{h}(\boldsymbol{\sigma}_n, \boldsymbol{\zeta}_n; T_n) + \theta\boldsymbol{h}(\boldsymbol{\sigma}^{i}_{n+1}, \boldsymbol{\zeta}^{i}_{n+1}; T_{n+1})\right],$$

$$\boldsymbol{\varepsilon}^{cr^{i+1}}_{n+1} = \boldsymbol{\varepsilon}^{cr}_{n} + \Delta\boldsymbol{\varepsilon}^{cr^i}_{n}, \qquad \boldsymbol{\zeta}^{i+1}_{n+1} = \boldsymbol{\zeta}_{n} + \Delta\boldsymbol{\zeta}^{i}_{n},$$

if $|\boldsymbol{\varepsilon}^{cr^{i+1}}_{n+1} - \boldsymbol{\varepsilon}^{cr^{i}}_{n+1}| > \epsilon$ and $|\boldsymbol{\zeta}^{i+1}_{n+1} - \boldsymbol{\zeta}^{i}_{n+1}| > \epsilon$

then solve BVP

$$
\begin{aligned}
\boldsymbol{DED}^T\boldsymbol{u}^{i+1}_{n+1} &= -\bar{\boldsymbol{f}}_{n+1} + \boldsymbol{DE}\boldsymbol{\varepsilon}^{cr^{i+1}}_{n+1}, \\
\boldsymbol{\sigma}^{i+1}_{n+1} &= \boldsymbol{E}\left(\boldsymbol{D}^T\boldsymbol{u}^{i+1}_{n+1} - \boldsymbol{\varepsilon}^{cr^{i+1}}_{n+1}\right)
\end{aligned}
\qquad (4.2.30)
$$

set $i := i + 1$ go to 1

else

set $\boldsymbol{\varepsilon}^{cr}_{n+1} = \boldsymbol{\varepsilon}^{cr^{i+1}}_{n+1}$, $\boldsymbol{\zeta}_{n+1} = \boldsymbol{\zeta}^{i+1}_{n+1}$

The accuracy and the efficiency of the implicit method in connection with the introduced iteration scheme is now additionally dependent on the tolerance ϵ and the convergence rate of the fixed point iterations. The first iteration in the above introduced scheme is the forward difference predictor. Since the convergence rate of the fixed point iterations is highly dependent on the "quality" of the first iteration, the efficiency of this scheme is determined again by the time step size. If the desired accuracy ϵ is not reached within $3-4$ iterations the time step size should be decreased and the calculations repeated starting from the step 1. The slow convergence of the fixed point iterations is the drawback of the proposed algorithm. However, in the case of creep-damage studies this algorithm is more efficient in comparison with the explicit forward method. Some examples are discussed in [22, 230]. Furthermore, it is possible to combine the implicit time integration scheme with the Newton-

Raphson iteration method or its modifications providing higher convergence rates. Examples can be found in [348].

Another widely used technique is to construct an explicit scheme based on the generalized trapezoidal rule (4.2.28), see e.g. [40, 254]. This can be accomplished by linearizing (4.2.29 with respect to $\dot{\boldsymbol{\xi}}_{n+1}$. For the sake of brevity let us assume that the functions $\boldsymbol{g}$ and $\boldsymbol{h}$ are independent from T. Then we can write

$$\begin{aligned} \dot{\boldsymbol{\varepsilon}}^{cr}_{n+1} &\cong \boldsymbol{g}(\boldsymbol{\sigma}_n, \boldsymbol{\xi}_n) + \boldsymbol{g}_{,\sigma}(\boldsymbol{\sigma}_n, \boldsymbol{\xi}_n)\Delta\boldsymbol{\sigma}_n + \boldsymbol{g}_{,\xi}(\boldsymbol{\sigma}_n, \boldsymbol{\xi}_n)\Delta\boldsymbol{\xi}_n, \\ \dot{\boldsymbol{\xi}}_{n+1} &\cong \boldsymbol{h}(\boldsymbol{\sigma}_n, \boldsymbol{\xi}_n) + \boldsymbol{h}_{,\sigma}(\boldsymbol{\sigma}_n, \boldsymbol{\xi}_n)\Delta\boldsymbol{\sigma}_n + \boldsymbol{h}_{,\xi}(\boldsymbol{\sigma}_n, \boldsymbol{\xi}_n)\Delta\boldsymbol{\xi}_n \end{aligned} \qquad (4.2.31)$$

with

$$\boldsymbol{g}_{,\sigma} = \frac{\partial \boldsymbol{g}}{\partial \boldsymbol{\sigma}}, \quad \boldsymbol{g}_{,\xi} = \frac{\partial \boldsymbol{g}}{\partial \boldsymbol{\xi}}, \quad \boldsymbol{h}_{,\sigma} = \frac{\partial \boldsymbol{h}}{\partial \boldsymbol{\sigma}}, \quad \boldsymbol{h}_{,\xi} = \frac{\partial \boldsymbol{h}}{\partial \boldsymbol{\xi}}$$

From (4.2.29) we obtain

$$\begin{aligned} \Delta\boldsymbol{\varepsilon}^{cr}_n &= \Delta t_n \left(\boldsymbol{g}_n + \theta \boldsymbol{g}_{n,\sigma}\Delta\boldsymbol{\sigma}_n + \theta \boldsymbol{g}_{n,\xi}\Delta\boldsymbol{\xi}_n\right), \\ \Delta\boldsymbol{\xi}_n &= \Delta t_n \left(\boldsymbol{h}_n + \theta \boldsymbol{h}_{n,\sigma}\Delta\boldsymbol{\sigma}_n + \theta \boldsymbol{h}_{n,\xi}\Delta\boldsymbol{\xi}_n\right), \end{aligned} \qquad (4.2.32)$$

where

$$\begin{aligned} \boldsymbol{g}_n &\equiv \boldsymbol{g}(\boldsymbol{\sigma}_n, \boldsymbol{\xi}_n), & \boldsymbol{h}_n &\equiv \boldsymbol{h}(\boldsymbol{\sigma}_n, \boldsymbol{\xi}_n), \\ \boldsymbol{g}_{n,\sigma} &\equiv \frac{\partial \boldsymbol{g}}{\partial \boldsymbol{\sigma}}(\boldsymbol{\sigma}_n, \boldsymbol{\xi}_n), & \boldsymbol{g}_{n,\xi} &\equiv \frac{\partial \boldsymbol{g}}{\partial \boldsymbol{\xi}}(\boldsymbol{\sigma}_n, \boldsymbol{\xi}_n), \\ \boldsymbol{h}_{n,\sigma} &\equiv \frac{\partial \boldsymbol{h}}{\partial \boldsymbol{\sigma}}(\boldsymbol{\sigma}_n, \boldsymbol{\xi}_n), & \boldsymbol{h}_{n,\xi} &\equiv \frac{\partial \boldsymbol{h}}{\partial \boldsymbol{\xi}}(\boldsymbol{\sigma}_n, \boldsymbol{\xi}_n) \end{aligned}$$

The second equation (4.2.32) can be rewritten as

$$\Delta\boldsymbol{\xi}_n = \Delta t_n \left[\boldsymbol{I} - \Delta t_n \theta \boldsymbol{h}_{n,\xi}\right]^{-1} \left[\boldsymbol{h}_n + \theta \boldsymbol{h}_{n,\sigma}\Delta\boldsymbol{\sigma}_n\right] \qquad (4.2.33)$$

Inserting this equation into the first equation (4.2.32) we obtain

$$\Delta\boldsymbol{\varepsilon}^{cr}_n = \Delta t_n (\boldsymbol{g}_n + \theta \boldsymbol{g}_{n,\sigma}\Delta t \boldsymbol{\sigma}_n) + \Delta t_n^2 \boldsymbol{h}_{n,\xi} \left[\boldsymbol{I} - \Delta t_n \theta \boldsymbol{h}_{n,\xi}\right]^{-1} \left[\boldsymbol{h}_n + \theta \boldsymbol{h}_{n,\sigma}\Delta\boldsymbol{\sigma}_n\right] \qquad (4.2.34)$$

Neglecting the last term in the right-hand side of (4.2.34), the first Eq. in (4.2.20) takes the form

$$\boldsymbol{D}\boldsymbol{E}\boldsymbol{D}^T \Delta\boldsymbol{u}_n = \boldsymbol{D}\boldsymbol{E}\Delta\boldsymbol{\varepsilon}^{cr}_n \cong \Delta t_n \boldsymbol{D}\boldsymbol{E}\left[\boldsymbol{g}_n + \theta \boldsymbol{g}_{n,\sigma}\Delta\boldsymbol{\sigma}_n\right] \qquad (4.2.35)$$

Here $\bar{\boldsymbol{f}} = const$ and $\boldsymbol{\varepsilon}^{th} = const$ are assumed. From (4.2.19) the increment of the stress vector can be computed as follows

$$\Delta\boldsymbol{\sigma}_n = \boldsymbol{E}\left[\boldsymbol{D}^T \Delta\boldsymbol{u}_n - \Delta t_n \boldsymbol{g}_n - \Delta t_n \theta \boldsymbol{g}_{n,\sigma}\Delta\boldsymbol{\sigma}_n\right],$$

or

$$\Delta\boldsymbol{\sigma}_n = \left[\boldsymbol{I} + \Delta t_n \theta \boldsymbol{E}\boldsymbol{g}_{n,\sigma}\right]^{-1} \boldsymbol{E}\boldsymbol{D}^T \Delta\boldsymbol{u}_n - \Delta t_n \left[\boldsymbol{I} + \Delta t_n \theta \boldsymbol{E}\boldsymbol{g}_{n,\sigma}\right]^{-1} \boldsymbol{E}\boldsymbol{g}_n \qquad (4.2.36)$$

After inserting into (4.2.35) we obtain

$$\begin{aligned} \boldsymbol{D}\left[\mathbf{E}-\mathbf{E}_n^*\right]\boldsymbol{D}^T\Delta\boldsymbol{u}_n &= \Delta t_n \boldsymbol{D}\mathbf{E}\boldsymbol{g}_n, \\ \mathbf{E}_n^* &= \Delta t_n \theta \mathbf{E}\boldsymbol{g}_{n,\sigma}\left[\boldsymbol{I}+\Delta t_n\theta\mathbf{E}\boldsymbol{g}_{n,\sigma}\right]^{-1}\mathbf{E} \end{aligned} \tag{4.2.37}$$

Based on the derived equations it is possible to formulate the following explicit one-step method:

set $n=0,\ \boldsymbol{\varepsilon}_0^{cr}=\mathbf{0},\quad \boldsymbol{\xi}_0=\mathbf{0}$

solve BVP $\boldsymbol{D}\mathbf{E}\boldsymbol{D}^T\boldsymbol{u}_0=-\bar{\boldsymbol{f}}$, calculate $\boldsymbol{\sigma}_0=\mathbf{E}\boldsymbol{D}^T\boldsymbol{u}_0$

1: calculate

$\Delta\boldsymbol{\varepsilon}_n^{cr}=\Delta t_n(\boldsymbol{g}_n+\theta\boldsymbol{g}_{n,\sigma}\Delta t\boldsymbol{\sigma}_n)$,

$\Delta\boldsymbol{\xi}_n=\Delta t_n\left[\boldsymbol{I}-\Delta t_n\theta\boldsymbol{h}_{n,\xi}\right]^{-1}\left[\boldsymbol{h}_n+\theta\boldsymbol{h}_{n,\sigma}\Delta\boldsymbol{\sigma}_n\right]$,

$\mathbf{E}_n^*=\Delta t_n\theta\mathbf{E}\boldsymbol{g}_{n,\sigma}\left[\boldsymbol{I}+\Delta t_n\theta\mathbf{E}\boldsymbol{g}_{n,\sigma}\right]^{-1}\mathbf{E}$

solve BVP

$$\begin{aligned} \boldsymbol{D}\left[\mathbf{E}-\mathbf{E}_n^*\right]\boldsymbol{D}^T\boldsymbol{\Delta u}_n &= \Delta t_n\boldsymbol{D}\mathbf{E}\boldsymbol{g}_n, \\ \Delta\boldsymbol{\sigma}_n &= \mathbf{E}(\boldsymbol{D}^T\boldsymbol{\Delta u}_n-\Delta\boldsymbol{\varepsilon}_n^{cr}) \end{aligned} \tag{4.2.38}$$

calculate

$$\begin{aligned} \boldsymbol{\varepsilon}_{n+1}^{cr} &= \boldsymbol{\varepsilon}_n^{cr}+\Delta\boldsymbol{\varepsilon}_n^{cr}, & \boldsymbol{\xi}_{n+1} &= \boldsymbol{\xi}_n+\Delta\boldsymbol{\xi}_n, \\ \boldsymbol{u}_{n+1} &= \boldsymbol{u}_n+\Delta\boldsymbol{u}_n, & \boldsymbol{\sigma}_{n+1} &= \boldsymbol{\sigma}_n+\Delta\boldsymbol{\sigma}_n \end{aligned}$$

if $t_{n+1}<t_N$ and $\omega_l<\omega_{l*},\quad l=1,\ldots,m$

then set $n:=n+1$ go to 1

else finish

For $\theta>0$ this method provides an accuracy of higher order if compared with that for the explicit one-step Euler method. For example, for $\theta=1/2$ the method has a second order accuracy while the explicit Euler method ($\theta=0$) provides a first order accuracy. Following this algorithm the fictitious force vector $\Delta t\boldsymbol{D}\mathbf{E}\boldsymbol{g}_n$ and the stiffness matrix $\mathbf{E}-\mathbf{E}_n^*$ must be computed at each time step. The modified stiffness leads to an additional effort in solving the boundary value problem (4.2.38). Furthermore, the matrix $\mathbf{E}-\mathbf{E}_n^*$ is non-symmetric.

4.2.3.2 Solution of Boundary Value Problems. According to the discussed time integration algorithms, linearized boundary value problems have to be solved at each time or iteration step. These problems include second order partial differential equations with respect to the unknown displacements $\boldsymbol{u}(\boldsymbol{x},t_n)$ or displacement increments $\Delta\boldsymbol{u}(\boldsymbol{x},t_n)$. The effect of the accumulated creep strain is considered by means of fictitious force vectors and/or complementary stiffness matrices. The accumulated creep strain is determined by the entire deformation history. Therefore, the known analytical methods from the theory of elasticity, e.g. the Fourier series approach [9] and the complex stress functions approach [129], are not applicable in the general case of creep with internal state variables. Only for some one-dimensional problems, e.g. for the Bernoulli-Euler type beam, analytical closed form solutions of the creep problems can be obtained [80, 208, 243]. These solutions are helpful for the verification of the general computational methods or general purpose solvers.

In what follows let us briefly discuss the numerical methods recently used in creep mechanics. These methods are:

- the finite difference method,
- the direct variational methods and
- the boundary element method.

Applying the finite difference method the partial differential operators are replaced by finite differences leading to the solution of algebraic equations instead of the partial differential ones. The utilization is mostly efficient for creep problems leading to ordinary differential equations. Examples include axi-symmetrically loaded shells of revolution and circular plates [20, 21, 33, 63, 64, 83, 85, 227, 262, 298].

The widely used approach is based on the variational formulations of the creep problem. Starting from appropriate variational functionals the following direct variational methods can be applied: the Ritz method, the Galerkin method and the Vlasov-Kantorovich method. We will briefly discuss the variational formulations and the classical variational methods in the next subsection. The most powerful variational method for the structural analysis is the finite element method [41, 348] which is the basis of commercial general purpose solvers, e.g., ABAQUS, ADINA, ANSYS, COSMOS, etc. The possibility to incorporate a creep material model with internal state variables is available in commercial codes. The implementation can be performed by writing a user-defined material subroutine.

The boundary element method is based on the transformation of the partial differential equations into boundary integral equations. In order to solve these equations the boundary of the domain is divided into finite elements. As a result a set of algebraic equations with respect to the vector of displacements (tractions) in the discretization points of the boundary can be obtained. In the case of creep an additional domain discretization is necessary in order to store the components of the creep strain vector [43]. For details of the boundary element technique we refer to [81, 133, 260].

4.2.3.3 Variational Formulations and Procedures. Variational formulations are widely used in several problems of solid mechanics. They are the basis for direct variational methods, e.g. the Ritz method, the Galerkin method, the finite element method. With respect to the type of the BVP, different variational functionals have been proposed. Here let us consider a variational functional in terms of the displacement vector. Let $\boldsymbol{u}(q^i, t)$ be the solution of the BVP (4.2.1) - (4.2.6) under given $\boldsymbol{\varepsilon}^{cr}$. Let $\delta\boldsymbol{u}$ be the vector of virtual displacements satisfying the kinematic boundary conditions (4.2.4). Starting from the equilibrium condition (4.2.3) we can write

$$\int_V (\nabla \cdot \boldsymbol{\sigma} + \rho \boldsymbol{f}) \cdot \delta \boldsymbol{u} dV = 0 \tag{4.2.39}$$

According to (B.2.2)

$$\int_V (\nabla \cdot \boldsymbol{\sigma}) \cdot \delta \boldsymbol{u} dV = \int_V \left[\nabla \cdot (\boldsymbol{\sigma} \cdot \delta \boldsymbol{u}) - \boldsymbol{\sigma} \cdot\cdot (\nabla \delta \boldsymbol{u})^T\right] dV \tag{4.2.40}$$

Applying the divergence theorem (see Sect. B.3) and the static boundary conditions (4.2.4) we obtain

$$\int_V \nabla \cdot (\boldsymbol{\sigma} \cdot \delta \boldsymbol{u}) dV = \int_A (\boldsymbol{\nu} \cdot \boldsymbol{\sigma}) \cdot \delta \boldsymbol{u} dA = \int_{A_p} \bar{\boldsymbol{p}} \cdot \delta \boldsymbol{u} dA \tag{4.2.41}$$

With $\boldsymbol{\sigma} \cdot\cdot (\nabla \delta \boldsymbol{u})^T = \boldsymbol{\sigma} \cdot\cdot \delta (\nabla \boldsymbol{u})^T = \boldsymbol{\sigma} \cdot\cdot \delta \boldsymbol{\varepsilon}$, (4.2.40) and (4.2.41), Eq. (4.2.39) takes the form

$$\int_V \boldsymbol{\sigma} \cdot\cdot \delta \boldsymbol{\varepsilon} dV - \int_V \rho \boldsymbol{f} \cdot \delta \boldsymbol{u} dV - \int_{A_p} \bar{\boldsymbol{p}} \cdot \delta \boldsymbol{u} dA = 0, \tag{4.2.42}$$

or

$$\delta W_i + \delta W_e = 0, \quad \delta W_i = -\int_V \boldsymbol{\sigma} \cdot\cdot \delta \boldsymbol{\varepsilon} dV, \quad \delta W_e = \int_V \bar{\boldsymbol{f}} \cdot \delta \boldsymbol{u} dV + \int_{A_p} \bar{\boldsymbol{p}} \cdot \delta \boldsymbol{u} dA \tag{4.2.43}$$

The principle of virtual displacements (4.2.43) states that if a deformable system is in equilibrium then the sum of the virtual work of external actions δW_e and the virtual work of internal forces δW_i is equal to zero, e.g., [9, 260, 329]. With the constitutive equation (4.2.6)

$$\begin{aligned} \boldsymbol{\sigma} \cdot\cdot \delta \boldsymbol{\varepsilon} &= \left({}^{(4)}\mathbf{C} \cdot\cdot (\boldsymbol{\varepsilon} - \boldsymbol{\varepsilon}^{cr} - \boldsymbol{\varepsilon}^{th})\right) \cdot\cdot \delta \boldsymbol{\varepsilon} \\ &= \frac{1}{2}\delta(\boldsymbol{\varepsilon} \cdot\cdot {}^{(4)}\mathbf{C} \cdot\cdot \boldsymbol{\varepsilon}) - (\boldsymbol{\varepsilon}^{cr} + \boldsymbol{\varepsilon}^{th}) \cdot\cdot {}^{(4)}\mathbf{C} \cdot\cdot \delta \boldsymbol{\varepsilon} \end{aligned}$$

the variational equation (4.2.43) can be formulated as follows

$$\delta \left[\frac{1}{2}\int_V \boldsymbol{\varepsilon} \cdot\cdot {}^{(4)}\mathbf{C} \cdot\cdot \boldsymbol{\varepsilon} dV - \int_V \bar{\boldsymbol{f}} \cdot \boldsymbol{u} dV - \int_{A_p} \bar{\boldsymbol{p}} \cdot \boldsymbol{u} dA \right. \\ \left. - \int_V (\boldsymbol{\varepsilon}^{cr} + \boldsymbol{\varepsilon}^{th}) \cdot\cdot {}^{(4)}\mathbf{C} \cdot\cdot \boldsymbol{\varepsilon} dV \right] = 0$$

or $\delta \Pi(\boldsymbol{u}) = 0$ with

$$\begin{aligned} \Pi(\boldsymbol{u}) &= \frac{1}{2}\int_V \boldsymbol{\varepsilon} \cdot\cdot {}^{(4)}\mathbf{C} \cdot\cdot \boldsymbol{\varepsilon} dV - \int_V \bar{\boldsymbol{f}} \cdot \boldsymbol{u} dV - \int_{A_p} \bar{\boldsymbol{p}} \cdot \boldsymbol{u} dA \\ &- \int_V (\boldsymbol{\varepsilon}^{cr} + \boldsymbol{\varepsilon}^{th}) \cdot\cdot {}^{(4)}\mathbf{C} \cdot\cdot \boldsymbol{\varepsilon} dV \end{aligned} \tag{4.2.44}$$

Applying the vector-matrix notation we can write

$$\begin{aligned}\Pi(\boldsymbol{u}) &= \frac{1}{2}\int\limits_V (\mathbf{D}^T\boldsymbol{u})^T\mathbf{E}\mathbf{D}^T\boldsymbol{u}\,dV - \int\limits_V \bar{\boldsymbol{f}}^T\boldsymbol{u}\,dV - \int\limits_{A_p} \bar{\boldsymbol{p}}^T\boldsymbol{u}\,dA \\ &- \int\limits_V \boldsymbol{\varepsilon}^{th}\mathbf{E}\mathbf{D}^T\boldsymbol{u}\,dV - \int\limits_V \boldsymbol{\varepsilon}^{cr}\mathbf{E}\mathbf{D}^T\boldsymbol{u}\,dV\end{aligned} \tag{4.2.45}$$

It is easy to verify that from the condition $\delta\Pi(\boldsymbol{u}) = 0$ follows the partial differential equation with respect to the displacement vector and the static boundary condition (4.2.20).

The variational functional (4.2.45) has been derived from the principle of virtual displacements. By analogy a variational functional in terms of stresses or stress functions can be formulated providing Eqs. (4.2.24) as Euler equations. Furthermore, a mixed variational formulation in terms of displacements and stresses can be convenient for numerous structural mechanics problems. In [21, 230] a mixed variational functional has been utilized for the solution of the von Kármán type plate equations. In [19, 23, 231] a mixed formulation has been applied to derive the first order shear deformation beam equations.

To solve the variational problem classical direct variational methods can be utilized. Let us illustrate the application of the Ritz method to the variational functional (4.2.45). The approximate solution for the displacement vector $\tilde{\boldsymbol{u}}$ is presented in the form of series

$$\tilde{u}_k = \sum_{i=1}^{N} a_{ki}\phi_{ki}(x_1, x_2, x_3), \quad k = 1,2,3 \tag{4.2.46}$$

(no summation over k) or

$$\tilde{\boldsymbol{u}} \equiv \begin{bmatrix} \tilde{u}_1 \\ \tilde{u}_2 \\ \tilde{u}_3 \end{bmatrix} = \begin{bmatrix} \boldsymbol{a}_1^T\boldsymbol{\phi}_1 \\ \boldsymbol{a}_2^T\boldsymbol{\phi}_2 \\ \boldsymbol{a}_3^T\boldsymbol{\phi}_3 \end{bmatrix} = \begin{bmatrix} \boldsymbol{\phi}_1 & \mathbf{0} & \mathbf{0} \\ \mathbf{0} & \boldsymbol{\phi}_2 & \mathbf{0} \\ \mathbf{0} & \mathbf{0} & \boldsymbol{\phi}_3 \end{bmatrix}^T \begin{bmatrix} \boldsymbol{a}_1 \\ \boldsymbol{a}_2 \\ \boldsymbol{a}_3 \end{bmatrix} = \mathbf{G}^T\boldsymbol{a}, \tag{4.2.47}$$

where $\boldsymbol{\phi}_k$ are vectors of the trial (basis or shape) functions which should be specified a priori and $\boldsymbol{a}_k$ are vectors of unknown (free) parameters. The functions ϕ_{ki} in (4.2.46) must be linearly independent and satisfy the kinematical boundary conditions. Furthermore, the set of these functions must be complete in order to provide the convergence of $\tilde{\boldsymbol{u}}$ as $N \to \infty$. Inserting the approximate solution $\tilde{\boldsymbol{u}}$ into the variational functional (4.2.45) we can obtain for the time step t_n

$$\begin{aligned}\tilde{\Pi}_n(\tilde{\boldsymbol{u}}) &= \boldsymbol{a}^T\left(\frac{1}{2}\int\limits_V (\mathbf{D}^T\mathbf{G})^T\mathbf{E}\mathbf{D}^T\mathbf{G}\,dV\right)\boldsymbol{a} - \boldsymbol{a}^T\int\limits_V \mathbf{G}\bar{\boldsymbol{f}}\,dV - \boldsymbol{a}^T\int\limits_{A_p} \mathbf{G}\bar{\boldsymbol{p}}\,dA \\ &- \boldsymbol{a}^T\left(\int\limits_V (\mathbf{D}^T\mathbf{G})^T\mathbf{E}\boldsymbol{\varepsilon}^{th^T}dV + \int\limits_V (\mathbf{D}^T\mathbf{G})^T\mathbf{E}\boldsymbol{\varepsilon}_n^{cr^T}dV\right) \\ &= \frac{1}{2}\boldsymbol{a}^T\boldsymbol{K}\boldsymbol{a} - \boldsymbol{a}^T(\boldsymbol{f} + \boldsymbol{f}^{th} + \boldsymbol{f}_n^{cr}) = \tilde{\Pi}_n(\boldsymbol{a})\end{aligned} \tag{4.2.48}$$

with

$$\boldsymbol{K} = \int\limits_V (\mathbf{D}^T\mathbf{G})^T \mathbf{E}\mathbf{D}^T\mathbf{G} dV = \int\limits_V \mathbf{B}^T\mathbf{E}\mathbf{B} dV, \quad \mathbf{B} = \mathbf{D}^T\mathbf{G},$$

$$\boldsymbol{f} = \int\limits_V \mathbf{G}\bar{\boldsymbol{f}} dV - \int\limits_{A_p} \mathbf{G}\bar{\boldsymbol{p}} dA,$$

$$\boldsymbol{f}^{th} = \int\limits_V \mathbf{B}^T\mathbf{E}\boldsymbol{\varepsilon}^{th^T} dV, \quad \boldsymbol{f}^{cr}_n = \int\limits_V \mathbf{B}^T\mathbf{E}\boldsymbol{\varepsilon}^{cr^T}_n dV$$

From the condition $\delta\tilde{\Pi}_n(\boldsymbol{a}) = 0$ follows the set of linear algebraic equations

$$\boldsymbol{K}\boldsymbol{a} = \boldsymbol{f} + \boldsymbol{f}^{th} + \boldsymbol{f}^{cr}_n \tag{4.2.49}$$

After the solution of (4.2.49) the displacements can be computed from (4.2.47) and the stresses from (4.2.19). With the Ritz method and the explicit time integration procedure the step-by-step solution of a creep problem can be utilized as follows:

set $n = 0,\ \boldsymbol{\varepsilon}^{cr}_0 = \mathbf{0},\quad \boldsymbol{\xi}_0 = \mathbf{0}$

calculate

$$\boldsymbol{K} = \int\limits_V \mathbf{B}^T\mathbf{E}\mathbf{B} dV, \quad \boldsymbol{f} = \int\limits_V \mathbf{G}\bar{\boldsymbol{f}} dV - \int\limits_{A_p} \mathbf{G}\bar{\boldsymbol{p}} dA, \quad \boldsymbol{f}^{th} = \int\limits_V (\mathbf{D}^T\mathbf{G})^T\mathbf{E}\boldsymbol{\varepsilon}^{th^T} dV$$

solve BVP $\boldsymbol{K}\boldsymbol{a}_0 = \boldsymbol{f} + \boldsymbol{f}^{th}$ calculate $\tilde{\boldsymbol{u}}_0 = \mathbf{G}^T\boldsymbol{a}_0,\ \boldsymbol{\sigma}_0 = \mathbf{E}\mathbf{D}^T\tilde{\boldsymbol{u}}_0$

1: calculate

$$\Delta\boldsymbol{\varepsilon}^{cr}_n = \Delta t_n \boldsymbol{g}(\boldsymbol{\sigma}_n, \boldsymbol{\xi}_n, T_n), \qquad \Delta\boldsymbol{\xi}_n = \Delta t_n \boldsymbol{h}(\boldsymbol{\sigma}_n, \boldsymbol{\xi}_n, T_n)$$

$$\boldsymbol{\varepsilon}^{cr}_{n+1} = \boldsymbol{\varepsilon}^{cr}_n + \Delta\boldsymbol{\varepsilon}^{cr}_n, \qquad \boldsymbol{\xi}_{n+1} = \boldsymbol{\xi}_n + \Delta\boldsymbol{\xi}_n,$$

calculate

$$\boldsymbol{f}^{cr}_{n+1} = \int\limits_V (\mathbf{D}^T\mathbf{G})^T\mathbf{E}\boldsymbol{\varepsilon}^{cr}_{n+1}{}^T dV$$

solve $\boldsymbol{K}\boldsymbol{a}_{n+1} = \boldsymbol{f} + \boldsymbol{f}^{th} + \boldsymbol{f}^{cr}_{n+1}$

calculate $\tilde{\boldsymbol{u}}_{n+1} = \mathbf{G}^T\boldsymbol{a}_{n+1},\ \boldsymbol{\sigma}_{n+1} = \mathbf{E}(\mathbf{D}^T\tilde{\boldsymbol{u}}_{n+1} - \boldsymbol{\varepsilon}^{cr}_{n+1})$

if $t_{n+1} < t_N$ and $\omega_l < \omega_{l*}, l = 1, \ldots, m$

then set $n := n + 1$ go to 1

else finish

The vector $\boldsymbol{f}^{cr}_n$ must be computed at each time step through a numerical integration. Therefore, the domain discretization is required to store the vectors $\boldsymbol{\varepsilon}^{cr}$ and $\boldsymbol{\xi}$. The accuracy of the solution by the Ritz method depends on the "quality" and the number of trial functions. For special problems with simple geometry, homogeneous boundary conditions, etc., trial functions can be formulated in terms of elementary functions (e.g. orthogonal polynomials, trigonometric or hyperbolic functions etc.) defined within the whole domain, e.g. [9]. Examples for such problems include beams [17, 80, 231] and plates [21, 230]. The Ritz method is simple in utilization and provides an approximate analytical solution.

In the general case of complex geometry, a powerful tool is the finite element method. The domain is subdivided into finite elements and the piecewise trial functions (polynomials) are defined within the elements. For details of finite element techniques we refer to the textbooks [41, 60, 334, 348]. By analogy with the Ritz method the finite element procedure results in a set of algebraic equations of the type

$$\boldsymbol{K}\boldsymbol{\delta}_n = \boldsymbol{f} + \boldsymbol{f}^{th} + \boldsymbol{f}_n^{cr}, \tag{4.2.50}$$

where $\boldsymbol{K}$ is the overall stiffness matrix, $\boldsymbol{\delta}_n$ is the vector of unknown nodal displacements and $\boldsymbol{f}$, $\boldsymbol{f}^{th}$ and $\boldsymbol{f}_n^{cr}$ are the nodal force vectors computed from given loads, thermal strains as well as creep strains at the time or iteration step. The commercial codes usually include more sophisticated time integration methods allowing the automatic time step size control. The vector $\boldsymbol{f}_n^{cr}$ depends on the distribution of creep strains at the current time step. The creep strains are determined by the constitutive model and a variety of constitutive models can be applied depending on the material type, type of loading, available experimental data, etc. Therefore the possibility to incorporate a user-defined material law is usually available in commercial codes.

4.3 Beams

Beams are widely discussed in monographs and textbooks on creep mechanics [80, 142, 155, 178, 207, 208, 241, 257, 299]. The presented examples are, however, limited to the classical Bernoulli-Euler beam theory and Norton-Bailey constitutive equations of steady-state creep. The objective of this section is to analyze time dependent behavior of beams under creep-damage conditions. For this purpose we apply the classical beam theory and a refined theory which includes the effect of transverse shear deformation (Timoshenko-type theory). Based on several examples we compare both theories as they describe creep-damage processes in beams. Furthermore, we develop and solve several benchmark problems. The reference solutions obtained by the Ritz method are applied to verify user-defined creep-damage material subroutines and the general purpose finite element codes.

4.3.1 Classical Beam Theory

Let us consider a straight homogeneous beam in the Cartesian coordinate system x, y, z as shown in Fig. 4.1. For the sake of brevity we limit to the case of symmetrical bending in the plane spanned on the x and z coordinate lines. Furthermore we introduce geometrically linear equations. Their validity is restricted to the case of infinitesimal strains, displacements and cross section rotations. The governing equations can be summarized as follows

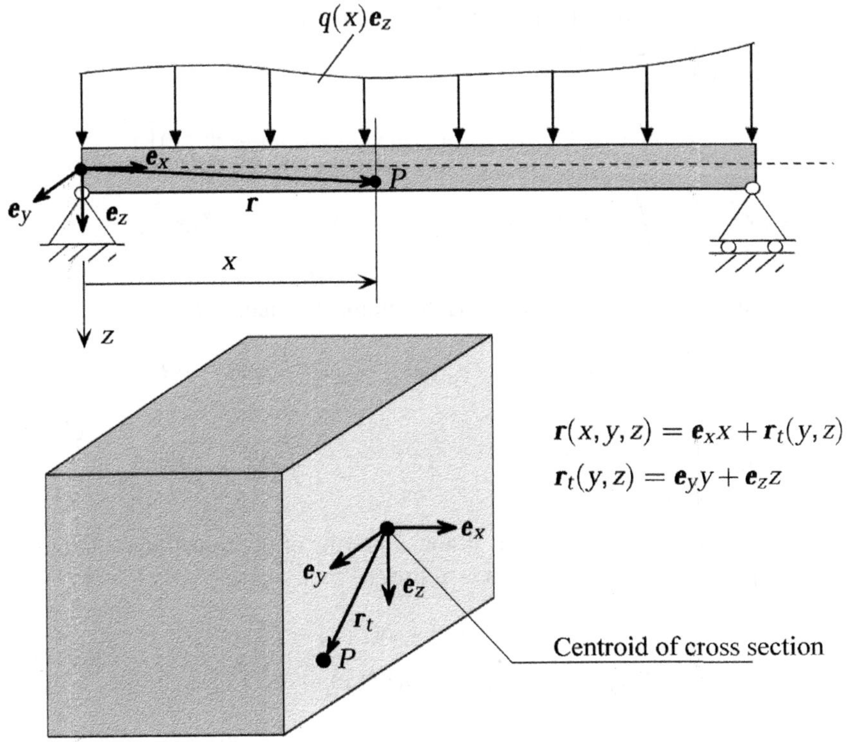

Fig. 4.1 Beam with a rectangular cross section. Geometry, loading and coordinates

- Kinematical equations

$$\begin{aligned} u(x,z) &= u_0(x) + \varphi(x)z, \quad \varphi(x) = -w'(x), \\ \varepsilon_x(x,z) &= u_0'(x) + \varphi'(x)z, \end{aligned} \tag{4.3.1}$$

  where $u(x,z)$ is the axial displacement, $u_0(x)$ is the axial displacement of the beam centerline, $\varphi(x)$ is the angle of the cross section rotation, $w(x)$ is the transverse displacement (deflection) and prime denotes the derivative with respect to the coordinate x.
- Equilibrium conditions

$$N'(x) = 0, \quad Q'(x) + q(x) = 0, \quad M'(x) = Q(x), \tag{4.3.2}$$

  where $N(x)$ is the normal force, $Q(x)$ is the shear force, $M(x)$ is the bending moment and $q(x)$ is the given distributed load.
- Constitutive equations
  - normal stress

$$\begin{aligned} \sigma_x(x,z) &= E[\varepsilon_x(x,z) - \alpha_T \Delta T(x,z) - \varepsilon_x^{cr}(x,z)] \\ &= E[\varepsilon_0(x) + \chi(x)z - \alpha_T(x,z)\Delta T - \varepsilon_x^{cr}(x,z)], \end{aligned} \tag{4.3.3}$$

where $\varepsilon_0 = u_0'$ is the strain of the beam centerline and $\chi = -w''$ is the curvature of the beam centerline.

- stress resultants

$$
\begin{aligned}
N(x) &= \int_A \sigma_x dA &= EA\left[\varepsilon_0(x) - \varepsilon_0^{cr}(x) - \varepsilon_0^{th}(x)\right],\\
M(x) &= \int_A \sigma_x z dA &= EI\left[\chi(x) - \chi^{cr}(x) - \chi^{th}(x)\right],
\end{aligned}
\tag{4.3.4}
$$

where A is the cross section area, I is the moment of inertia and

$$
\begin{aligned}
\varepsilon_0^{cr}(x) &= \frac{1}{A}\int_A \varepsilon_x^{cr}(x,z)dA, \quad \varepsilon_0^{th}(x) = \alpha_T\frac{1}{A}\int_A \Delta T(x,z)dA,\\
\chi^{cr}(x) &= \frac{1}{I}\int_A \varepsilon_x^{cr}(x,z)zdA, \quad \chi^{th}(x) = \alpha_T\frac{1}{I}\int_A \Delta T(x,z)zdA
\end{aligned}
\tag{4.3.5}
$$

are averages of thermal and creep strains. In terms of fictitious forces and moments Eqs (4.3.4) can be rewritten as follows

$$
\begin{aligned}
N(x) &= EA\varepsilon_0(x) - N^{cr}(x) - N^{th}(x),\\
M(x) &= EI\chi(x) - M^{cr}(x) - M^{th}(x)
\end{aligned}
\tag{4.3.6}
$$

with

$$
\begin{aligned}
N^{cr}(x) &= E\int_A \varepsilon_x^{cr}(x,z)dA, \quad N^{th}(x) = E\alpha_T\int_A \Delta T(x,z)dA,\\
M^{cr}(x) &= E\int_A \varepsilon_x^{cr}(x,z)zdA, \quad M^{th}(x) = E\alpha_T\int_A \Delta T(x,z)zdA
\end{aligned}
\tag{4.3.7}
$$

- creep-damage constitutive and evolution equations (see Sect. 2.4.1.1)

$$
\begin{aligned}
\dot{\varepsilon}_x^{cr} &= \frac{a|\sigma_x|^{n-1}\sigma_x}{(1-\omega)^n}, \quad \dot{\omega} = \frac{b\sigma_{eq}^k}{(1-\omega)^l},\\
\sigma_{eq} &= \alpha\frac{|\sigma_x| + \sigma_x}{2} + (1-\alpha)|\sigma_x|
\end{aligned}
\tag{4.3.8}
$$

The boundary conditions at $x = 0$ and $x = l$ (l is the beam length) must be formulated with respect to the kinematical quantities w, φ and/or the dual static quantities Q, M. The initial conditions at $t = 0$ are $\varepsilon_x^{cr} = 0$ and $\omega = 0$.

4.3.2 Closed Form Solution

Assuming idealized creep behavior with the secondary creep stage only (see Sect. 1.2) a steady state exists, for which the bending stress and the deflection rate in a beam are constant. Constitutive equation for secondary creep follows from (4.3.8) by setting $b = 0$. For the sake of brevity let us neglect the thermal strains. The constitutive equation for the bending stress (4.3.6) takes the form

$$\sigma_x(x,z) = E\left[\chi(x)z - \varepsilon_x^{cr}(x,z)\right] \tag{4.3.9}$$

In the following derivations let us drop the arguments. Taking the time derivative of (4.3.9) and applying the constitutive equation (4.3.8) we obtain

$$\dot{\sigma}_x = E(\dot{\chi}z - a|\sigma_x|^{n-1}\sigma_x) \tag{4.3.10}$$

Equation (4.3.10) describes the stress redistribution in a beam. The steady state solution follows from (4.3.10) by setting $\dot{\sigma}_x = 0$

$$\sigma_x = \left(\frac{1}{a}\right)^{\frac{1}{n}} |\dot{\chi}z|^{\frac{1}{n}-1}\dot{\chi}z \tag{4.3.11}$$

The bending moment in the steady state can be calculated as follows

$$M = \int\limits_A \sigma_x z dA = \left(\frac{1}{a}\right)^{\frac{1}{n}} I_n|\dot{\chi}|^{\frac{1}{n}-1}\dot{\chi}, \tag{4.3.12}$$

where

$$I_n = \int\limits_A |z|^{\frac{1}{n}-1}z^2 dA \tag{4.3.13}$$

is the generalized moment of inertia.

As an example let us consider a simply supported beam subjected to the uniformly distributed load q. In this statically determinate case the bending moment is $M(x) = qx(l-x)/2$. From (4.3.12) follows the differential equation for the deflection rate

$$\dot{w}(x)'' = -\frac{a}{I_n^n}\frac{q^n}{2^n}x^n(l-x)^n, \quad 0 \leq x \leq l \tag{4.3.14}$$

For integer values of the power n the solution is

$$\dot{w}(x) = \frac{a}{I_n^n}\frac{q^n}{2^n}x\sum_{k=0}^{n}\alpha_k(l^{n+k+1} - x^{n+k+1}) \tag{4.3.15}$$

with

$$\alpha_k = (-1)^k \frac{n!}{k!(n-k)!}\frac{l^{n-k}}{(n+k+1)(n+k+2)}$$

The reference elastic deflection is

$$w(x) = \frac{q}{24EI}x(x-l)(x^2 - lx - l^2)$$

Let us note that the closed form solution for the steady state deflection rate (4.3.15) is a polynomial of the order $2n+2$. Therefore, if the creep problem is numerically solved applying variational methods (see Sect. 4.2.3.3), the trial functions for the deflection or deflection rate should contain polynomial terms of the order $2n+2$ instead of 4 in the elastic case. The order of the polynomial terms of the creep solution is material-dependent since n is the creep exponent in the Norton-Bailey creep law. Furthermore, for the analysis of steady-state creep, an accurate solution cannot be obtained applying approximations justified from the elastic solution. Closed form solutions for steady-state creep in beams with various types of boundary conditions and loading are presented in [80, 208, 241].

4.3.3 Variational Formulation and the Ritz Method

The variational formulations and the Ritz method are discussed in Sect. 4.2.3.3 for three-dimensional solids. To derive the variational functional for the beam we start from the principle of virtual displacements (4.2.43). Applying the kinematical relations (4.3.1) and the constitutive equations (4.3.3) we can write

$$
\begin{aligned}
\int_V \sigma_x \delta\varepsilon_x dV &= EI\int_0^l w''\delta w'' dx + EA\int_0^l u_0'\delta u_0' dx \\
&+ \int_0^l M^{cr}\delta w'' dx - \int_0^l N^{cr}\delta u_0' dx \\
&= \int_0^l q\delta w dx
\end{aligned}
\tag{4.3.16}
$$

To simplify the solution, in Eq. (4.3.16) the fictitious force and moment associated with the thermal strain are neglected. Assuming the creep strain to be known function of the coordinates x and z for the fixed time t, the following functional can be formulated

$$
\begin{aligned}
\Pi_t(w, u_0) &= \frac{1}{2}EI\int_0^l w''^2 dx + \frac{1}{2}EA\int_0^l u_0'^2 dx \\
&+ \int_0^l M^{cr} w'' dx - \int_0^l N^{cr} u_0' dx - \int_0^l q w dx
\end{aligned}
$$

The problem is to find such functions w and u_0 that yield an extremum of the functional for the given values of M^{cr}, N^{cr} and q. The approximate solutions can be represented in the form of series

$$
w(x,t) = a_0^w(t)\varphi_0^w(x) + \sum_{i=1}^{\mathcal{N}} a_i^w(t)\varphi_i^w(x), \quad u_0(x) = \sum_{i=1}^{\mathcal{M}} a_i^u(t)\varphi_i^u(x) \tag{4.3.17}
$$

The trial functions should be formulated according to the types of constraints and the loading conditions. As an example let us consider the simply supported and uniformly loaded beam. In this case the following trial functions can be applied. $\varphi_0^w(x) = x(x-l)(x^2 - lx - l^2)$ is the first approximation which follows from the solution for the elastic deflection. $\varphi_i^w(x)$ are the polynomials satisfying the boundary conditions for the deflection $w = 0$ and for the bending moment $M = 0$ at $x = 0$ and $x = l$

$$\varphi_i^w(x) = x^{i+2}(l-x)^{i+2} \tag{4.3.18}$$

Assuming that $u_0 = 0$ at $x = 0$ the functions $\varphi_i^u(x) = x^i$ can be utilized. After collecting the unknown constant coefficients into the vector $\boldsymbol{a}^T = [\boldsymbol{a}^{w^T} \quad \boldsymbol{a}^{u^T}]$ with $\boldsymbol{a}^{w^T} = [a_0^w \quad a_i^w]$, $i = 1, \ldots, \mathcal{N}$ and $\boldsymbol{a}^{u^T} = [a_i^u]$, $i = 1, \ldots, \mathcal{M}$, the Ritz method yields a set of linear algebraic equations

$$\frac{\partial \Pi_t}{\partial a_k} = 0, \quad \Rightarrow \quad \boldsymbol{R}^{ww}\boldsymbol{a}^w = \boldsymbol{f}^w, \quad \boldsymbol{R}^{uu}\boldsymbol{a}^u = \boldsymbol{f}^u \tag{4.3.19}$$

with

$$R_{kj}^{ww} = EI\int_0^l \varphi_k^{w\prime\prime}\varphi_j^{w\prime\prime}dx, \qquad R_{kj}^{uu} = EA\int_0^l \varphi_k^{u\prime}\varphi_j^{u\prime}dx,$$
$$f_k^w = q\int_0^l \varphi_k^w dx - \int_0^l M^{cr}\varphi_k^{w\prime\prime}dx, \qquad f_k^u = \int_0^l N^{cr}\varphi_k^{u\prime}dx,$$
$$k = 0, \ldots, \mathcal{M}, \qquad j = 0, \ldots, \mathcal{M}$$

After solving (4.3.19) the displacements can be computed from (4.3.17). With (4.3.3) the stress $\sigma_x(x,z,t)$ can be calculated as follows

$$\sigma_x(x,z,t)) = E\left[u_0'(x,t) - zw''(x,t) - \varepsilon_x^{cr}(x,z,t)\right]$$

For the known values of the stress and the damage parameter the constitutive model (4.3.8) yields the rates of creep strain and damage for the fixed time t. The corresponding values for the time $t + \Delta t$ are calculated using the implicit time integration procedure (see Sect. 4.2.3.1)

$$\varepsilon^{cr}(x,z,t+\Delta t) = \varepsilon^{cr}(x,z,t) + \frac{\Delta t}{2}[\dot{\varepsilon}^{cr}(x,z,t) + \dot{\varepsilon}^{cr}(x,z,t+\Delta t)],$$
$$\omega(x,z,t+\Delta t) = \omega(x,z,t) + \frac{\Delta t}{2}[\dot{\omega}(x,z,t) + \dot{\omega}(x,z,t+\Delta t)],$$
$$\varepsilon^{cr}(x,z,0) = 0, \quad \omega(x,z,0) = 0, \quad \omega(x,z,t) < \omega_*$$

To compute the fictitious creep force N^{cr} and the creep moment M^{cr} defined in (4.3.7) the Simpson quadrature rule with $\mathcal{N}_h$ integration points in the thickness direction is used. To obtain the matrices $\boldsymbol{R}^{ww}$ and $\boldsymbol{R}^{uu}$ as well as the vectors $\boldsymbol{f}^w$ and $\boldsymbol{f}^u$ in (4.3.19) the Simpson quadrature rule with $\mathcal{N}_l$ integration points along the beam length axis x is applied. The values of the creep strain and damage parameter at the current time step are stored in all integration points. They are used for the calculations at the next time step.

4.3.4 Examples

The solution accuracy of the time dependent creep problem is primarily determined by the solution accuracy of the BVP at each time or iteration step. By utilizing the Ritz method the solution accuracy depends on the type and the number of trial functions in (4.3.17). By applying the finite element method the type of finite elements and the mesh density are most responsible for the accuracy. The following example is selected to examine the convergence of time dependent solutions with respect to different types of approximations including the series (4.3.17) and the finite element discretization.

Let us consider a beam with the length l and the rectangular cross section $g \times h$, where g is the width and h is the height. The beam is simply supported and subjected to the uniformly distributed load q. Two cases of the material behavior will be discussed. In the first case we assume the idealized creep having the secondary stage only. The Norton-Bailey creep constitutive equation will be assumed. The numerical results will be compared with the closed form solution presented in Sect. 4.3.2. The second case is related to the creep-damage analysis. The Kachanov-Rabotnov model is applied to take into account tertiary creep. The reference numerical solution for the creep-damage behavior of the beam will be obtained by the procedure presented in Sect. 4.3.3. This solution will be used to verify the capability of finite element codes for the creep-damage analysis. In particular, we present the results obtained by ABAQUS and ANSYS finite element codes and user-defined creep-damage material subroutines.

For the numerical analysis we set

$$q = 60\ \text{N/mm}, \quad l = 10^3\ \text{mm}, \quad g = 30\ \text{mm}, \quad h = 80\ \text{mm} \tag{4.3.20}$$

Furthermore, we apply the creep-damage material model (3.1.11) with the material constants (3.1.12) identified for the aluminium alloy BS 1472 (see Sect. 3.1.4).

4.3.4.1 Solutions by the Ritz Method.

Norton-Bailey Creep Law. First let us simplify the material behavior by neglecting the damage process. This can be accomplished by setting $b = 0$ in the damage evolution equation (3.1.11). The creep constitutive equation (3.1.11) simplifies to the Norton-Bailey law of steady-state creep. The creep rates are the same for the tensile and compressive loading. Therefore the fictitious normal force N^{cr} is zero. Consequently Eqs. (4.3.19) can be simplified to

$$\boldsymbol{R}^{ww}\boldsymbol{a}^{w} = \boldsymbol{f}^{w}, \quad \boldsymbol{f}^{u} = \boldsymbol{0}, \quad \boldsymbol{a}^{u} = \boldsymbol{0} \tag{4.3.21}$$

The accuracy of the numerical solution depends on the number of trial functions in Eq. (4.3.17), and on the number of integration points $\mathcal{N}_l \times \mathcal{N}_h$. In the following example we fix the number of integration points to $N_l = 41$ and $N_h = 11$ and analyze the accuracy of the series approximation (4.3.17).

Figure 4.2 illustrates the time variation of the maximum deflection in the midpoint of the beam. We observe, that after a certain period of time the deflection

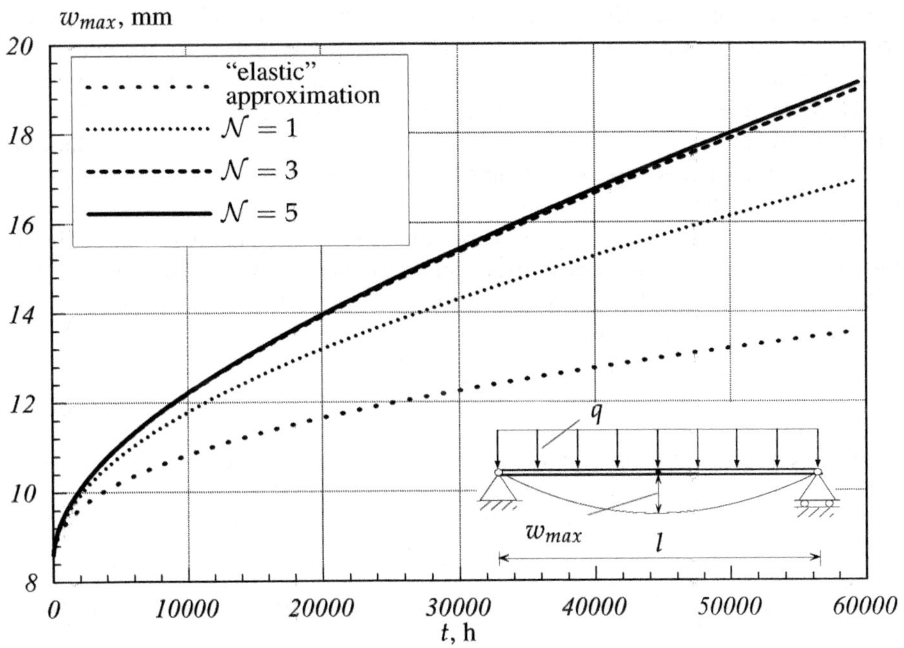

Fig. 4.2 Maximum deflection vs. time based on the Ritz method with different number of trial functions in (4.3.17)

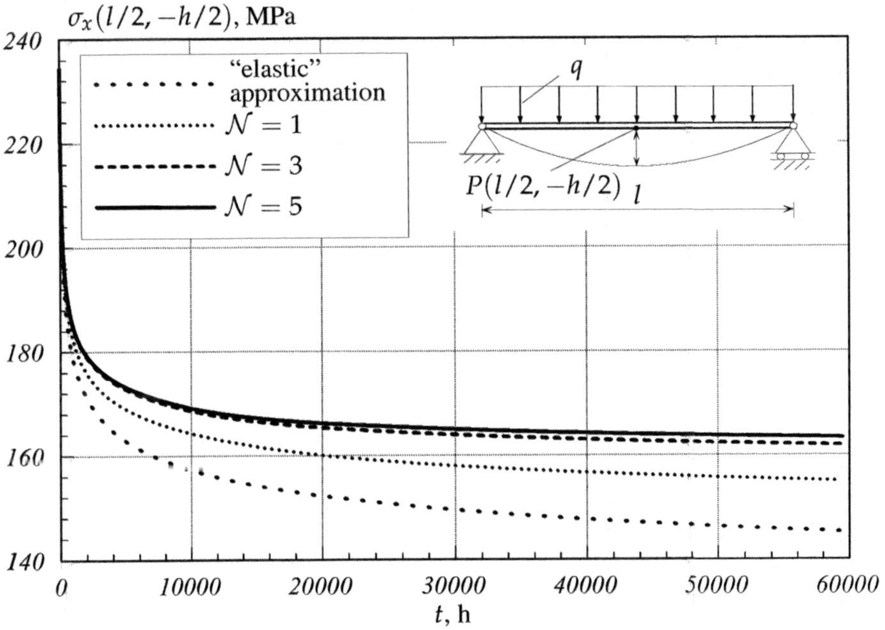

Fig. 4.3 Normal stress vs. time based on the Ritz method with different number of trial functions in (4.3.17)

rate becomes constant. The steady state deflection rate depends significantly on the number of trial functions in (4.3.17). The first approximation for the deflection $w(x,t) = a_0^w(t)x(x-l)(x^2 - lx - l^2)$ is exact for the reference elastic state with $a_0^w(0) = q/(24EI)$. However, it is hardly sufficient to represent the creep behavior. The convergent solution can only be obtained with $\mathcal{N} = 5$. Let us note, that the number of series terms required depends on the material behavior and, in particular, on the value of the creep exponent n (see Sect. 4.3.2). Figure 4.3 shows the time variation of the normal stress σ_x in the bottom layer of the middle cross section of the beam. To verify the numerical results let us compare the steady state stress value with the closed form solution given in Sect. 4.3.2. From Eqs. (4.3.11) and (4.3.12) the normal stress can be computed as follows

$$\sigma_x(x,z) = \frac{M(x)}{I_n}|z|^{\frac{1}{n}-1}z$$

For the rectangular cross section the generalized moment of inertia (4.3.13) reads

$$I_n = \frac{2gn}{n+1}\left(\frac{h}{2}\right)^{\frac{1}{n}+2}$$

With $M(l/2) = ql^2/8$ the normal stress in the middle cross section takes the form

$$\sigma(l/2,z) = \frac{ql^2}{16}\frac{2n+1}{gn}\left(\frac{2}{h}\right)^{\frac{1}{n}+2}|z|^{\frac{1}{n}-1}z \qquad (4.3.22)$$

Figure 4.4 shows the distribution of the normal stress over the thickness direction in the steady-state creep range. The solid line is the plot of Eq. (4.3.22) by taking into

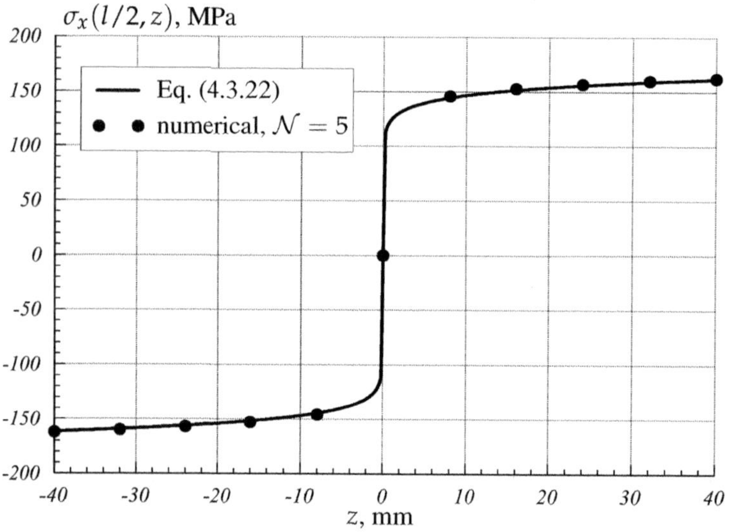

Fig. 4.4 Distribution of the normal stress over the thickness direction in the middle cross section

account (4.3.20) and $n = 14.37$. The circles denote the numerical stress values in the integration points. We observe that the numerical solution agrees well with the closed form one. According to Eq. (4.3.22) the stress value in the bottom layer of the middle cross section is $\sigma_x(l/2, -h/2) = 161.7$ MPa, which is in good agreement with the obtained numerical solution.

Kachanov-Rabotnov Creep Law. The constitutive model (3.1.11) suggests that the damage evolution is controlled by the von Mises equivalent stress. By setting $\alpha = 0$ in Eq. (4.3.8) we obtain $\sigma_{eq} = |\sigma_x|$. Therefore, the damage rate will be the same for tensile and compressive layers of the beam. The distribution of $|\sigma_x|$ will be symmetrical with respect to the beam centerline. In this case one may apply the simplified equation (4.3.21). The time step solutions were performed until the critical damage is achieved in one of the integration points. The condition of termination is $\omega(x_f, z_f, t_*) > 0.9$, where the integration point $P(x_f, z_f)$ can be specified as a point of failure initiation and the time t_* as the time to failure initiation. Let us note, that more integration points over the thickness direction are required for the the creep damage analysis if compared with the steady-state creep analysis. Our convergence studies suggest that 17 integration points provide enough accuracy for the deformations, stresses and the time to failure initiation.

Figures 4.5 and 4.6 illustrate the maximum deflection and the stress $\sigma_x(l/2, -h/2)$ as functions of time. The results have been obtained with differ-

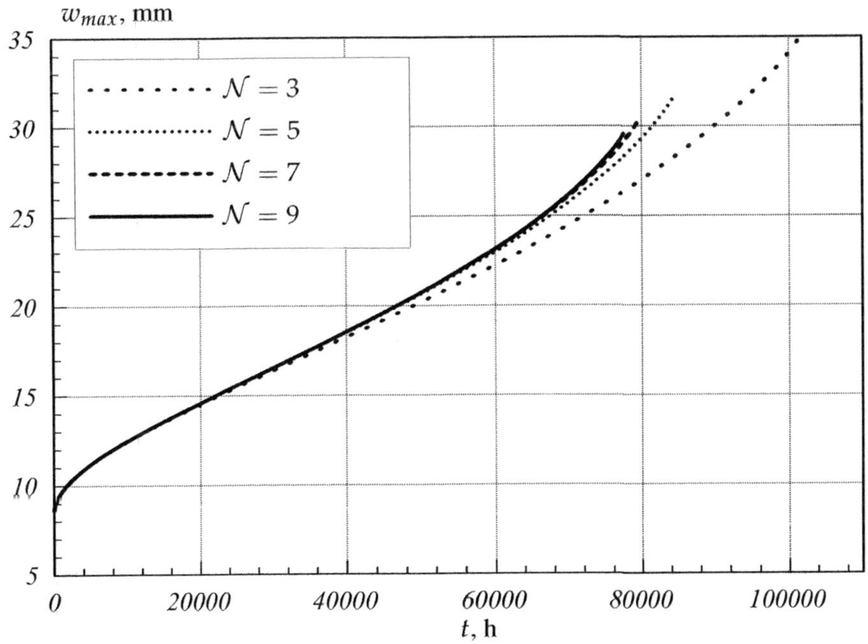

Fig. 4.5 Maximum deflection vs. time based on the Ritz method with different number of trial functions in (4.3.17)

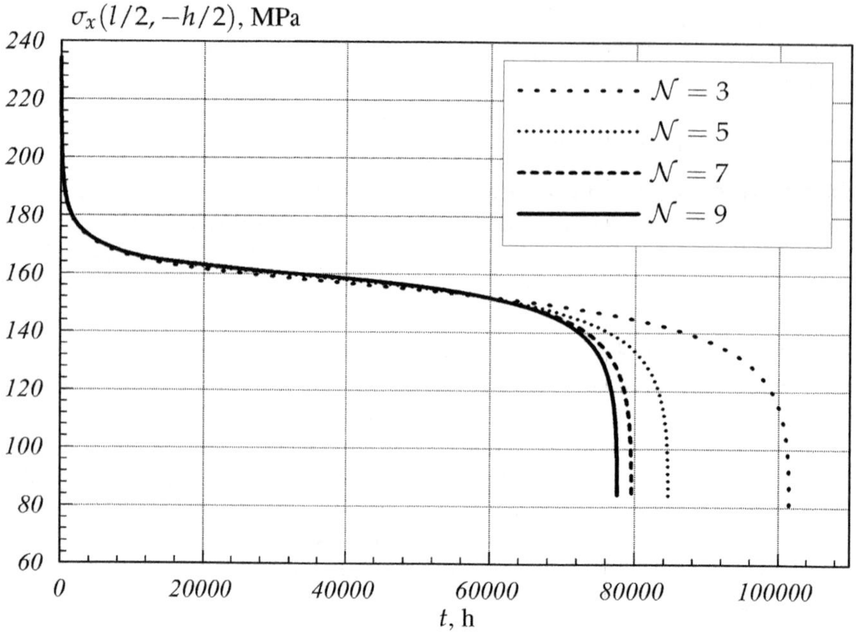

Fig. 4.6 Normal stress vs. time based on the Ritz method with different number of trial functions in (4.3.17)

ent number of trial functions in (4.3.17). All applied approximations to the deflection function provide the same result for the reference elastic state. However, the results for creep are quite different and depend essentially on the number of polynomials. From Figs. 4.5 and 4.6 we observe that the solutions converge against the accurate solution with increasing number of trial functions. By analogy with the uniaxial creep curve three creep stages of the beam can be recognized. The "primary" stage is characterized by the decrease in the deflection rate and significant stress relaxation. The "secondary" stage can be identified by slow changes in the rates of deflection growth and stress relaxation. During the "tertiary" stage the rates rapidly increase. The critical damage $\omega_* = 0.9$ appears in the bottom layer of the middle cross section. In all cases presented in Figs. 4.5 and 4.6 the applied approximations provide almost the same solutions for the primary and secondary creep stages. The results differ only in the final stage. Therefore, we may conclude that the consideration of damage needs an increased order of approximation in comparison with the steady-state creep analysis.

4.3.4.2 Finite Element Solutions.

Norton-Bailey Creep Law. Creep analysis under the assumption of the Norton-Bailey creep law is standard in the commercial finite element codes. Many benchmarks exist which allow to verify the capability of the finite element method for the steady-state creep analysis, e.g. [1]. They are based on the available closed form solutions for simple structures in the steady state creep range. The next example is

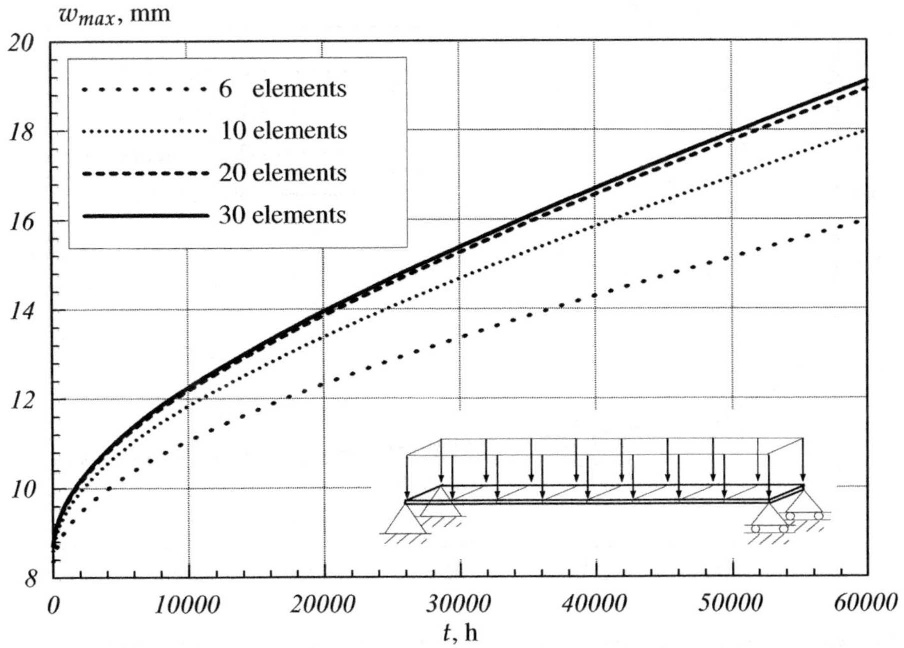

Fig. 4.7 Maximum deflection vs. time using the ABAQUS code with shell elements S4R5

selected to examine the accuracy of the finite element solution for the simply supported beam. The beam is modeled as a plate strip with shell type elements. The solution is performed by the ABAQUS code. The midplane of the strip is divided by 4-node shell elements S4R5 as shown in Fig. 4.7. The Simpson quadrature rule with 11 integration points through the thickness of the cross section is selected. The automatic time step feature with the minimum time step size of 0.01 h, the maximum time step size of 1000 h and the creep strain error tolerance of 10^{-6} is applied for the time integration. Figures 4.7 and 4.8 illustrate the results for time variations of the maximum deflection and the normal stress $\sigma_x(l/2, -h/2)$ obtained with different number of finite elements. We observe that all of the used meshes provide the same result for the reference elastic state. However, the time variations of the maximum deflection and the normal stress are sensitive to the mesh density. A mesh adjusted to the convergent solution of the linear elasticity problem (6 elements) is not fine enough for the creep analysis. The best result has been obtained with 30 elements. The corresponding plots for the time variations of the deflection and the stress agree well with those, previously obtained by the Ritz method (cp. Figs. 4.2 and 4.3). By analogy with the convergence of the series approximation (4.3.17) we may conclude, that the number of finite elements required for an accurate solution of the creep problem depends on the material material behavior, and in particular, on the value of the creep exponent n.

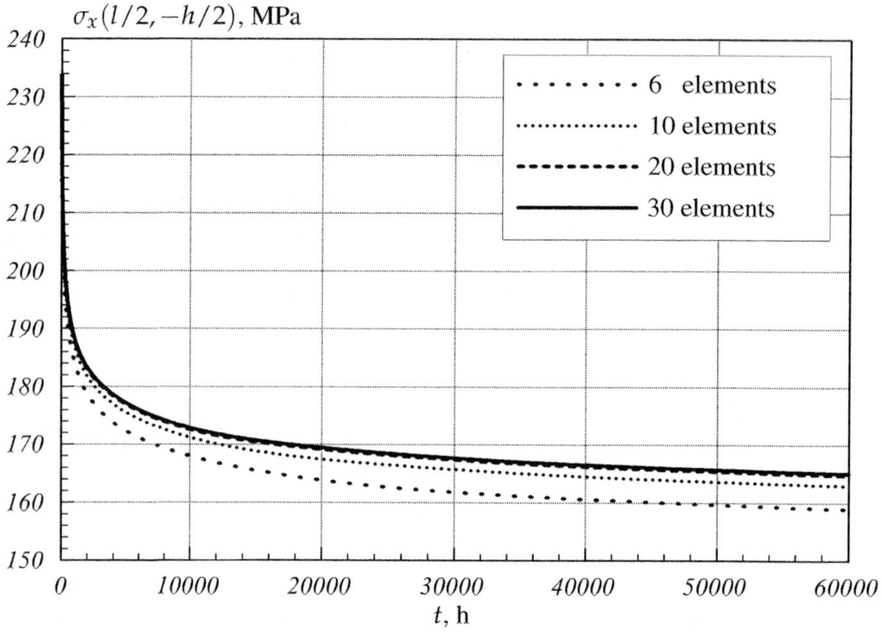

Fig. 4.8 Normal stress vs. time using the ABAQUS code with shell elements S4R5

Kachanov-Rabotnov Creep Law. Creep constitutive models with damage state variables are not available in the general purpose finite element codes. However, a specific constitutive model with internal state variables can be incorporated in a general purpose code by writing a user-defined material subroutine. To verify the developed subroutine, the results of the finite element modeling must be compared with reference solutions of benchmark problems. Let us note that the closed form solutions to creep-damage problems are only available for the case of homogeneous stress (strain) states. Examples include a bar subjected to the uniform tension (see Sect. 2.4.1.1), a thin-walled tube subjected to the axial force and torque (see Sect. 1.1.2), etc. Such benchmarks can be applied to assure that the developed subroutine is correctly coded and implemented. To analyze how the discretization parameters e.g., finite element type, mesh density, time step size time step control affect the solution, additional benchmark problems which involve non-homogeneous stress (strain) states are required.

The constitutive model (3.1.3) is incorporated into the ABAQUS and ANSYS finite element codes by means of the user-defined creep material subroutines. For details of the User Programmable Features provided by ANSYS an ABAQUS as well as the utilized time integration methods we refer to [2, 35]. To verify the subroutines we select the simply supported beam as a benchmark problem. This problem will be solved applying the shell and the plane stress finite elements. Furthermore, we perform the mesh convergence study illustrating basic features of the finite element solution of creep-damage problems. The results will be compared with those ob-

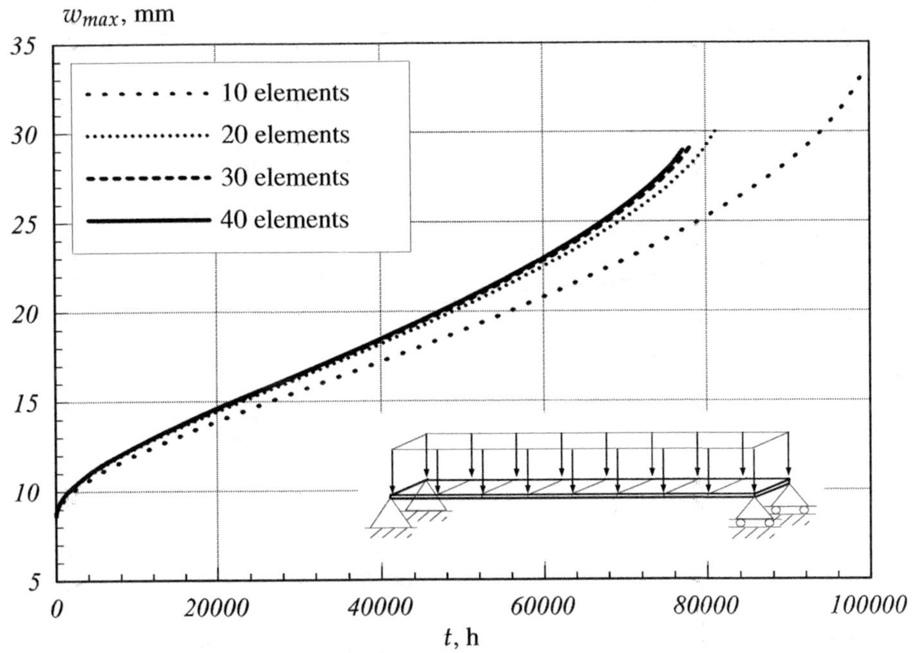

Fig. 4.9 Maximum deflection vs. time using the ABAQUS code with shell elements S4R5

tained by the Ritz method. In the case of the shell type elements the beam will be considered as a plate strip. In addition, the beam will be modeled as a "wall" by use of the plane stress type finite elements.

In the next example we apply the 4-node shell elements S4R5 with 17 integration points through the thickness. The settings for time step method are the same as in the previous example. Figures 4.9 and 4.10 illustrate time variations for the maximum defection and the normal stress in the bottom layer of the middle cross section. The results have been obtained with different number of elements. We observe that all of the used meshes provide the same solutions for the reference elastic state. Furthermore, the meshes with 20, 30 and 40 elements lead to almost the same results in the "primary" and "secondary" creep ranges of the beam. The results differ only in the final stage before the failure initiation. However, such a difference is not essential if we take into account the scatter of material data and the inaccuracy of the material behavior description. In this sense the mesh with 30 elements adjusted to the convergent solution in the steady-state creep range, cp. Figs. 4.7 and 4.8, is fine enough for the numerical life-time predictions.

When studying creep-damage in structures with complex geometry it is difficult to perform the mesh convergence studies due to large computational time. From our experience we may recommend to adjust the mesh to the convergent solution in the steady-state creep range. With such meshes the accuracy of long term predictions is not less than the accuracy of the material data involved in computations.

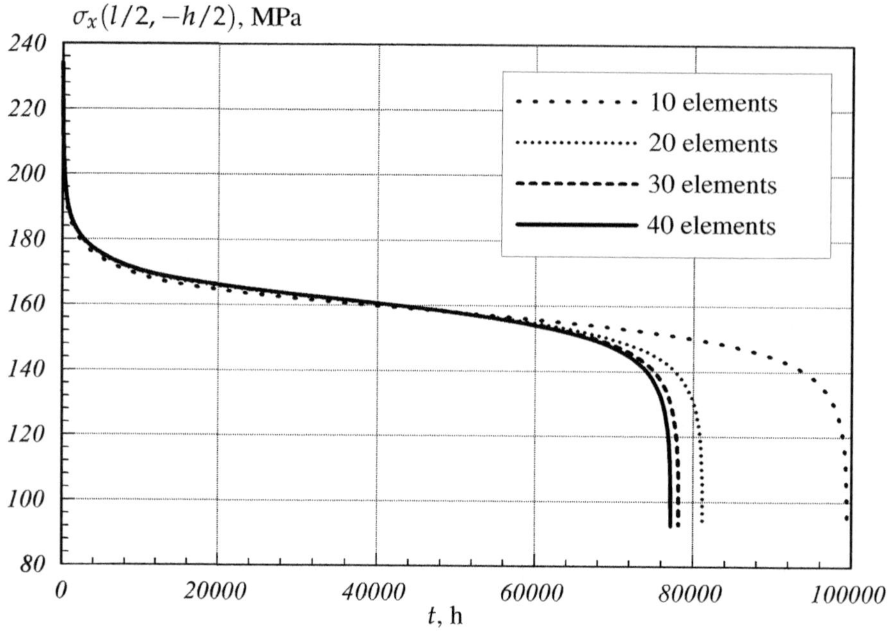

Fig. 4.10 Normal stress vs. time using the ABAQUS code with shell elements S4R5

In [17, 18] several benchmark problems for beams and rectangular plates solved by the Ritz method are presented. Finite element solutions for the same problems have been performed by ANSYS code applying shell, plane stress and solid type elements. The results illustrate the correctness of the developed subroutine over the wide range of element types as well as the convergence behavior of solutions.

To complete the analysis of the simply supported beam let us compare the results obtained by the Ritz method with those of ABAQUS and ANSYS finite element codes. Shell and plane stress type finite elements elements were applied. Table 4.1 provides a summary of element types, the number of finite elements as well as the number of integration points through the thickness. Figures 4.11 and 4.12 illustrate the time variations of maximum deflection and the normal stress obtained by the Ritz method and the finite element method with shell and plane stress type elements. We observe that the results are in a very good agreement. The exception

Table 4.1 Element types and discretization parameters

Element Type	Element Abbreviation	Number of integration points	Mesh
Shell (ABAQUS)	S4R5	17 (Simpson)	40
Shell (ANSYS)	SHELL 43	5 (Gauss)	40
Plane Stress (ABAQUS)	CPS4R	-	100×8
Plane Stress (ANSYS)	PLANE 42	-	100×8

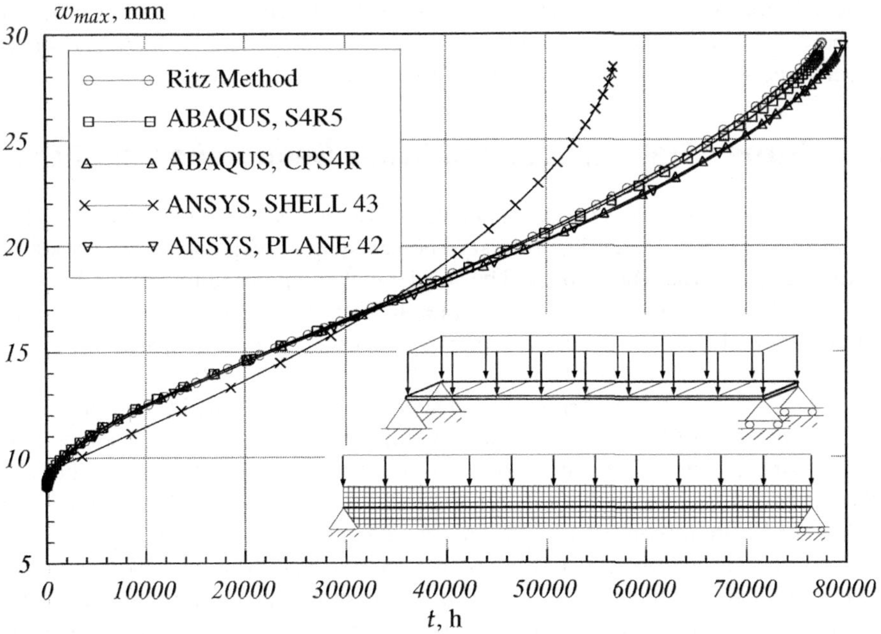

Fig. 4.11 Maximum deflection vs. time using the Ritz method and the finite element codes

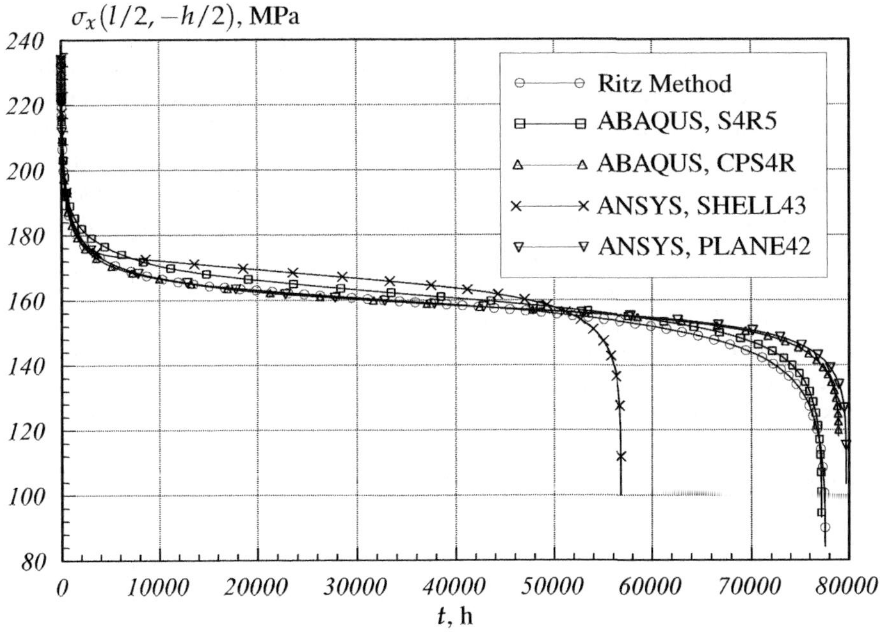

Fig. 4.12 Normal stress vs. time using the Ritz method and the finite element codes

is the solution based on the shell element SHELL 43, which is due to insufficient number of integration points. Finally let us note that the results of the plane stress model agree well with those based on the elementary beam theory.

4.3.5 Stress State Effects and Cross Section Assumptions

For many materials stress state dependent tertiary creep has been observed in multi-axial tests (see Sect. 1.1.2). The primary and secondary creep rates are dominantly controlled by the von Mises stress. The accelerated creep is additionally influenced by the kind of the stress state. For example, different tertiary creep rates and times to fracture can be obtained from creep tests under uni-axial tension with the stress σ and under torsion with the shear stress $\sqrt{3}\tau = \sigma$, e.g. [174]. Figure 4.13a shows creep curves for tensile, compressive and shear stresses simulated according to the constitutive model (3.1.1), (3.1.2) and the material constants (3.1.3) for type 316 stainless steel at 650°C . The selected stress values provide the same value of the von Mises stress. It is obvious that the tertiary creep rate is significantly dependent on the kind of loading. Figure 4.13b presents creep curves calculated by the combined action of the normal and shear stresses. We observe that even a small superposed shear stress can significantly influence the axial strain response and decrease the fracture time. Furthermore, combined tension-shear and compression-shear loadings with the same stress magnitudes lead to quite different creep responses. The change of the sign of the normal stress influences both the normal and the shear creep rates.

The considered loading case is typical for transversely loaded beams, plates and shells. For beams the local stress state is characterized by normal (bending) stress and small superposed transverse shear stress. Transverse shear stress and transverse shear deformation are neglected within the classical theory of beams. The considered example indicates that small shear stress can significantly influence the material response and cause significant shear strains. Furthermore, the dependence of creep on the sign of the normal stress can lead to non-classical thickness distributions of the displacement, strain and stress fields. For example, the concept of the neutral stress-free plane fails and the distribution of the transverse shear stresses is non-symmetrical with respect to the midplane.

Cross section assumptions are usually the basis for different refined models of beams, plates and shells developed within the theory of elasticity. Below we apply the first order shear deformation theory (Timoshenko-type theory) to creep analysis. For a beam with a rectangular cross section we compare the results based on different structural mechanics models (classical beam, shear deformable beam and plane stress model).

4.3.6 First Order Shear Deformation Theory

The classical beam theory is based on the assumption that beam cross sections remain plane and perpendicular to the beam axis during the deformation. The cross

a

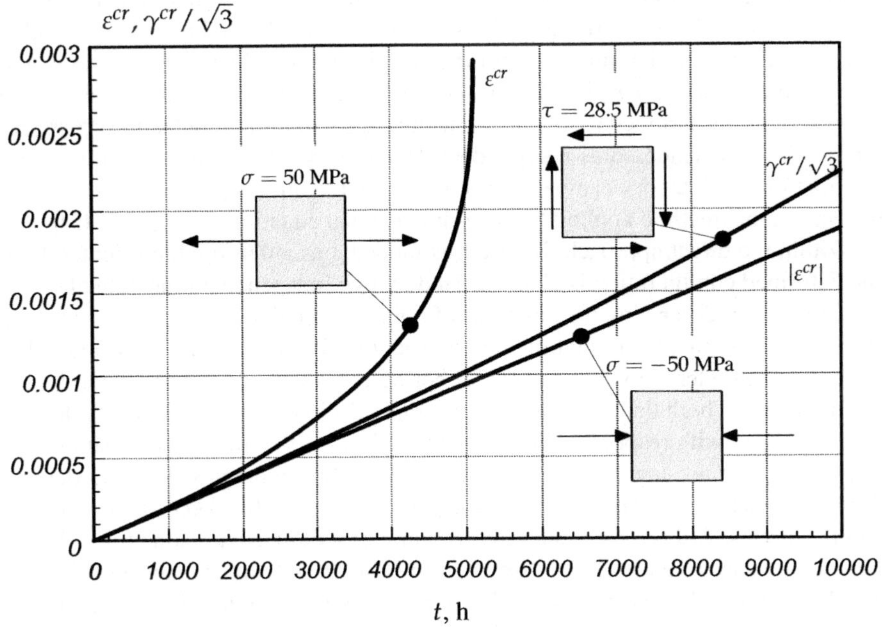

b

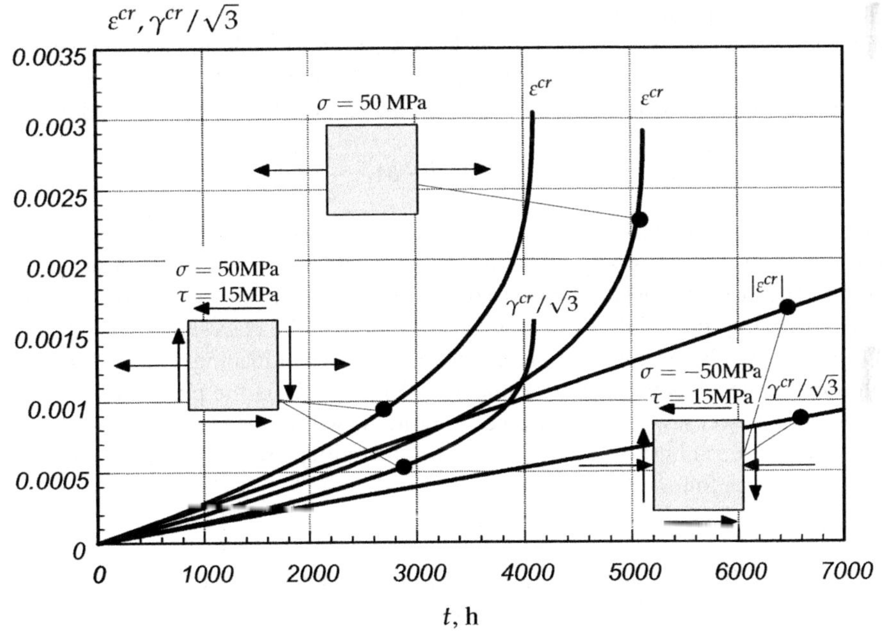

Fig. 4.13 Creep responses for various stress states computed using Eqs (3.1.1) – (3.1.3). **a** Responses by tension, torsion and compression, **b** responses by combined tension (compression) and torsion

section rotation is related to the first derivative of the deflection function. The transverse shear strain is therefore zero. The shear force is not defined by the constitutive equation. It must be determined from the equilibrium condition. In this sense the classical model is shear rigid. Within the first order shear deformation theory the cross sections are assumed to remain plane during the deformation but the cross section rotation is considered as independent degree of freedom. The first order shear deformation beam theory can be derived either by the direct approach, e.g. [36], or by a variational method applied to three-dimensional equations, e.g. [23, 143].

Within the direct approach the beam is modeled as a deformable oriented line. The deformed configuration is characterized by two independent kinematical quantities: the vector describing the positions of points on the line and the rotation tensor or vector describing the orientation of cross sections. Furthermore, it is assumed that the mechanical interaction between neighboring cross sections is only due to forces and moments. The balance equations are applied directly to the deformable line and formulated with respect to the beam quantities, i.e. the line mass density (mass density per unit arc length), the vectors of forces and moments, the line density of internal energy, etc. The constitutive equations connect the forces and moments with the strains. A direct approach to formulate constitutive equations for rods and shells in the case of elasticity is discussed in [26]. Despite the elegance of this approach several problems arise in application to creep mechanics. The creep constitutive equations must be formulated for inelastic parts of beam like strains (tensile, transverse shear and bending strains). By analogy to the creep theories discussed in Chapt. 2 the creep potential should be constructed as a function of the force and the moment vectors. For example, let $\boldsymbol{T}$ be the force vector and $\boldsymbol{M}$ the moment vector. Following the classical creep theory (see Sect. 2.2.1.1) an equivalent stress for the deformable line can be formulated as a quadratic form with respect to $\boldsymbol{T}$ and $\boldsymbol{M}$

$$t_{eq}^2 = \frac{1}{2}\boldsymbol{T}\cdot\boldsymbol{A}\cdot\boldsymbol{T} + \boldsymbol{T}\cdot\boldsymbol{B}\cdot\boldsymbol{M} + \frac{1}{2}\boldsymbol{M}\cdot\boldsymbol{C}\cdot\boldsymbol{M}$$

The structure of second rank material tensors $\boldsymbol{A}$, $\boldsymbol{B}$ and $\boldsymbol{C}$ must be established according to the material symmetries and geometrical symmetries of the beam cross section. The material constants have to be identified either from creep tests on beams or by comparing the solutions of beam equations with the corresponding solutions of three- or two-dimensional problems for special cases of loading. Only a few such solutions are available in creep mechanics. An example is the pure bending of a beam under power law secondary creep condition (see Sect. 4.3.2). In this case the steady-state creep constitutive equation for the bending strain rate can be obtained from (4.3.12) as follows

$$\dot{\chi} = \frac{a}{I_n^n}|M|^{n-1}M$$

Alternatively the beam equations may be derived in the sense of approximate solution of two- or three-dimensional equations. First, through-the-thickness approximations of displacements and/or stresses are specified. Then, the two- or three-dimensional boundary value problem is reduced to ordinary differential equations by means of a variational principle. In order to discuss this approach let us consider

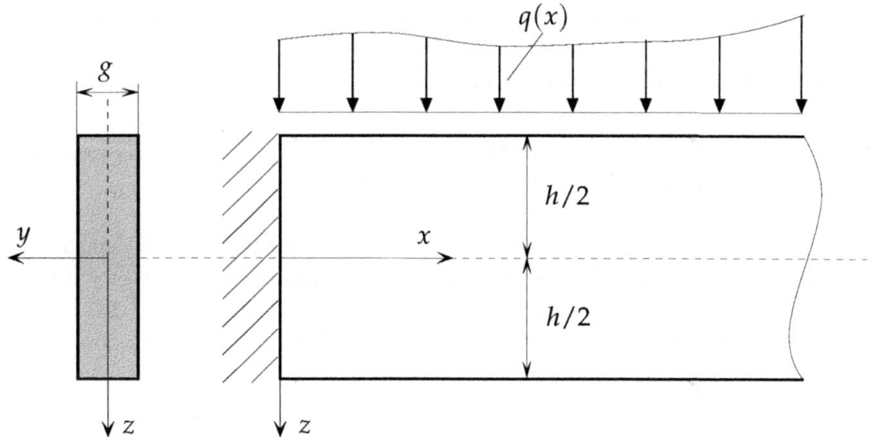

Fig. 4.14 Straight beam with a rectangular cross-section

a beam with a rectangular cross-section, Fig. 4.14. The governing two-dimensional equations for this case can be derived from (4.2.1) - (4.2.9) under the assumption of the plane stress state, i.e. $\boldsymbol{\sigma} \cdot \boldsymbol{e}_y = \mathbf{0}$. The principle of virtual displacements (4.2.43) yields

$$\frac{gh}{2}\int_0^l\int_{-1}^1(\sigma_x\delta\varepsilon_x + \tau_{xz}\delta\gamma_{xz} + \sigma_z\delta\varepsilon_z)d\zeta dx = \int_0^l q(x)\delta w(x,-h/2)dx \quad (4.3.23)$$

For the sake of brevity, here we assume that the virtual work of tractions on the edges $x = 0$ and $x = l$ is zero. In (4.3.23) l denotes the beam length, $\sigma_x, \sigma_z, \tau_{xz}$ and $\varepsilon_x, \varepsilon_z, \gamma_{xz}$ are the Cartesian components of the stress and strain tensors, respectively, w is the beam deflection and $\zeta = 2z/h$ is the dimensionless thickness coordinate. In the following derivations we use the abbreviations

$$(\ldots)_{,x} \equiv \frac{\partial}{\partial x}(\ldots), \quad (\ldots)_{,z} \equiv \frac{\partial}{\partial z}(\ldots), \quad (\ldots)' \equiv \frac{d}{dx}(\ldots),$$
$$(\ldots)^{\bullet} \equiv \frac{d}{d\zeta}(\ldots), \quad (\dot{\ldots}) \equiv \frac{d}{dt}(\ldots)$$

Specifying through-the-thickness approximations for the axial displacement u and the deflection w, various one-dimensional displacement based beam theories can be derived [275]. The classical Bernoulli-Euler beam theory is based on the following displacement approximations

$$u(x,z) = u_0(x) - w_0'(x)\frac{h}{2}\zeta, \quad w(x,z) = w_0(x), \quad (4.3.24)$$

where u_0, w_0 are the displacements of the beam centerline. The refined assumption

$$u(x,z) = u_0(x) + \varphi(x)\frac{h}{2}\zeta, \quad (4.3.25)$$

where φ denotes the independent cross-section rotation, provides the first order shear deformation (Timoshenko-type) beam theory. Another refined displacement based beam model can be obtained with

$$\begin{aligned} u(x,\zeta) &= u_0(x) + \varphi(x)\frac{h}{2}\zeta + u_1(x)\Phi(\zeta), \\ w(x,\zeta) &= w_0(x) + w_1(x)\Omega(\zeta), \end{aligned} \tag{4.3.26}$$

where u_0 and w_0 are the displacements of the beam centerline, $\Phi(\zeta)$ and $\Omega(\zeta)$ are distribution functions, which should be specified, and $u_1(x)$ and $w_1(x)$ are unknown functions of the x-coordinate. The assumptions $\Phi(\zeta) = (\zeta h/2)^3, \Omega(z) = 0$ result in a Levinson-Reddy type theory [192, 274]. From the boundary conditions $\gamma_{xz}(x, \pm 1) = 0$ it follows

$$\begin{aligned} u(x,\zeta) &= u_0(x) + \varphi(x)\frac{h}{2}\zeta - [w_0'(x) + \varphi(x)]\frac{h}{6}\zeta^3, \\ w(x,\zeta) &= w_0(x) \end{aligned}$$

and

$$\left.\frac{d\Phi}{d\zeta}\right|_{\zeta=-1} = \left.\frac{d\Phi}{d\zeta}\right|_{\zeta=1}$$

The next possibility is the use of stress based approximations. For example, the solution of the Bernoulli-Euler beam equations in the linear-elastic range leads to the following stress distributions

$$\begin{aligned} \sigma_x &= \frac{6M(x)}{gh^2}\zeta, \\ \tau_{xz} &= \frac{3Q(x)}{2gh}\left(1 - \zeta^2\right), \\ \sigma_z &= \frac{3q(x)}{4g}\left(-\frac{2}{3} + \zeta - \frac{1}{3}\zeta^3\right) \end{aligned} \tag{4.3.27}$$

Applying the stress approximations, equations for an elastic shear deformable plate have been derived by E. Reissner [277] by means of a mixed variational principle. The displacement approximations (4.3.26) neglecting the terms $u_1\Phi$ and $w_1\Omega$ or the stress approximations (4.3.27) lead to the first order shear deformation beam theory. The stress approximations (4.3.27) are not applicable to creep problems because the normal stress σ_x is a non-linear function of the thickness coordinate even in the case of steady-state creep (see Sect. 4.3.2). To derive the beam equations for the creep analysis the following approximations for the transverse shear and normal stresses were proposed in [23]

$$\begin{aligned} \tau_{xz} &= \frac{2Q(x)}{gh}\frac{\psi^{\bullet}(\zeta)}{\psi_0}, \\ \sigma_z &= \frac{q(x)}{g}\frac{\psi(\zeta) - \psi(1)}{\psi_0}, \quad \psi_0 = \psi(1) - \psi(-1), \end{aligned} \tag{4.3.28}$$

where $\psi(\zeta)$ is a function of distribution satisfying the boundary conditions $\psi^\bullet(\pm 1) = 0$. Furthermore, the linear through-the-thickness approximation of the axial displacement $u(x,\zeta) = u_0(x) + \zeta\varphi(x)h/2$ was assumed. Applying a mixed variational principle the following beam equations were derived in [23]

– equilibrium conditions

$$N' = 0, \quad M' - Q = 0, \quad Q' + q = 0, \tag{4.3.29}$$

– constitutive equation for the shear force

$$Q = GAk(\varphi + \tilde{w}' - \tilde{\gamma}^{cr}), \tag{4.3.30}$$

where G is the shear modulus and

$$\begin{aligned} \frac{1}{k} &= \frac{2}{\psi_0^2}\int_{-1}^{1} \psi^{\bullet 2}(\zeta)d\zeta, \quad \tilde{w}(x) = \frac{1}{\psi_0}\int_{-1}^{1} w(x,\zeta)\psi^\bullet(\zeta)d\zeta, \\ \tilde{\gamma}^{cr}(x) &= \frac{1}{\psi_0}\int_{-1}^{1} \gamma_{xz}^{cr}(x,\zeta)\psi^\bullet(\zeta)d\zeta \end{aligned} \tag{4.3.31}$$

By setting $GAk \to \infty$ and $\tilde{\gamma}^{cr} = 0$ in (4.3.30) the classical beam equations can be obtained. In this case $\varphi = -\tilde{w}'$ (the straight normal hypothesis). Let us note that Eqs. (4.3.29) and (4.3.30) can be derived applying the direct approach. For plates and shells this way is shown in [29]. However, in this case the meaning of the quantities GAk and $\tilde{\gamma}^{cr}$ is different. The shear stiffness GAk plays the role of the beam like material constant and must be determined either from tests or by comparison of results according to the beam theory with solutions of three-dimensional equations of elasto-statics or -dynamics. For a review of different estimates of the shear correction factor k we refer to [5, 143, 157]. Furthermore, the direct approach would require a constitutive equation for the rate of transverse shear strain $\dot{\tilde{\gamma}}^{cr}$. Within the applied variational procedure, Eqs. (4.3.29) and (4.3.30) represent an approximate solution of the plane stress problem under special trial functions (4.3.28). Therefore k and $\tilde{\gamma}^{cr}$ appear in (4.3.31) as numerical quantities and depend on the choice of the function $\psi(\zeta)$. For example, setting $\psi(\zeta) = \zeta$ we obtain

$$k = 1, \quad \tilde{\gamma}^{cr}(x) = \frac{1}{2}\int_{-1}^{1} \gamma_{xz}^{cr}(x,\zeta)d\zeta \tag{4.3.32}$$

With

$$\psi(\zeta) = \zeta - \zeta^3/3$$

we obtain the Reissner type approximation (4.3.27) and

$$k = 5/6, \quad \tilde{\gamma}^{cr}(x) = \frac{3}{4}\int_{-1}^{1} \gamma_{xz}^{cr}(x,\zeta)(1-\zeta^2)d\zeta \tag{4.3.33}$$

As the next choice let us consider the steady-state creep solution of a Bernoulli-Euler beam (see Sect. 4.3.2). According to (4.3.11) and (4.3.12) the bending stress σ_x can be expressed as

$$\sigma_x(x,\zeta) = \frac{M(x)}{gh^2}\frac{2(2n+1)}{n}|\zeta|^{(1/n)-1}\zeta$$

After inserting this equation into the equilibrium condition

$$\sigma_{x,x} + \frac{2}{h}\tau_{xz,\zeta} = 0 \tag{4.3.34}$$

and the integration with respect to ζ we obtain the transverse shear stress

$$\tau_{xz} = \frac{Q(x)}{gh}\frac{2n+1}{n+1}(1-\zeta^2|\zeta|^{(1/n)-1})$$

With the trial functions

$$\psi^\bullet(\zeta) = 1-\zeta^2|\zeta|^{(1/n)-1}, \quad \psi(\zeta) = \zeta - \frac{n}{2n+1}\zeta|\zeta|^{\frac{1}{n}+1} \tag{4.3.35}$$

from (4.3.31) follows

$$k = \frac{3n+2}{4n+2}, \quad \tilde{\gamma}^{cr}(x) = \frac{2n+1}{2n+2}\int_{-1}^{1}\gamma_{xz}^{cr}(x,\zeta)(1-\zeta^2|\zeta|^{(1/n)-1})d\zeta \tag{4.3.36}$$

By setting $n = 1$ in (4.3.36) we obtain (4.3.33). The value of n usually varies between 3 and 10 for metallic materials. For example, if $n = 3; 10, k = 11/14; 16/21$, respectively. It can be observed that with increasing creep exponent and consequently with increasing creep rate the value of k decreases (for $n \to \infty$ we obtain $k_\infty = 3/4$). The effect of damage is connected with the increase of the creep rate. Therefore a decrease of the value of k can be expected if damage evolution is taken into account. In addition, if the damage rate differs for tensile and compressive stresses, the thickness distribution of the transverse shear stress will be non-symmetrical. In this case the function $\psi^\bullet$ cannot be selected a priori.

4.3.7 Example: Refined vs. Classical Beam Theory

In [231] the first order shear deformation equations are solved by the use of the Ritz method and a time step integration procedure. At a current time step the transverse shear stress is recovered by an approximate solution of (4.3.34). The proposed numerical procedure allows to modify the trial functions as well as k and $\tilde{\gamma}^{cr}$ according to the time dependent redistribution of τ_{xz}.

Figure 4.15 presents the results for the uniformly loaded beam with clamped edges. The calculations have been performed with $l = 1000$ mm, $g = 50$ mm, $h = 100$ mm and $q_0 = 50$ N/mm. The constitutive model (3.1.1) and the material

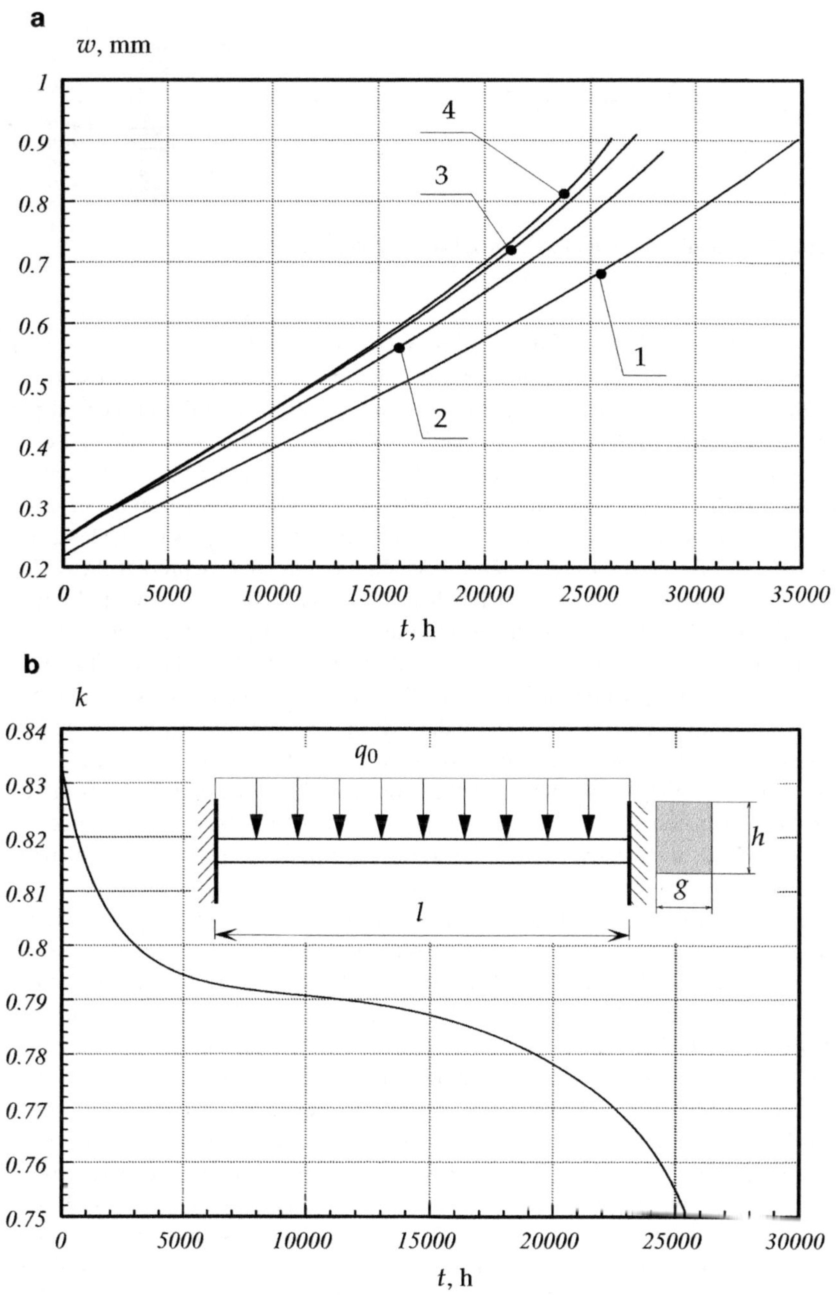

Fig. 4.15 Time-dependent solutions for a clamped beam. **a** Maximum deflection vs. time, **b** shear correction factor vs. time, 1 – Bernoulli–Euler beam theory, 2 – first order shear deformation theory with parabolic shear stress distribution, 3 – first order shear deformation theory with modified shear stress distribution, 4 – plane stress solution using the ANSYS code with PLANE 42 elements

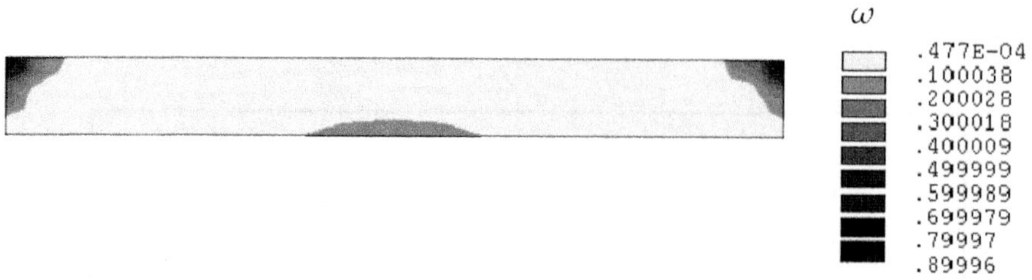

Fig. 4.16 Damage distribution in a beam at last time step

constants for the type 316 stainless steel at 650°C (3.1.3) were applied. Curve 1 in Fig. 4.15a is the time dependent maximum deflection calculated by the use of the Bernoulli-Euler beam theory. The corresponding equations and the numerical procedure are presented in Sect. 4.3.1. Curve 2 is obtained by the use of the first order shear deformation theory with the approximations (4.3.27) and (4.3.33). Curve 3 is the solution of the same equations but with the modified trial functions. Curve 4 is the ANSYS code solution of the plane stress problem with elements PLANE 42. It is obvious that the Bernoulli-Euler beam theory cannot adequately predict the deflection growth. Furthermore, the first order shear deformation equations with the fixed trial functions underestimate the deflection particularly in the tertiary creep range. The best agreement with the plane stress solution is obtained if the trial functions are modified according to redistribution of the transverse shear stress. In this case the shear correction factor is time-dependent, Fig. 4.15b. With decreasing value of k we can conclude that the influence of the shear correction terms increases.

The results for the beam show that the modified shear stress influences the deflection growth in the creep-damage process. On the other hand if we neglect the damage evolution, the steady-state creep solution provides the shear stress distribution close to the parabolic one, see Eq. (4.3.35).

Figure 4.16 shows the distribution of the damage parameter at the last step of calculation. The damage evolution is controlled by the maximum tensile stress, see Eqs. (3.1.1) and (3.1.3). Therefore the zones of the dominant damage are tensile layers of the clamped edges. Figure 4.17a presents the results for τ_{xz} obtained by ANSYS code with PLANE 42 elements. It can be observed that in the neighborhood of the beam edges, where the maximum damage occurs, the distribution of the transverse shear stress is non-symmetrical with respect to the beam midplane. Figures 4.17b and 4.17c show the solution for the transverse shear stress according to the derived beam equations. The transverse shear stress is calculated as a product of the shear force, the distribution function $\psi^{\bullet}$ and a constant factor. For the considered beam the shear force $Q(x) = q(l/2 - x)$ remains constant during the creep process. Therefore, the time redistribution of the transverse shear stress is only determined by the time-dependence of the function $\psi^{\bullet}$. Figure 4.17c illustrates $\psi^{\bullet}$ for different time steps.

The presented example shows that transverse shear deformation and transverse shear stress cannot be ignored in creep-damage analysis of beams. The first order

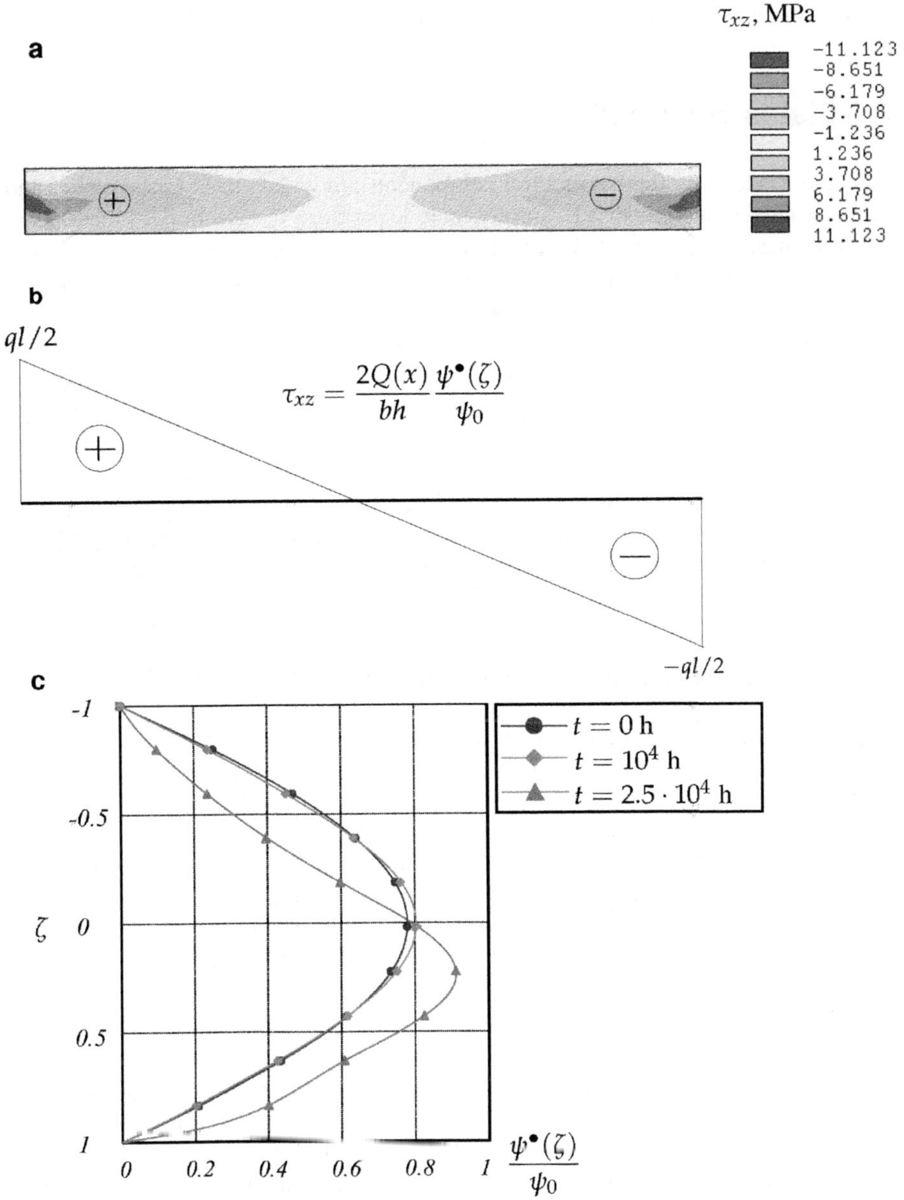

Fig. 4.17 Time-dependent solutions of a clamped beam. **a** Transverse shear stresses at last time step, solution with PLANE 42 elements, **b** shear force according to the beam equations, **c** function of the transverse shear stress distribution for different time steps

shear deformation theory provides satisfactory results if compared to the results of the plane stress model. Further investigations are required to establish the constitutive equations and material constants for beams with arbitrary cross sections.

4.4 Plates and Shells

Thin and moderately thick structural components are widely used in the power and petrochemical plant industry. Examples include pressure vessels, boiler tubes, steam transfer lines, etc. At elevated temperature the load carrying capacity and the lifetime of a structure are limited by the development of irreversible creep strains and damage processes. The failure modes under creep-damage conditions may include unacceptable changes of the components shape, creep buckling and loss of the material strength, e.g. [282]. The first two modes are associated with excessive creep deformations and stress redistributions. Local changes of shape of the component may lead to the loss of functionality of the whole structure. Creep buckling may occur if external loading leads to compressive stresses. A thinwalled structure designed against spontaneous "elastic" buckling may fail after a certain critical time as a consequence of stress redistribution. The degradation of material strength is the result of damage processes including creep cavitation, thermal ageing, oxidation, etc.

4.4.1 Approaches to the Analysis of Plates and Shells

To discuss available results of creep in plate and shell structures let us categorize the recent studies according to the problem statement, the type of the constitutive model and the type of the structural mechanics model involved in the analysis. Creep problems for thin and moderately thick plates and shells are summarized in Table 4.2. Constitutive equations of creep under multi-axial stress states were discussed in Chapt. 2. Table 4.3 provides an overview of several constitutive models recently applied to the analysis of plates and shells. The corresponding structural mechanics models are given in Table 4.4. The overviews presented in Tables 4.2 - 4.4 lead to a conclusion that the type and the order of complexity of the applied structural mechanics models are connected with the problem statement and with the type of the material behavior description.

The early works were primarily concerned with the analysis of steady-state creep in plates and shells. The creep behavior was assumed to have only primary and secondary creep stages and the Norton-Bailey-Odqvist creep constitutive equation sometimes extended by strain or time hardening functions was applied. The structural mechanics models were those of the Kirchhoff plate and the Kirchhoff-Love shell. In [196] the biharmonic equation describing a deflection surface of the Kirchhoff plate taking into account the given distribution of creep strains has been derived. It is shown that the deflection of the plate can be computed by applying additional fictitious lateral loads on the plate face and additional fictitious moments on the plate edges. In many cases this equation can be solved by special numerical methods, e.g. the finite difference method [262] or direct variational methods [22, 80]. The results obtained by special methods are useful in verifying general purpose finite element codes and creep material subroutines. Creep equations for axisymmetrically loaded shells of revolution were derived in [256] by the use of the Kirchhoff-Love hypotheses. The influence of creep is expressed in terms of fictitious membrane forces and bending moments. In [85, 226, 227] problems of stress redistribution from the reference state of elastic deformation up to the steady creep state were solved for axisymmetrically loaded shells of revolution by means of the finite difference method.

The classical models of Kirchhoff plate or Kirchhoff-Love shell are based on geometrically linear equations. Because the development of creep strains may lead to significant changes of the components shape, geometrically nonlinear terms should be taken into account in the kinematical equations and as well as in the equilibrium conditions. For elastic plates the governing equations (finite deflection model) were originally proposed by von Kármán [327]. Geometrically non-linear equations for creep in membranes and plates have been derived by Odqvist [240]. Problems of long term stability and long term strength have required the use of refined geometrically-nonlinear structural mechanics models. Creep buckling analysis of cylindrical shells under internal pressure and compressive force has been performed in [65, 215, 216] (see also references cited therein). The governing equations correspond to the Kirchhoff-Love type shell with geometrical non-linearities in the von Kármán's sense. In [20, 21, 22, 33] we applied a geometrically-nonlinear theory to the creep-damage analysis of rectangular plates and cylindrical shells. We demonstrated that the effect of geometrical non-linearity may be associated with "structural hardening", i.e. an increase in the structural resistance to time dependent deformations. Furthermore, we have shown that even in the case of moderate bending, the classical geometrically-linear theory leads to a significant underestimation of the life-time and overestimation of the deformation.

Table 4.2 Problem statements for creep in plates and shells

References	Type of Problem
[85, 226, 227]	Shells of revolution, steady-state creep
[230, 336]	Plates, primary and secondary creep
[20, 63, 65]	Shells of revolution, plates, finite deflections, creep buckling
[14, 83, 84]	Shells of revolution, dynamic creep, long term strength
[21, 22]	Shells of revolution, shallow shells, plates, creep-damage
[313, 314]	Moderately thick and layered shells, steady state creep
[8, 33]	Shells of revolution, plates, creep-damage
[73, 121]	Plates, thermo-mechanical coupling, creep-damage
[19, 34]	Moderately thick plates, curved shells, creep-damage
[16, 67, 123]	Moderately thick plates, damage induced anisotropy
[117, 169, 182]	Moderately thick shells, creep-damage

Table 4.3 Constitutive models applied to analysis of plates and shells

	Type of Constitutive Model for Creep Stages				
	Primary	Secondary	Tertiary		
References	Time or Strain Hardening	Power or sinh law	Kachanov-Rabotnov-Hayhurst	Mecha-nism-based	Damage Tensors
[85, 226, 227]		x			
[230, 336]		x			
[20, 63, 65]	x	x			
[14, 83, 84]		x	x		
[21, 22]		x	x		
[313, 314]	x	x			
[8, 33]	x	x	x	x	
[73, 121]	x	x	x		
[19, 34]		x	x		
[16, 67, 123]		x			x
[117, 169, 182]	x	x		x	

Table 4.4 Structural mechanics models of plates and shells (FOSDT – first order shear deformation theory)

	Type of Structural Mechanics Model				Verification
References	Kirchhoff-Love Type	FOSDT	Geomet-rical Non-linearities	3-D Models	Experi-mental Analysis
[85, 226, 227]	x				
[230, 336]	x		x		x
[20, 63, 65]	x		x		
[14, 83, 84]	x				x
[21, 22]	x		x		
[313, 314]		x			
[8, 33]	x		x		
[73, 121]		x			
[19, 34]		x		x	
[16, 67, 123]		x		x	
[117, 169, 182]				x	x

A first order shear deformation shell theory has been firstly applied in [313] to analyze primary and secondary creep of simply supported cylindrical shells under internal pressure. The initial-boundary value problem is solved by the use of the finite difference method. Time dependent distributions of displacements and stress resultants are compared with those according to the Kirchhoff-Love type theory. It is demonstrated that the results agree well only for thin shells. In the case of moderately thick shells the difference between the results is essential and increases with time. Reissner type plate equations were applied in [121, 123] for the creep-damage analysis of a simply supported circular plate considering thermo-mechanical couplings. The derived plate equations as well as the equations of the three-dimensional theory were solved by means of the finite difference method. It is demonstrated that in the tertiary creep range the thickness distribution of the transverse shear stress differs from the parabolic one. Similar effects have been illustrated for beams in Sect. 4.3.7.

Let us note that unlike the Kirchhoff-Love type theories, the first order shear deformation theories have been found more convenient for the finite element implementations due to C^0 continuity [348]. They are standard in commercial finite element codes, e.g. [35]. Examples of creep-damage analysis of plates and shells by the use of ANSYS code are presented in [17, 19].

Numerous refined finite element techniques were designed to solve non-linear problems of shells. For reviews we refer to [334, 338]. One feature of the refined theories of plates and shells is that except special types of boundary conditions (e.g. simple support) they describe additional edge zone effects. The use of the finite element or the finite difference method to solve refined equations of plates and shells requires advanced numerical techniques to represent the rapidly varying behavior in the edge zones. Several closed form and approximate analytical solutions of the first order shear deformation plate equations in the linear elastic range illustrate edge zone effects for different types of boundary conditions, e.g. [234, 346]. Similar solutions in the case of creep-damage in plates and shells are not available. Further investigations should be made to formulate corresponding benchmark problems and to assess the validity of different available shell and solid type finite elements in problems of creep mechanics.

4.4.2 Examples

4.4.2.1 Edge Effects in a Moderately Thick Plate. An important step in the creep analysis of plates and shells is to select a suitable structural mechanics model. One way is the "three-dimensional approach" which is based on three-dimensional equations of continuum mechanics. This approach seems more preferable for creep-damage analysis since the existing constitutive models of creep-damage are developed with respect to the Cauchy stress and strain (rate) tensors and the proposed measures of damage (scalars or tensors of different rank) are defined in the three-dimensional space. Another way is the use of the classical two dimensional structural mechanics equations of beams, plates and shells and the balance equations formulated in terms of force and moment tensors. This approach often finds application because of the simplicity of the model creation, smaller effort in solving non-linear initial-boundary value problems of creep, and easily interpretable results.

The governing mechanical equations describing creep in three-dimensional solids are summarized in Sect. 4.2. Various approaches to derive a shell theory have been recently applied within the assumption of elastic or viscoelastic material behavior. As far as we know, a "closed form" shell theory in the case of creep does not exist at present. The principal problem lies in establishing the constitutive equations of creep with respect to the shell type strain measures, i.e. the membrane strains, changes of curvature and transverse shear strains. Although, a general structure of such equations can be found based on the direct approach, e.g. [26, 29], the open question is the introduction of appropriate damage measures as well as the identification of damage mechanisms under the shell type stress states, i.e. under bending and twisting moments, membrane and transverse shear forces, or their interactions.

Here we apply the standard approach which can be summarized as follows:

1. Assume that equations (4.2.1) - (4.2.9) are applicable to the analysis of creep-damage in a moderately thick plate.
2. Formulate a variational equation of statics (e.g., based on the principle of virtual displacements) with the known tensor $\boldsymbol{\varepsilon}^{cr}$ for a fixed time (time step).
3. Specify cross-section approximations for the functions to be varied (e.g., the displacement vector $\boldsymbol{u}$).
4. Formulate and solve the two-dimensional version of Eqs. (4.2.1) – (4.2.9).
5. Recover the three-dimensional stress field $\boldsymbol{\sigma}$ from the two-dimensional solution.
6. Insert $\boldsymbol{\sigma}$ into constitutive equations (4.2.9) in order to calculate the time increments of $\boldsymbol{\varepsilon}^{cr}$ and ω.
7. Update the tensor $\boldsymbol{\varepsilon}^{cr}$ for the next time step and repeat the cycle from step 2.

Depending on the type of the applied variational equation (e.g., displacement type or mixed type) and the type of incorporated cross-section assumptions, different two-dimensional versions of Eqs. (4.2.1) - (4.2.9) with a different order of complexity can be obtained (i.e. models with forces and moments or models with higher order stress resultants). In the case of linear-elastic plates a huge number of such kind plate theories has been proposed, e.g., [200, 213, 274]. Note that the steps 2 and 3 can be performed numerically applying, e.g., the Galerkin method to Eqs. (4.2.1) – (4.2.9). Various types of finite elements which were developed for the inelastic analysis of shells are reviewed in [334]. Let us note that if studying the creep behavior coupled with damage, the type of assumed cross-section approximations may have a significant influence on the result. For example, if we use a mixed type variational equation and approximate both the displacements and stresses, a parabolic through-the-thickness approximation for the transverse shear stress or a linear approximation for the in-plane stresses is in general not suitable for the creep-damage estimations [23]. In what follows we compare finite element solutions based on the three-dimensional approach and a two-dimensional plate model and discuss the possibilities and limitations of each approach in connection with the creep-damage analysis.

Consider a square plate with $l_x = l_y = 1000$ mm, $h = 100$ mm, loaded by a pressure $q = 2$ MPa uniformly distributed on the top surface as shown in Fig. 4.18. The edges $x = 0$ and $x = l_x$ are simply supported (hard hinged support) and the edges $y = 0$ and $y = l_y$ are clamped. According to the first order

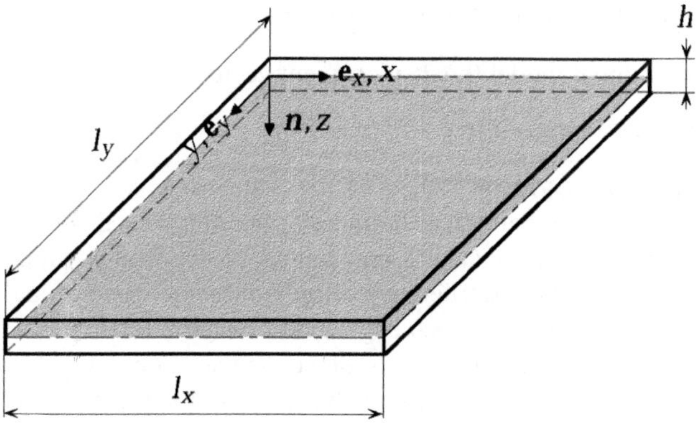

Fig. 4.18 Rectangular plate

shear deformation plate model we can specify the vectors of "plate displacements" $\boldsymbol{u}_p(x,y) = \boldsymbol{u}_0(x,y) + w(x,y)\boldsymbol{n}$, $\boldsymbol{u}_0 \cdot \boldsymbol{n} = 0$ and cross-section rotations $\boldsymbol{\varphi}(x,y)$ on the lines $x = const$ or $y = const$, Fig. 4.19. Applying such a model and assuming infinitesimal cross-section rotations the displacement vector $\boldsymbol{u}(x,y,z)$ is usually formulated as follows

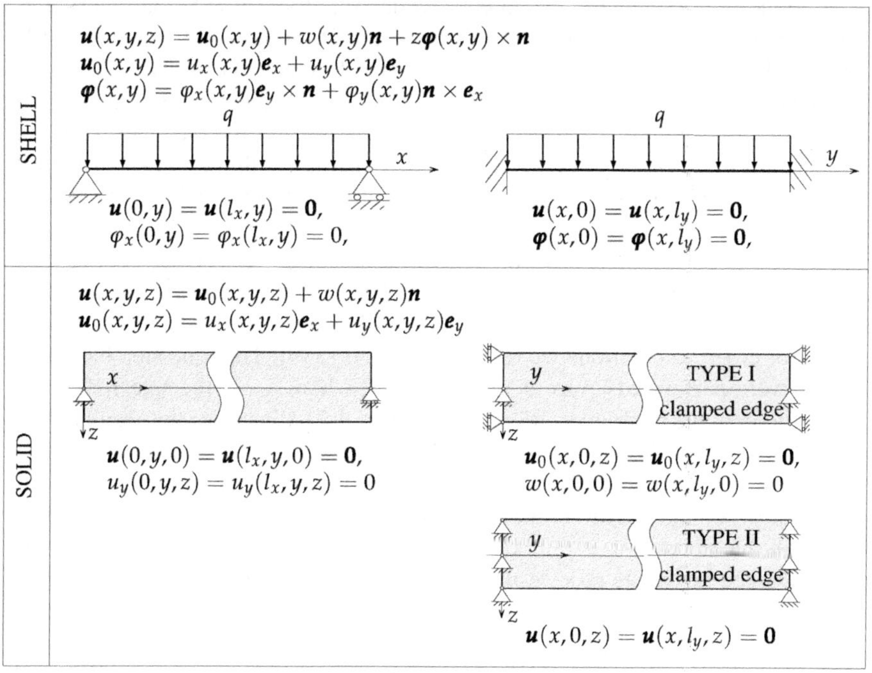

Fig. 4.19 Types of kinematical boundary conditions

$$\boldsymbol{u}(x,y,z) \approx \boldsymbol{u}_p(x,y) + z\boldsymbol{\varphi}(x,y) \times \boldsymbol{n}$$

In the case of the three-dimensional model the displacement vector

$$\boldsymbol{u}(x,y,z) = \boldsymbol{u}_0(x,y,z) + w(x,y,z)\boldsymbol{n}$$

can be prescribed on the planes x_c, y, z or x, y_c, z of the plate edges $x = x_c$ or $y = y_c$. Figure 4.19 illustrates the kinematical boundary conditions used for the shell and the solid models. Let us note that different boundary conditions which correspond to the clamped edge can be specified if we apply the three-dimensional model. Here we discuss two types of the clamped edge conditions. For the first type (TYPE I), see Fig. 4.19, we assume the vector of in-plane displacements $\boldsymbol{u}_0$ to be zero. The deflection w is zero only in the points of the plate mid-surface. In the second type (TYPE II) the whole displacement vector $\boldsymbol{u}$ is assumed to be zero in all points which belong to the plate edges. The TYPE II boundary conditions are the simplest possibility with respect to the effort in the model creation on the computer and the preprocessing since all nodal displacements can be simultaneously set to zero on the whole surfaces of the edges $x = const$ and $y = const$.

The analysis has been performed using the ANSYS finite element code after incorporating the material model (3.1.1) with the help of the user-defined creep-damage material subroutine. In Sects. 4.3.4 and 4.3.7 we discussed various examples for beams, which verify the developed subroutine. Similar examples for plates may be found in [17]. The finite elements available in the ANSYS code for plasticity and creep analysis were applied as follows: the 20-nodes solid element SOLID 95 and the 4-nodes shell element SHELL 43. 30×15 elements were used for a half of the plate in the case of the shell model and $30 \times 15 \times 3$ elements in the case of the solid model. The meshes have been justified based on the elasticity solutions and the steady-state creep solutions neglecting damage. With these meshes the reference stress distributions as well as the distributions of the von Mises stresses in the steady creep state were approximately the same for both the solid and the shell elements and did not change anymore by further re-meshing. The automatical time stepping feature with a minimum time step 0.1 h has been applied. For details of the used elements, the time integration and equilibrium iteration methods used in ANSYS for creep calculations we refer to [35] and [348]. The time step based calculations were performed up to $\omega = \omega_* = 0.9$, where ω_* is the selected critical value of the damage parameter. Figures 4.20 and 4.21 illustrates the results of the computations, where the maximum deflection and the maximum value of the damage parameter are plotted as functions of time. From Fig. 4.20 we observe that the starting values of maximum deflection as well as the starting rates of the deflection growth due to creep are approximately the same for the shell and the two solid models. Consequently the type of the elements (shell or solid) and the type of the applied boundary conditions in the case of the solid elements has a small influence on the description of the steady-state creep process. However, the three used models lead to quite different life time predictions. The difference can be seen in Fig. 4.21. The shell model overestimates the time to failure, while the result based on the solid model depends significantly on the type of the clamped edge boundary conditions.

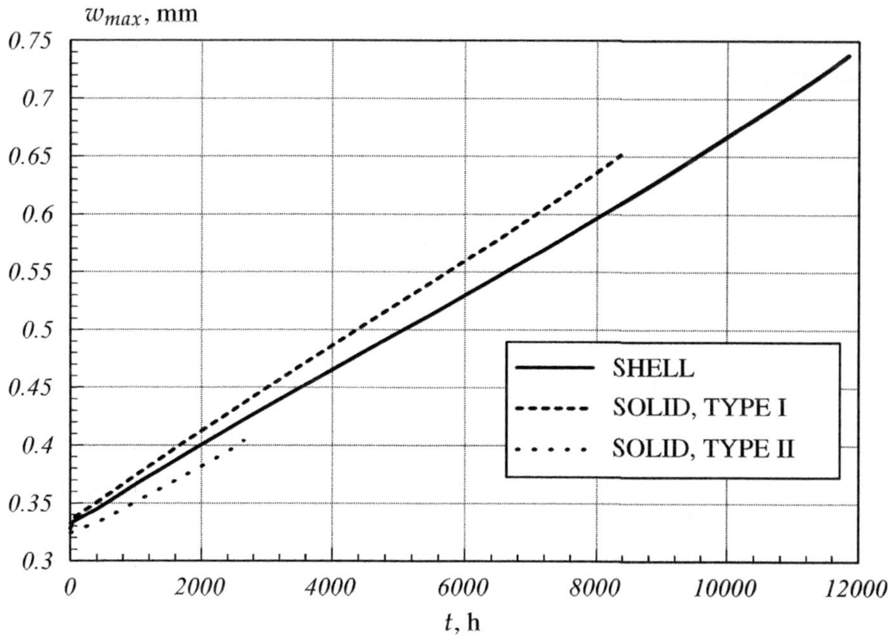

Fig. 4.20 Time variations of the maximum deflection

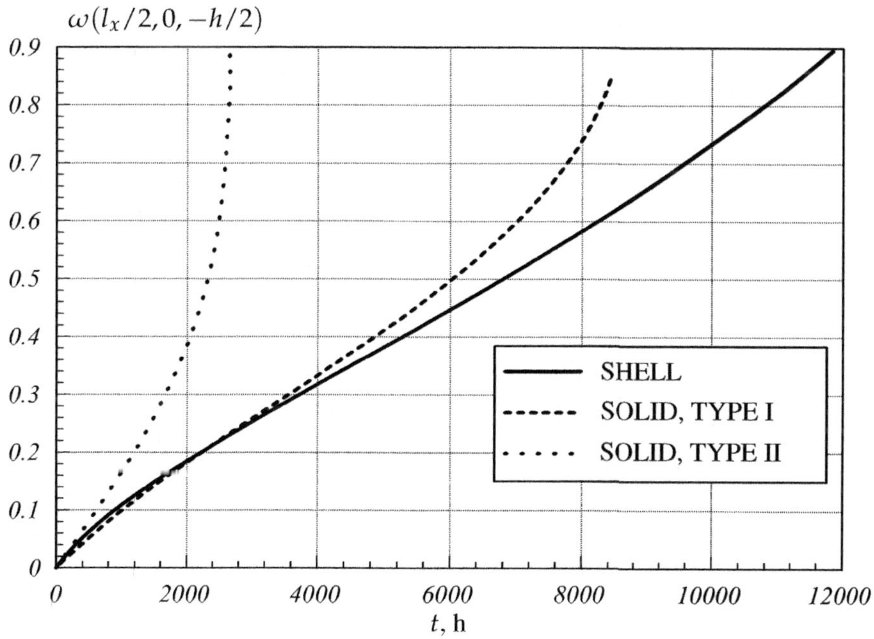

Fig. 4.21 Time variations of the damage parameter

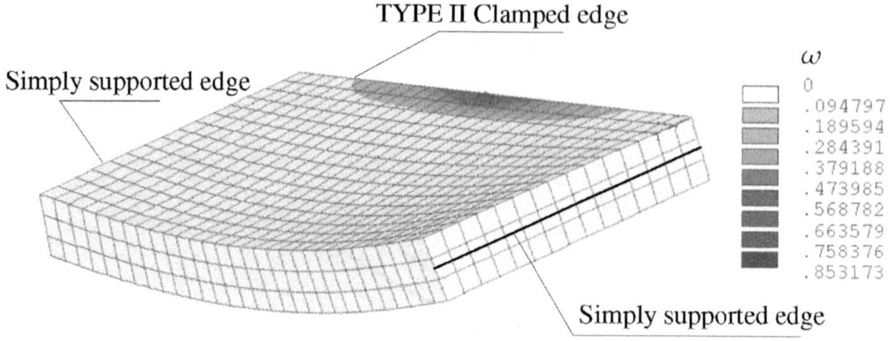

Fig. 4.22 Deformed shape of a half of the plate and distribution of the damage parameter in the zone of a clamped edge (SOLID elements, TYPE II boundary conditions, last time step)

In the case of the TYPE II clamped edge much more accelerated damage growth is obtained. The corresponding time to failure is approximately four times shorter compared to those based on the TYPE I clamped edge. All considered models predict the zone of maximum damage to be in the midpoint of the clamped edge on the plate top surface, as shown in Fig. 4.22.

The creep response of a structure is connected with the time-dependent stress redistributions. If the applied load and the boundary conditions are assumed to be constant and the effect of tertiary creep is ignored, than an asymptotic stress state exists, which is known as the state of stationary or steady creep (see Sect. 1.2). If tertiary creep is considered, then stresses change with time up to the critical damage state. It is obvious that the damage growth and the tertiary creep behavior of the considered plate are controlled by the local stress state in the vicinity of the clamped edges. Figure 4.23 illustrates the stress states in the midpoint of the clamped edge with the coordinates $x = l_x/2, y = 0$. Four components of the stress tensor (the two remaining components are zero due to symmetry conditions) are plotted as functions of the normalized thickness coordinate. The starting elastic distributions (solid lines) as well as the creep solutions at the last time step (dotted lines) are presented. The maximum starting stresses obtained by the use of the three considered models are the normal in-plane stresses σ_{yy} and σ_{xx} (the stresses which results in the maximum bending and twisting moments in the clamped edges), Fig. 4.23. These in-plane stresses remain dominant during the whole creep process for the used shell and solid elements. Therefore, all the applied models predict the damage evolution in the zone of the clamped edge on the plate top side. However, the influence of the "second order" stresses (stresses which are usually neglected in the plate theories) is different and depends on the type of the boundary conditions. For the TYPE I clamped edge the effect of the transverse normal stress σ_{zz} decreases with time and has negligible influence on the stress state. In contrast, for the TYPE II clamped edge the initial transverse normal stress σ_{zz} remains approximately constant, while σ_{yy}

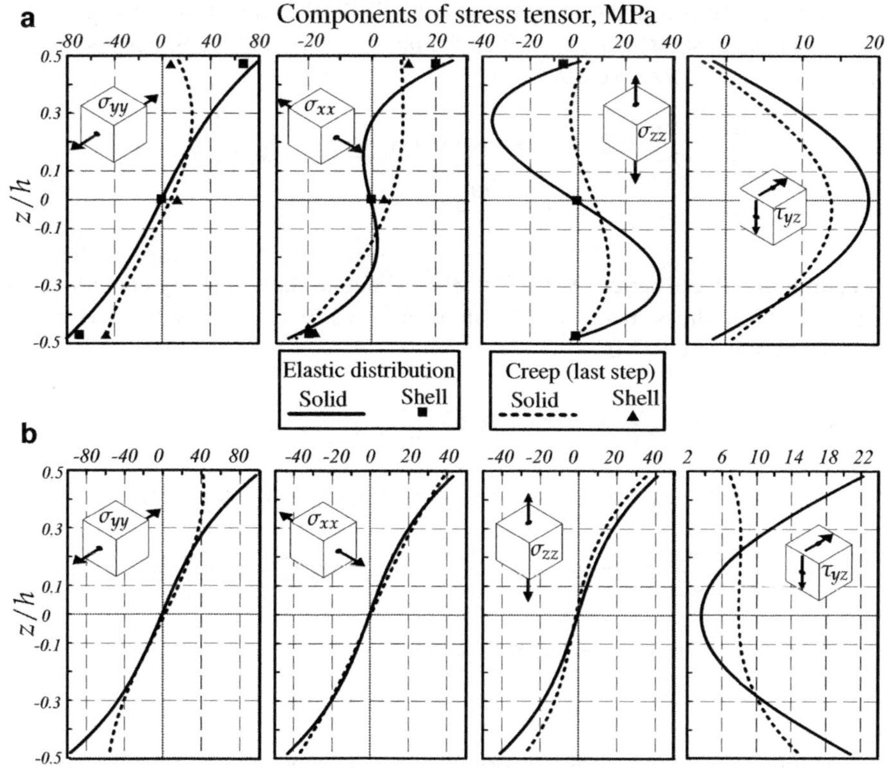

Fig. 4.23 Local stress state in a midpoint of the clamped edge vs. thickness coordinate (last time step). **a** TYPE I clamped edge, **b** TYPE II clamped edge

relaxes with time as the consequence of creep. The transverse normal stress becomes comparable with the bending stress and cannot be considered as the "second order" effect anymore.

In order to explain the difference in life-time predictions let us compare the stress states in the critical zone for the considered models. With respect to the transverse normal and transverse shear stresses, the TYPE I and TYPE II boundary conditions lead to different results. For the TYPE I clamped edge the transverse normal stress σ_{zz} has the value of the applied transverse load q on the top plate face and remains constant during the creep process. The transverse shear stress τ_{xz} is zero due to the applied boundary conditions. The stress state on the top side of the plate is primarily determined by two in-plane stresses σ_{xx} and σ_{yy}, Fig. 4.23. Such a stress state with dominant in-plane stresses and small transverse normal and shear stresses can be obtained applying the first order shear deformation plate theory. In contrast, if applying the TYPE II boundary conditions the results show the considerable value of the transverse normal stress σ_{zz} which remains approximately constant during the creep process.

Now let us estimate the stress state for the TYPE II clamped edge $y = y_c$. In this case we have to set $\boldsymbol{u}(x, y_c, z) = \boldsymbol{0}$ on the plane x, y_c, z, Fig. 4.19. For $0 < x < l_x$ and $-h/2 < z < h/2$ we can write

$$\begin{aligned} \frac{\partial \boldsymbol{u}}{\partial x} &= \frac{\partial \boldsymbol{u}}{\partial z} = \boldsymbol{0} \qquad \Rightarrow \qquad \boldsymbol{\nabla u}(x, y_c, z) = \boldsymbol{e}_y \otimes \frac{\partial \boldsymbol{u}}{\partial y}, \\ \mathrm{tr}\boldsymbol{\varepsilon}(x, y_c, z) &= \boldsymbol{\nabla} \cdot \boldsymbol{u} = \frac{\partial u_y}{\partial y} \end{aligned} \tag{4.4.1}$$

In addition, we can set $\boldsymbol{e}_x \cdot \boldsymbol{u}(l_x/2, y, z) = 0$ due to the symmetry condition. The starting elastic stress state at $t = 0$ can be obtained from the constitutive equations (4.2.6) by setting $\boldsymbol{\varepsilon}^{cr} = \boldsymbol{0}$

$$\begin{aligned} \sigma_m|_{t=0} &= \frac{1}{3}\frac{1+\nu}{1-2\nu}\sigma_0, \qquad \sigma_0 = 2G\frac{\partial u_y}{\partial y}|_{t=0}, \qquad \tau_0 = G\frac{\partial w}{\partial y}|_{t=0}, \\ \boldsymbol{s}|_{t=0} &= \frac{1}{3}\sigma_0\left[2\boldsymbol{e}_y \otimes \boldsymbol{e}_y - (\boldsymbol{I} - \boldsymbol{e}_y \otimes \boldsymbol{e}_y)\right] + \tau_0(\boldsymbol{e}_y \otimes \boldsymbol{n} + \boldsymbol{n} \otimes \boldsymbol{e}_y), \\ \boldsymbol{\sigma}|_{t=0} &= \frac{1-\nu}{1-2\nu}\sigma_0\left[\boldsymbol{e}_y \otimes \boldsymbol{e}_y + \frac{\nu}{1-\nu}(\boldsymbol{I} - \boldsymbol{e}_y \otimes \boldsymbol{e}_y)\right] + \tau_0(\boldsymbol{e}_y \otimes \boldsymbol{n} + \boldsymbol{n} \otimes \boldsymbol{e}_y) \end{aligned} \tag{4.4.2}$$

From the last equation in (4.4.2) we see that

$$\sigma_{zz} = \sigma_{xx} = \sigma_{yy}\nu/(1-\nu)$$

This well known result of the theory of linear isotropic elasticity agrees with the obtained finite element solution for $\nu = 0.314$, Fig. 4.23b (solid lines).

Let us estimate the stress redistribution in the TYPE II clamped edge as a consequence of creep. For this purpose we neglect the damage evolution by setting $\omega = 0$ in (3.1.1). Because the boundary conditions and the applied pressure are independent of time, we can estimate the type of the stress state under stationary state creep by setting $\dot{\boldsymbol{\varepsilon}} \approx \dot{\boldsymbol{\varepsilon}}^{cr}$, $\dot{\varepsilon}_V \approx 0$ or

$$\frac{1}{2}\left(\boldsymbol{e}_y \otimes \frac{\partial \dot{\boldsymbol{u}}}{\partial y} + \frac{\partial \dot{\boldsymbol{u}}}{\partial y} \otimes \boldsymbol{e}_y\right) \approx \dot{\boldsymbol{\varepsilon}}^{cr} = \frac{3}{2}a\sigma_{vM}^{n-1}\boldsymbol{s}, \qquad \boldsymbol{\nabla} \cdot \dot{\boldsymbol{u}} \approx 0 \tag{4.4.3}$$

Consequently

$$\frac{1}{2}\frac{\partial \dot{w}}{\partial y}(\boldsymbol{e}_y \otimes \boldsymbol{n} + \boldsymbol{n} \otimes \boldsymbol{e}_y) \approx \frac{3}{2}a\sigma_{vM}^{n-1}\boldsymbol{s} \tag{4.4.4}$$

From Eq. (4.4.4) we observe that the stress deviator in the steady-state creep has the form $\boldsymbol{s} \approx \tau(\boldsymbol{e}_y \otimes \boldsymbol{n} + \boldsymbol{n} \otimes \boldsymbol{e}_y)$ and is completely determined by the transverse shear stress. The mean stress σ_m cannot be determined from the constitutive equation, it must be found from the equilibrium conditions (4.2.3). The stress state in the zone of the clamped edge $(l_x/2, y, z)$ is then of the type

$$\boldsymbol{\sigma} \approx \sigma_m \boldsymbol{I} + \tau(\boldsymbol{e}_y \otimes \boldsymbol{n} + \boldsymbol{n} \otimes \boldsymbol{e}_y)$$

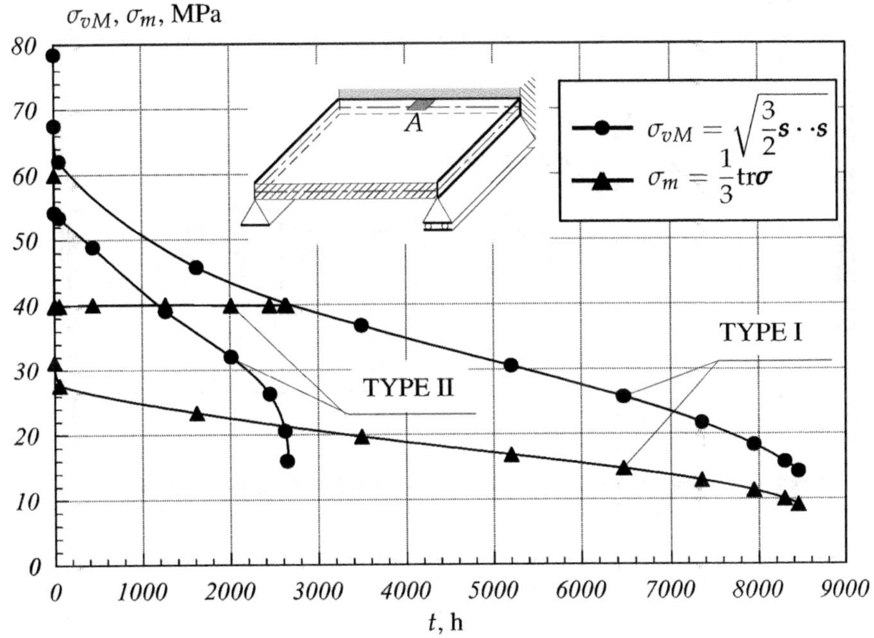

Fig. 4.24 Time variations of the von Mises equivalent stress and the hydrostatic stress in element A of the clamped edge

We observe that $\sigma_{zz} \approx \sigma_{yy} \approx \sigma_{xx} \approx \sigma_m$ after the transient stress redistribution. This estimation agrees again with the obtained finite element solution Fig. 4.23b (dotted lines). The transverse normal stress is approximately equal to the in-plane stresses and cannot be neglected.

Let us compare the finite element results for the mean stress and the von Mises equivalent stress. Figure 4.24 shows the corresponding time variations in the element A of the solid model for the TYPE I and TYPE II boundary conditions. We observe that the TYPE II boundary condition leads to a lower starting value of the von Mises stress and a higher starting value of the mean stress when compared with those for the TYPE I boundary condition. In addition, for the TYPE II clamped edge we observe that the mean stress rapidly decreases within the short transition time and after that remains constant while the von Mises stress relaxes during the whole creep process. With the relaxation of σ_{vM} the stress state tends to $\boldsymbol{\sigma} = \sigma_m \boldsymbol{I}$. The relatively high constant value of σ_m is the reason for the obtained increase of damage and much shorter time to fracture in the case of the TYPE II clamped edge (see Fig. 4.21). The above effect of the mean stress has a local character and is observed only in the neighborhood of the edge. As Fig. 4.25 shows the value of the transverse normal stress decreases rapidly with increased distance from the boundary.

We discussed the possibilities of creep-damage behavior modeling in moderately thick structural elements. The selected constitutive model of creep is based on

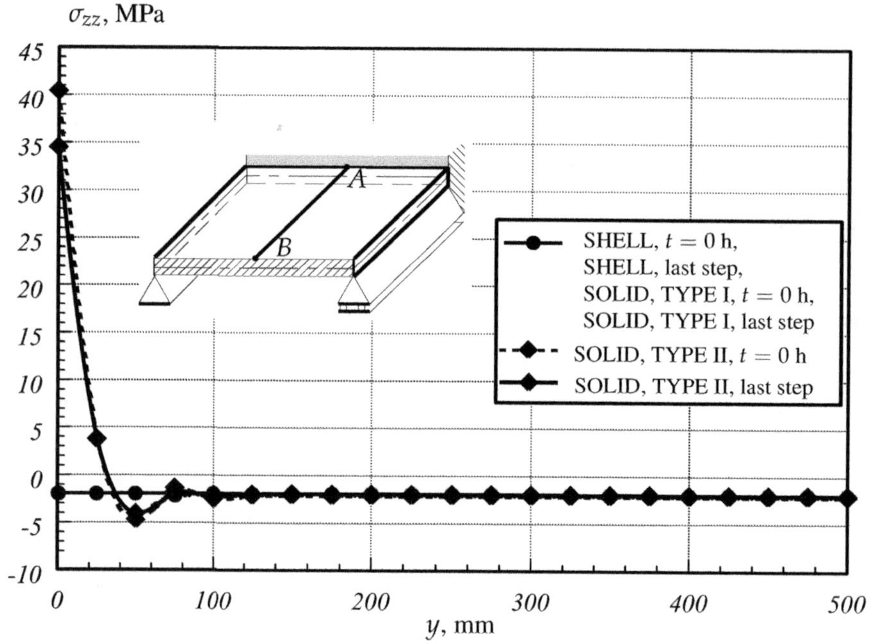

Fig. 4.25 Distribution of the transverse normal stress σ_{zz} in elements along the line AB

the assumption that the secondary creep strain rate is determined by the deviatoric part of the stress tensor and the von Mises equivalent stress, while the increase of the creep rate in the tertiary range is due to isotropic damage evolution which is controlled by the mean stress, the first principal stress and the von Mises equivalent stress. The use of this model in connection with long term predictions of structural elements has motivated a numerical comparative study of two approaches: the three-dimensional approach and the approach based on the first order shear deformation type plate theory. The finite element results as well as some simplified estimates have shown that the approaches based on standard solid and shell finite elements provide quite different predictions. The model based on the shell elements overestimates the fracture time. The reason for the obtained differences is the local stress response in the zone of the clamped edge. In the case of linear isotropic elasticity, the transverse normal and shear stresses in the zone of the clamped edge can be assumed to be the second order quantities in comparison to the dominant in-plane stresses. In the case of steady-state creep, the transverse normal and shear stresses are comparable with the in-plane stresses due to the stress redistribution. If studying the creep behavior coupled with damage, the influence of these factors cannot be ignored.

If a shell or a plate theory is considered to be an approximate version of the three-dimensional equations (4.2.1) – (4.2.9) then we can conclude that "more accurate" cross-section approximations for the transverse normal and shear stresses have to be used in the case of creep. In this sense it is more reliable to solve the three-dimensional equations (4.2.1) – (4.2.9) which are "free" from ad hoc assumptions for the displacements and stresses.

4.4.2.2 Long Term Strength Analysis of a Steam Transfer Line. From the practice of power and petrochemical plants it is well known that pipe bends are the most critical structural components with respect to possible creep failures, e.g. [186, 202]. An example for a steam transfer line between a header and a desuperheater of a boiler is presented in [186]. The pipeline from steel 1Cr0.5Mo (13CrMo4-5) had operated under the temperature in the range $500 - 550°C$ and the internal pressure 11.8 MPa. After a service life of 77000 h rupture occurred along the outer radius of a pipe bend. A metallographic analysis of a section cut from the bend close to the main crack has shown typical creep damage due to microvoids and microcracks on grain boundaries. Several incidents of pipe bend failures in different power plants are reported in [131]. Inspection techniques were developed to examine the state of creep damage during the service. However, as noted in [186], any inspection must be conducted at exactly the critical position, or the presence of damage may not be detected.

Many investigations have addressed the analysis of mechanical behavior of pressurized curved tubes. Problems of elastic and elasto-platic deformation and stability are reviewed in [68, 194]. Creep behavior of curved tubes subjected to in-plane bending moment was firstly analyzed by Spence [305] (see also [80]). Based on Hill's variational estimates he found the relationship between the rate of cross section rotation and the applied bending moment in the steady-state creep range. He defined the creep flexibility factor as the ratio of the rate of cross section rotation of the pipe bend to that of a straight pipe subjected to the same bending moment. The obtained values of the flexibility factor as functions of characteristic dimensions of pipe bends and the creep exponent are useful for preliminary design of pipelines. Analysis of long term strength in pipe bends requires the consideration of damage processes in the constitutive model of creep. In [19] we analyzed creep-damage of a pipe bend subjected to internal pressure by use of the finite element method. The results show that the predicted life time and the location of critical damage zones highly depend on the types of the applied boundary conditions. In studying the long term behavior of real pipe bends one should take into account the initial ovality of the cross section and non-uniformity of the wall thickness induced by the manufacturing process [150, 330]. In [150] the finite element based creep-damage analysis of pressurized pipe bends is performed by taking into account the initial ovality. The results are compared with those for pipe bends with circular cross section. It is shown that the introduction of ovality results in a large reduction in the creep failure time.

The aforementioned studies were concerned with the analysis of a single pipe bend subjected to special loading conditions, i.e. in-plane bending moments and internal pressure. In the following example we attempt to analyze the behavior of pipe bends in a real spatial pipeline. Figure 4.26 shows the reference geometry of the structure which includes three straight pipe segments (I, III and V) and two pipe

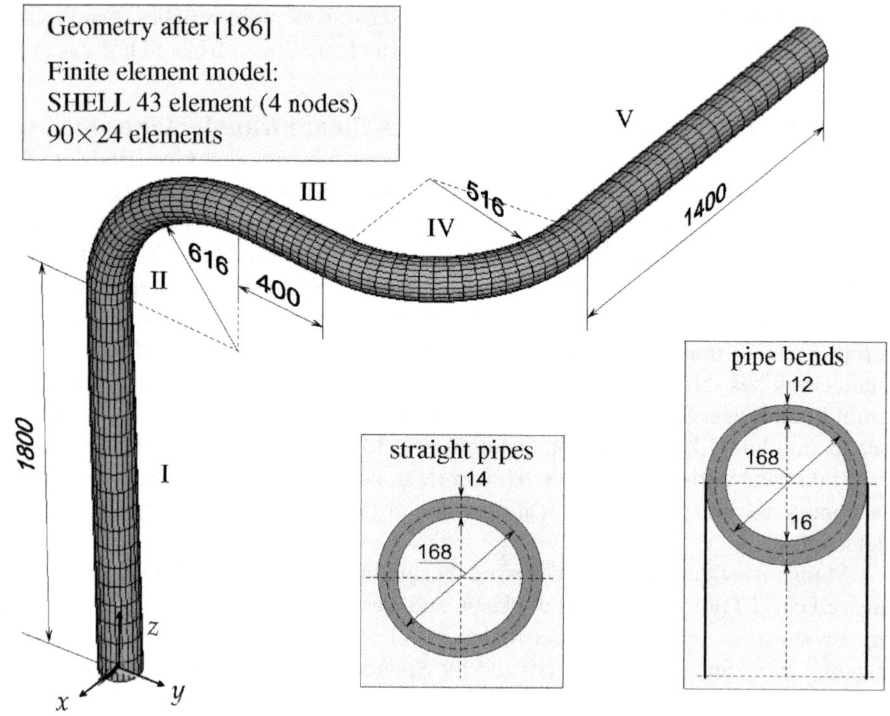

Fig. 4.26 Pipeline. Geometry and finite element mesh

bends (II and IV). The lengths of the pipe segments, the mean diameter of the cross section and the wall thickness correspond to the data given in [186]. In addition, we take into account the non-uniformity of the wall thickness in the pipe bends as a result of processing by induction bending. The circumferential thickness distribution is selected according to standard tolerances presented in [45, 118, 202]. The flanges of the pipeline are clamped. The internal pressure and the temperature are assumed to be constant during the creep process. The corresponding values are presented in Fig. 4.27. The constitutive model and the material constants for steel 13CrMo4-5 at 550°C are taken from [296] (see Sect. 3.1.2).

Figure 4.27 illustrates the deformed shape and the distribution of the magnitude of the displacement vector in the reference state. Figure 4.28a shows the corresponding distribution of the von Mises equivalent stress. From the results we may conclude that the pipe bends are subjected to complex spatial loading and deformation conditions as a result of internal pressure and uniform heating.

Time dependent changes in the deformation and stress states are illustrated in Figs. 4.28b and 4.29. In addition, the values of the von Mises equivalent stress in three points of the pipe bend IV are plotted as functions of time. According to the

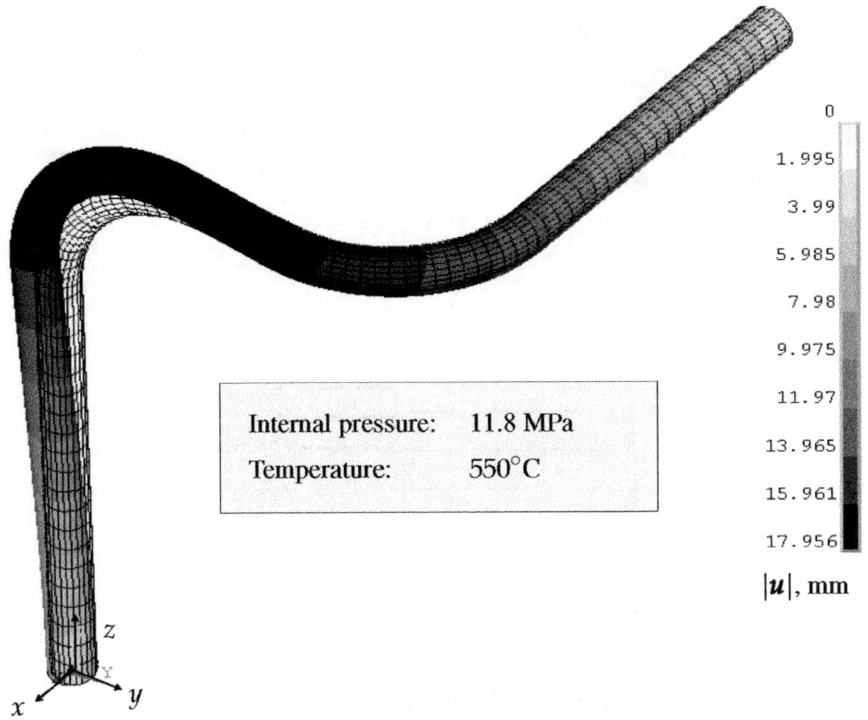

Fig. 4.27 Deformed shape and magnitude of the displacement vector in the reference elastic state

results the creep process of the pipeline may be divided into three stages. During the first stage (approximately 50% of the total live) significant stress redistributions occur leading to quite different stress state in the pipeline (cp. Fig. 4.28 and Fig. 4.29a). The second stage (approximately 45% of the total live) is characterized by slow changes in the stress state. In the final stage (approximately 5% of the total live) we observe additional stress redistributions, Fig. 4.29b. The distribution of the damage parameter at the final time step is shown in Fig. 4.30. According to the results the critical position of possible creep failure is the point A of the pipe bend IV. This result agrees well with the data presented in [186], where the creep failure has been detected at the same position.

Similar processes have been already discussed in Sects. 4.3.4, 4.3.7 and 4.4.2.1. One feature of the example considered here is that the final creep stage is not only the result of the local material deterioration but is additionally governed by the flattening (ovalisation) of the pipe bend cross section. Let us note that some parameters of the reference pipe bend geometries were not given in [186] and have been assumed in the presented calculation. Furthermore, many additional details including

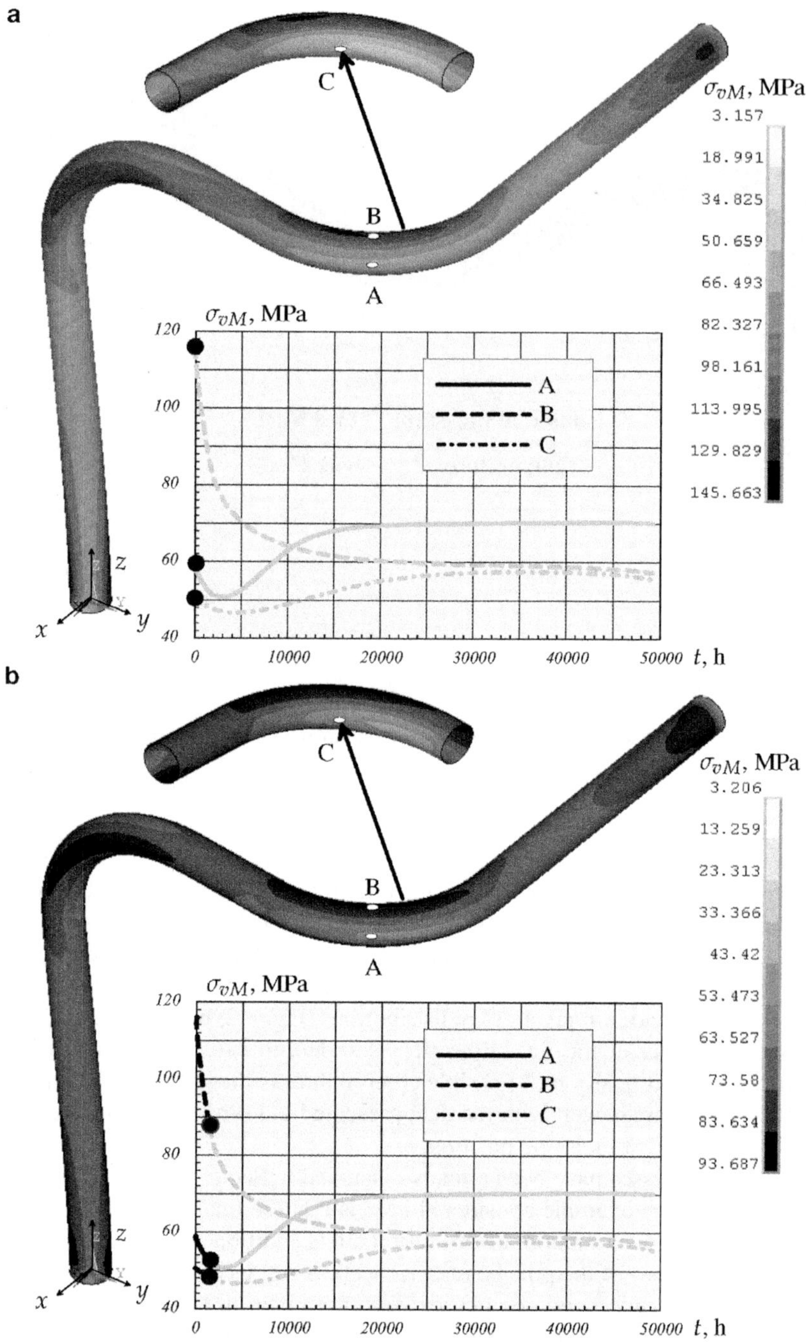

Fig. 4.28 Distribution of the von Mises equivalent stress and corresponding time variations in three points of the pipe bend. **a** Reference elastic state, **b** $t = 2000$ h

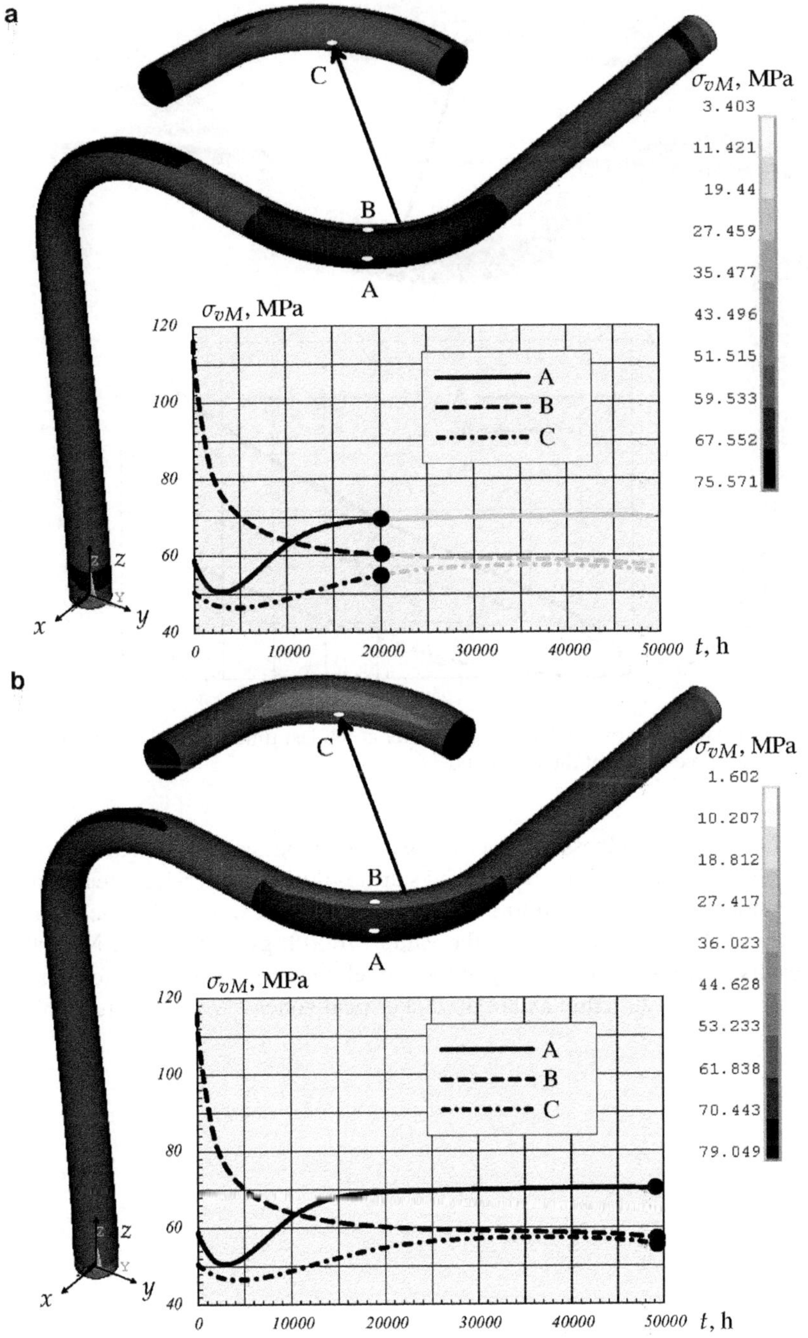

Fig. 4.29 Distribution of the von Mises equivalent stress and corresponding time variations in three points of the pipe bend. **a** $t = 20000$ h, **b** last time step

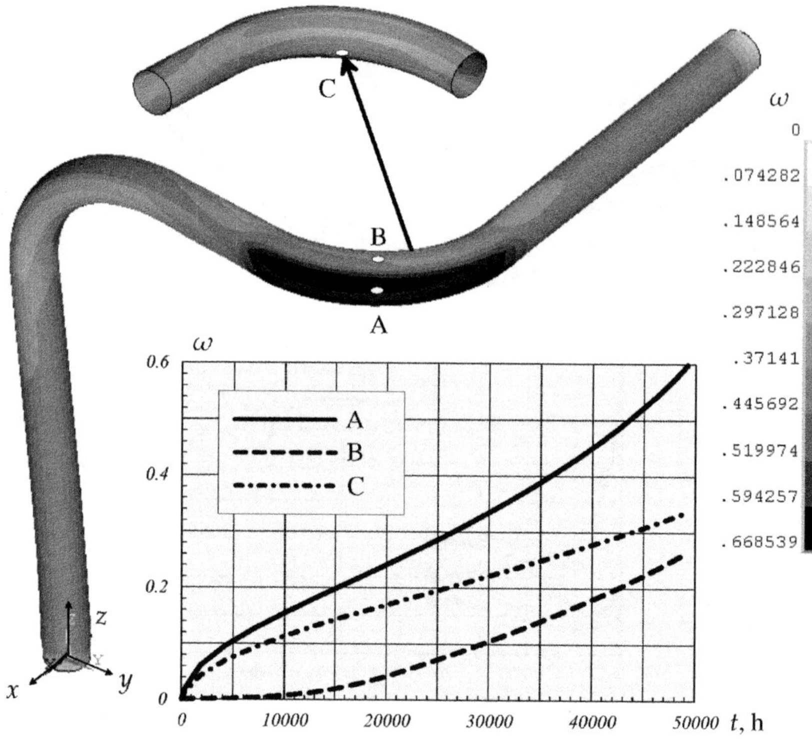

Fig. 4.30 Distribution of the damage parameter at the last time step and corresponding time variations in three points of the pipe bend

the initial out of roundness of the cross section, inhomogeneous material properties as a result of processing, shutdowns and startups during the service, are not included in the presented model. Therefore the obtained numerical result for the failure time (49000 h) "slightly" differs from the value 77000 h given in [186]. Nevertheless, the results demonstrate the ability of the modeling to represent basic features of the creep process in a structure and to predict critical zones of possible creep failure.

A Basic Operations of Tensor Algebra

The tensor calculus is a powerful tool for the description of the fundamentals in continuum mechanics and the derivation of the governing equations for applied problems. In general, there are two possibilities for the representation of the tensors and the tensorial equations:

- the direct (symbolic, coordinate-free) notation and
- the index (component) notation

The direct notation operates with scalars, vectors and tensors as physical objects defined in the three-dimensional space (in this book we are limit ourselves to this case). A vector (first rank tensor) $\boldsymbol{a}$ is considered as a directed line segment rather than a triple of numbers (coordinates). A second rank tensor $\boldsymbol{A}$ is any finite sum of ordered vector pairs $\boldsymbol{A} = \boldsymbol{a} \otimes \boldsymbol{b} + \ldots + \boldsymbol{c} \otimes \boldsymbol{d}$. The scalars, vectors and tensors are handled as invariant (independent from the choice of the coordinate system) quantities. This is the reason for the use of the direct notation in the modern literature of mechanics and rheology, e.g. [32, 36, 53, 126, 134, 205, 253, 321, 343] among others. The basics of the direct tensor calculus are given in the classical textbooks of Wilson (founded upon the lecture notes of Gibbs) [331] and Lagally [183].

The index notation deals with components or coordinates of vectors and tensors. For a selected basis, e.g. $\boldsymbol{g}_i$, $i = 1, 2, 3$ one can write

$$\boldsymbol{a} = a^i \boldsymbol{g}_i, \quad \boldsymbol{A} = \left(a^i b^j + \ldots + c^i d^j \right) \boldsymbol{g}_i \otimes \boldsymbol{g}_j$$

Here the Einstein's summation convention is used: in one expression the twice repeated indices are summed up from 1 to 3, e.g.

$$a^k \boldsymbol{g}_k \equiv \sum_{k=1}^{3} a^k \boldsymbol{g}_k, \quad A^{ik} b_k \equiv \sum_{k=1}^{3} A^{ik} b_k$$

In the above examples k is a so-called dummy index. Within the index notation the basic operations with tensors are defined with respect to their coordinates, e. g. the sum of two vectors is computed as the sum of their coordinates $c^i = a^i + b^i$. The introduced basis remains in the background. It must be noted that a change of the coordinate system leads to the change of the components of tensors.

In this book we prefer the direct tensor notation over the index one. When solving applied problems the tensor equations can be "translated into the language" of matrices for a specified coordinate system. The purpose of this Appendix is to give a brief guide to notations and rules of the tensor calculus applied throughout this book. For more comprehensive overviews on tensor calculus we recommend [58, 99, 126, 197, 205, 319, 343]. The calculus of matrices is presented in [44, 114, 350], for example. Section A provides a summary of basic algebraic operations with vectors and second rank tensors. Several rules from tensor analysis are given in Sect. B. Basic sets of invariants for different groups of symmetry transformation are presented in Sect. C, where a novel approach to find the functional basis is discussed.

A.1 Polar and Axial Vectors

A vector in the three-dimensional Euclidean space is defined as a directed line segment with specified scalar-valued magnitude and direction. The magnitude (the length) of a vector $\boldsymbol{a}$ is denoted by $|\boldsymbol{a}|$. Two vectors $\boldsymbol{a}$ and $\boldsymbol{b}$ are equal if they have the same direction and the same magnitude. The zero vector $\mathbf{0}$ has a magnitude equal to zero. In mechanics two types of vectors can be introduced. The vectors of the first type are directed line segments. These vectors are associated with translations in the three-dimensional space. Examples for polar vectors include the force, the displacement, the velocity, the acceleration, the momentum, etc. The second type is used to characterize spinor motions and related quantities, i.e. the moment, the angular velocity, the angular momentum, etc. Figure A.1a shows the so-called spin vector $\boldsymbol{a}_*$ which represents a rotation about the given axis. The direction of rotation is specified by the circular arrow and the "magnitude" of rotation is the corresponding length. For the given spin vector $\boldsymbol{a}_*$ the directed line segment $\boldsymbol{a}$ is introduced according to the following rules [343]:

1. the vector $\boldsymbol{a}$ is placed on the axis of the spin vector,
2. the magnitude of $\boldsymbol{a}$ is equal to the magnitude of $\boldsymbol{a}_*$,
3. the vector $\boldsymbol{a}$ is directed according to the right-handed screw, Fig. A.1b, or the left-handed screw, Fig. A.1c

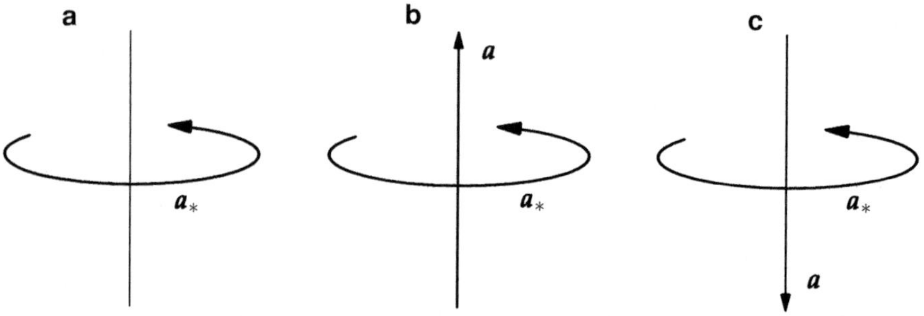

Fig. A.1 Spin vector and its representation by an axial vector. **a** Spin vector, **b** axial vector in the right-screw oriented reference frame, **c** axial vector in the left-screw oriented reference frame

The selection of one of the two cases in of the third item corresponds to the convention of orientation of the reference frame [343] (it should be not confused with the right- or left-handed triples of vectors or coordinate systems). The directed line segment is called a polar vector if it does not change by changing the orientation of the reference frame. The vector is called to be axial if it changes the sign by changing the orientation of the reference frame. The above definitions are valid for scalars and tensors of any rank too. The axial vectors (and tensors) are widely used in the rigid body dynamics, e.g. [342], in the theories of rods, plates and shells, e.g. [28], in the asymmetric theory of elasticity, e.g. [238], as well as in dynamics of micro-polar media, e.g. [111]. When dealing with polar and axial vectors it should be remembered that they have different physical meanings. Therefore, a sum of a polar and an axial vector has no sense.

A.2 Operations with Vectors

A.2.1 Addition

For a given pair of vectors $\boldsymbol{a}$ and $\boldsymbol{b}$ of the same type the sum $\boldsymbol{c} = \boldsymbol{a} + \boldsymbol{b}$ is defined according to one of the rules in Fig. A.2. The sum has the following properties

- $\boldsymbol{a} + \boldsymbol{b} = \boldsymbol{b} + \boldsymbol{a}$ (commutativity),
- $(\boldsymbol{a} + \boldsymbol{b}) + \boldsymbol{c} = \boldsymbol{a} + (\boldsymbol{b} + \boldsymbol{c})$ (associativity),
- $\boldsymbol{a} + \boldsymbol{0} = \boldsymbol{a}$

A.2.2 Multiplication by a Scalar

For any vector $\boldsymbol{a}$ and for any scalar α a vector $\boldsymbol{b} = \alpha\boldsymbol{a}$ is defined in such a way that

- $|\boldsymbol{b}| = |\alpha||\boldsymbol{a}|$,
- for $\alpha > 0$ the direction of $\boldsymbol{b}$ coincides with that of $\boldsymbol{a}$,
- for $\alpha < 0$ the direction of $\boldsymbol{b}$ is opposite to that of $\boldsymbol{a}$.

For $\alpha = 0$ the product yields the zero vector, i.e. $\boldsymbol{0} = 0\boldsymbol{a}$. It is easy to verify that

$$\alpha(\boldsymbol{a} + \boldsymbol{b}) = \alpha\boldsymbol{a} + \alpha\boldsymbol{b},$$
$$(\alpha + \beta)\boldsymbol{a} = \alpha\boldsymbol{a} + \beta\boldsymbol{a}$$

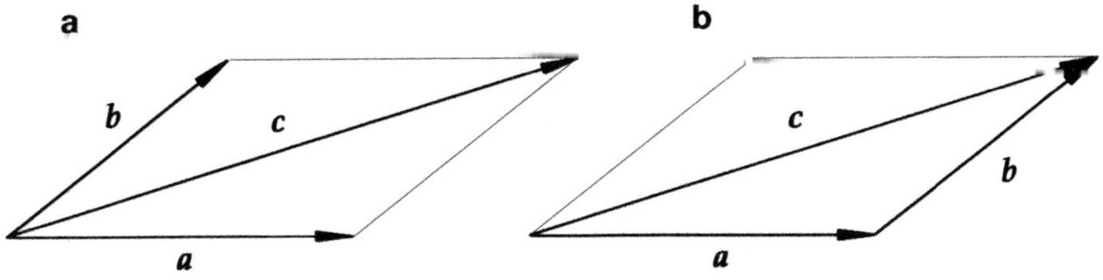

Fig. A.2 Addition of two vectors. **a** Parallelogram rule, **b** triangle rule

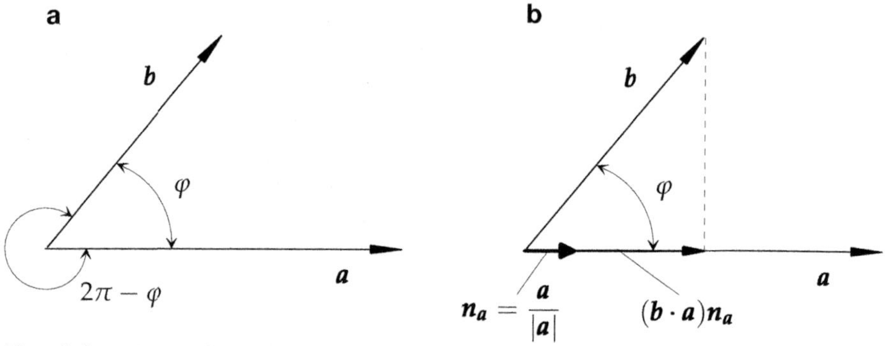

Fig. A.3 Scalar product of two vectors. **a** Angles between two vectors, **b** unit vector and projection

A.2.3 Scalar (Dot) Product of Two Vectors

For any pair of vectors $\boldsymbol{a}$ and $\boldsymbol{b}$ a scalar α is defined by

$$\alpha = \boldsymbol{a} \cdot \boldsymbol{b} = |\boldsymbol{a}||\boldsymbol{b}| \cos \varphi,$$

where φ is the angle between the vectors $\boldsymbol{a}$ and $\boldsymbol{b}$. As φ one can use any of the two angles between the vectors, Fig. A.3a. The properties of the scalar product are

- $\boldsymbol{a} \cdot \boldsymbol{b} = \boldsymbol{b} \cdot \boldsymbol{a}$ (commutativity),
- $\boldsymbol{a} \cdot (\boldsymbol{b} + \boldsymbol{c}) = \boldsymbol{a} \cdot \boldsymbol{b} + \boldsymbol{a} \cdot \boldsymbol{c}$ (distributivity)

Two nonzero vectors are said to be orthogonal if their scalar product is zero. The unit vector directed along the vector $\boldsymbol{a}$ is defined by (see Fig. A.3b)

$$\boldsymbol{n}_a = \frac{\boldsymbol{a}}{|\boldsymbol{a}|}$$

The projection of the vector $\boldsymbol{b}$ onto the vector $\boldsymbol{a}$ is the vector $(\boldsymbol{b} \cdot \boldsymbol{a})\boldsymbol{n}_a$, Fig. A.3b. The length of the projection is $|\boldsymbol{b}||\cos \varphi|$.

A.2.4 Vector (Cross) Product of Two Vectors

For the ordered pair of vectors $\boldsymbol{a}$ and $\boldsymbol{b}$ the vector $\boldsymbol{c} = \boldsymbol{a} \times \boldsymbol{b}$ is defined in two following steps [343]:

- the spin vector $\boldsymbol{c}_*$ is defined in such a way that
  - ⋆ the axis is orthogonal to the plane spanned on $\boldsymbol{a}$ and $\boldsymbol{b}$, Fig. A.4a,
  - ⋆ the circular arrow shows the direction of the "shortest" rotation from $\boldsymbol{a}$ to $\boldsymbol{b}$, Fig. A.4b,
  - ⋆ the length is $|\boldsymbol{a}||\boldsymbol{b}| \sin \varphi$, where φ is the angle of the "shortest" rotation from $\boldsymbol{a}$ to $\boldsymbol{b}$,

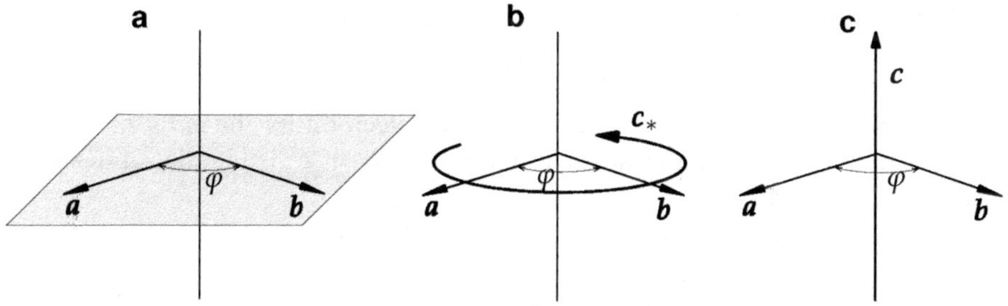

Fig. A.4 Vector product of two vectors. **a** Plane spanned on two vectors, **b** spin vector, **c** axial vector in the right-screw oriented reference frame

- from the resulting spin vector the directed line segment $\boldsymbol{c}$ is constructed according to one of the rules listed in Sect. A.1.

The properties of the vector product are

$$\boldsymbol{a} \times \boldsymbol{b} = -\boldsymbol{b} \times \boldsymbol{a},$$
$$\boldsymbol{a} \times (\boldsymbol{b} + \boldsymbol{c}) = \boldsymbol{a} \times \boldsymbol{b} + \boldsymbol{a} \times \boldsymbol{c}$$

The type of the vector $\boldsymbol{c} = \boldsymbol{a} \times \boldsymbol{b}$ can be established for the known types of the vectors $\boldsymbol{a}$ and $\boldsymbol{b}$, [343]. If $\boldsymbol{a}$ and $\boldsymbol{b}$ are polar vectors the result of the cross product will be the axial vector. An example is the moment of momentum for a mass point m defined by $\boldsymbol{r} \times (m\dot{\boldsymbol{v}})$, where $\boldsymbol{r}$ is the position of the mass point and $\boldsymbol{v}$ is the velocity of the mass point. The next example is the formula for the distribution of velocities in a rigid body $\boldsymbol{v} = \boldsymbol{\omega} \times \boldsymbol{r}$. Here the cross product of the axial vector $\boldsymbol{\omega}$ (angular velocity) with the polar vector $\boldsymbol{r}$ (position vector) results in the polar vector $\boldsymbol{v}$.

The mixed product of three vectors $\boldsymbol{a}, \boldsymbol{b}$ and $\boldsymbol{c}$ is defined by $(\boldsymbol{a} \times \boldsymbol{b}) \cdot \boldsymbol{c}$. The result is a scalar. For the mixed product the following identities are valid

$$\boldsymbol{a} \cdot (\boldsymbol{b} \times \boldsymbol{c}) = \boldsymbol{b} \cdot (\boldsymbol{c} \times \boldsymbol{a}) = \boldsymbol{c} \cdot (\boldsymbol{a} \times \boldsymbol{b}) \tag{A.2.1}$$

If the cross product is applied twice, the first operation must be set in parentheses, e.g., $\boldsymbol{a} \times (\boldsymbol{b} \times \boldsymbol{c})$. The result of this operation is a vector. The following relation can be applied

$$\boldsymbol{a} \times (\boldsymbol{b} \times \boldsymbol{c}) = \boldsymbol{b}(\boldsymbol{a} \cdot \boldsymbol{c}) - \boldsymbol{c}(\boldsymbol{a} \cdot \boldsymbol{b}) \tag{A.2.2}$$

By use of (A.2.1) and (A.2.2) one can calculate

$$\begin{aligned} (\boldsymbol{a} \times \boldsymbol{b}) \cdot (\boldsymbol{c} \times \boldsymbol{d}) &= \boldsymbol{a} \cdot [\boldsymbol{b} \times (\boldsymbol{c} \times \boldsymbol{d})] \\ &= \boldsymbol{a} \cdot (\boldsymbol{c}\, \boldsymbol{b} \cdot \boldsymbol{d} - \boldsymbol{d}\, \boldsymbol{b} \cdot \boldsymbol{c}) \\ &= \boldsymbol{a} \cdot \boldsymbol{c}\, \boldsymbol{b} \cdot \boldsymbol{d} - \boldsymbol{a} \cdot \boldsymbol{d}\, \boldsymbol{b} \cdot \boldsymbol{c} \end{aligned} \tag{A.2.3}$$

A.3 Bases

Any triple of linear independent vectors $\boldsymbol{e}_1, \boldsymbol{e}_2, \boldsymbol{e}_3$ is called basis. A triple of vectors $\boldsymbol{e}_i$ is linear independent if and only if $\boldsymbol{e}_1 \cdot (\boldsymbol{e}_2 \times \boldsymbol{e}_3) \neq 0$.

For a given basis $\boldsymbol{e}_i$ any vector $\boldsymbol{a}$ can be represented as follows

$$\boldsymbol{a} = a^1\boldsymbol{e}_1 + a^2\boldsymbol{e}_2 + a^3\boldsymbol{e}_3 \equiv a^i\boldsymbol{e}_i$$

The numbers a^i are called the coordinates of the vector $\boldsymbol{a}$ for the basis $\boldsymbol{e}_i$. In order to compute the coordinates a^i the dual (reciprocal) basis $\boldsymbol{e}^k$ is introduced in such a way that

$$\boldsymbol{e}^k \cdot \boldsymbol{e}_i = \delta_i^k = \begin{cases} 1, & k = i, \\ 0, & k \neq i \end{cases}$$

δ_i^k is the Kronecker symbol. The coordinates a^i can be found by

$$\boldsymbol{e}^i \cdot \boldsymbol{a} = \boldsymbol{a} \cdot \boldsymbol{e}^i = a^m\boldsymbol{e}_m \cdot \boldsymbol{e}^i = a^m\delta_m^i = a^i$$

For the selected basis $\boldsymbol{e}_i$ the dual basis can be found from

$$\boldsymbol{e}^1 = \frac{\boldsymbol{e}_2 \times \boldsymbol{e}_3}{(\boldsymbol{e}_1 \times \boldsymbol{e}_2) \cdot \boldsymbol{e}_3}, \quad \boldsymbol{e}^2 = \frac{\boldsymbol{e}_3 \times \boldsymbol{e}_1}{(\boldsymbol{e}_1 \times \boldsymbol{e}_2) \cdot \boldsymbol{e}_3}, \quad \boldsymbol{e}^3 = \frac{\boldsymbol{e}_1 \times \boldsymbol{e}_2}{(\boldsymbol{e}_1 \times \boldsymbol{e}_2) \cdot \boldsymbol{e}_3} \tag{A.3.1}$$

By use of the dual basis a vector $\boldsymbol{a}$ can be represented as follows

$$\boldsymbol{a} = a_1\boldsymbol{e}^1 + a_2\boldsymbol{e}^2 + a_3\boldsymbol{e}^3 \equiv a_i\boldsymbol{e}^i, \quad a_m = \boldsymbol{a} \cdot \boldsymbol{e}_m, \quad a^m \neq a_m$$

In the special case of the orthonormal vectors $\boldsymbol{e}_i$, i.e. $|\boldsymbol{e}_i| = 1$ and $\boldsymbol{e}_i \cdot \boldsymbol{e}_k = 0$ for $i \neq k$, from (A.3.1) follows that $\boldsymbol{e}^k = \boldsymbol{e}_k$ and consequently $a_k = a^k$.

A.4 Operations with Second Rank Tensors

A second rank tensor is a finite sum of ordered vector pairs $\boldsymbol{A} = \boldsymbol{a} \otimes \boldsymbol{b} + \ldots + \boldsymbol{c} \otimes \boldsymbol{d}$ [343]. One ordered pair of vectors is called the dyad [331]. The symbol $\otimes$ is called the dyadic (tensor) product of two vectors. A single dyad or a sum of two dyads are special cases of the second rank tensor. Any finite sum of more than three dyads can be reduced to a sum of three dyads. For example, let

$$\boldsymbol{A} = \sum_{i=1}^{n} \boldsymbol{a}_{(i)} \otimes \boldsymbol{b}_{(i)}$$

be a second rank tensor. Introducing a basis $\boldsymbol{e}_k$ the vectors $\boldsymbol{a}_{(i)}$ can be represented by $\boldsymbol{a}_{(i)} = a_{(i)}^k\boldsymbol{e}_k$, where $a_{(i)}^k$ are coordinates of the vectors $\boldsymbol{a}_{(i)}$. Now we may write

$$\boldsymbol{A} = \sum_{i=1}^{n} a_{(i)}^k\boldsymbol{e}_k \otimes \boldsymbol{b}_{(i)} = \boldsymbol{e}_k \otimes \sum_{i=1}^{n} a_{(i)}^k\boldsymbol{b}_{(i)} = \boldsymbol{e}_k \otimes \boldsymbol{d}^k, \quad \boldsymbol{d}^k \equiv \sum_{i=1}^{n} a_{(i)}^k\boldsymbol{b}_{(i)}$$

A.4.1 Addition

The sum of two tensors is defined as the sum of the corresponding dyads. The sum has the properties of associativity and commutativity. In addition, the following operation can be introduced

$$\boldsymbol{a} \otimes (\boldsymbol{b} + \boldsymbol{c}) = \boldsymbol{a} \otimes \boldsymbol{b} + \boldsymbol{a} \otimes \boldsymbol{c}, \quad (\boldsymbol{a} + \boldsymbol{b}) \otimes \boldsymbol{c} = \boldsymbol{a} \otimes \boldsymbol{c} + \boldsymbol{b} \otimes \boldsymbol{c}$$

A.4.2 Multiplication by a Scalar

This operation is introduced first for one dyad. For any scalar α and any dyad $\boldsymbol{a} \otimes \boldsymbol{b}$

$$\begin{aligned} \alpha(\boldsymbol{a} \otimes \boldsymbol{b}) &= (\alpha \boldsymbol{a}) \otimes \boldsymbol{b} = \boldsymbol{a} \otimes (\alpha \boldsymbol{b}), \\ (\alpha + \beta)\boldsymbol{a} \otimes \boldsymbol{b} &= \alpha \boldsymbol{a} \otimes \boldsymbol{b} + \beta \boldsymbol{a} \otimes \boldsymbol{b} \end{aligned} \tag{A.4.1}$$

By setting $\alpha = 0$ in the first equation of (A.4.1) the zero dyad can be defined, i.e. $0(\boldsymbol{a} \otimes \boldsymbol{b}) = \boldsymbol{0} \otimes \boldsymbol{b} = \boldsymbol{a} \otimes \boldsymbol{0}$. The above operations can be generalized for any finite sum of dyads, i.e. for second rank tensors.

A.4.3 Inner Dot Product

For any two second rank tensors $\boldsymbol{A}$ and $\boldsymbol{B}$ the inner dot product is specified by $\boldsymbol{A} \cdot \boldsymbol{B}$. The rule and the result of this operation can be explained in the special case of two dyads, i.e. by setting $\boldsymbol{A} = \boldsymbol{a} \otimes \boldsymbol{b}$ and $\boldsymbol{B} = \boldsymbol{c} \otimes \boldsymbol{d}$

$$\boldsymbol{A} \cdot \boldsymbol{B} = \boldsymbol{a} \otimes \boldsymbol{b} \cdot \boldsymbol{c} \otimes \boldsymbol{d} = (\boldsymbol{b} \cdot \boldsymbol{c})\boldsymbol{a} \otimes \boldsymbol{d} = \alpha \boldsymbol{a} \otimes \boldsymbol{d}, \quad \alpha \equiv \boldsymbol{b} \cdot \boldsymbol{c}$$

Note that in general $\boldsymbol{A} \cdot \boldsymbol{B} \neq \boldsymbol{B} \cdot \boldsymbol{A}$. This can be again verified for two dyads. The operation can be generalized for two second rank tensors as follows

$$\begin{aligned} \boldsymbol{A} \cdot \boldsymbol{B} = \sum_{i=1}^{3} \boldsymbol{a}_{(i)} \otimes \boldsymbol{b}_{(i)} \cdot \sum_{k=1}^{3} \boldsymbol{c}_{(k)} \otimes \boldsymbol{d}_{(k)} &= \sum_{i=1}^{3} \sum_{k=1}^{3} (\boldsymbol{b}_{(i)} \cdot \boldsymbol{c}_{(k)}) \boldsymbol{a}_{(i)} \otimes \boldsymbol{d}_{(k)} \\ &= \sum_{i=1}^{3} \sum_{k=1}^{3} \alpha_{(ik)} \boldsymbol{a}_{(i)} \otimes \boldsymbol{d}_{(k)} \end{aligned}$$

with $\alpha_{(ik)} \equiv \boldsymbol{b}_{(i)} \cdot \boldsymbol{c}_{(k)}$. The result of this operation is a second rank tensor.

A.4.4 Transpose of a Second Rank Tensor

The transpose of a second rank tensor $\boldsymbol{A}$ is defined as follows

$$\boldsymbol{A}^T = \left(\sum_{i=1}^{3} \boldsymbol{a}_{(i)} \otimes \boldsymbol{b}_{(i)} \right)^T = \sum_{i=1}^{3} \boldsymbol{b}_{(i)} \otimes \boldsymbol{a}_{(i)}$$

A.4.5 Double Inner Dot Product

For any two second rank tensors $\boldsymbol{A}$ and $\boldsymbol{B}$ the double inner dot product is specified by $\boldsymbol{A} \cdot\cdot\, \boldsymbol{B}$. The result of this operation is a scalar. This operation can be explained for two dyads $\boldsymbol{A} = \boldsymbol{a} \otimes \boldsymbol{b}$ and $\boldsymbol{B} = \boldsymbol{c} \otimes \boldsymbol{d}$ as follows

$$\boldsymbol{A} \cdot\cdot\, \boldsymbol{B} = \boldsymbol{a} \otimes \boldsymbol{b} \cdot\cdot\, \boldsymbol{c} \otimes \boldsymbol{d} = (\boldsymbol{b} \cdot \boldsymbol{c})(\boldsymbol{a} \cdot \boldsymbol{d})$$

By analogy to the inner dot product one can generalize this operation for two second rank tensors. It can be verified that $\boldsymbol{A} \cdot\cdot \boldsymbol{B} = \boldsymbol{B} \cdot\cdot \boldsymbol{A}$ for arbitrary second rank tensors $\boldsymbol{A}$ and $\boldsymbol{B}$. For a second rank tensor $\boldsymbol{A}$ and for a dyad $\boldsymbol{a} \otimes \boldsymbol{b}$ the double inner dot product yields

$$\boldsymbol{A} \cdot\cdot \boldsymbol{a} \otimes \boldsymbol{b} = \boldsymbol{b} \cdot \boldsymbol{A} \cdot \boldsymbol{a} \tag{A.4.2}$$

A scalar product of two second rank tensors $\boldsymbol{A}$ and $\boldsymbol{B}$ is defined by

$$\alpha = \boldsymbol{A} \cdot\cdot \boldsymbol{B}^T$$

One can verify that

$$\boldsymbol{A} \cdot\cdot \boldsymbol{B}^T = \boldsymbol{B}^T \cdot\cdot \boldsymbol{A} = \boldsymbol{B} \cdot\cdot \boldsymbol{A}^T$$

A.4.6 Dot Products of a Second Rank Tensor and a Vector

The right dot product of a second rank tensor $\boldsymbol{A}$ and a vector $\boldsymbol{c}$ is defined by

$$\boldsymbol{A} \cdot \boldsymbol{c} = \left(\sum_{i=1}^{3} \boldsymbol{a}_{(i)} \otimes \boldsymbol{b}_{(i)}\right) \cdot \boldsymbol{c} = \sum_{i=1}^{3} (\boldsymbol{b}_{(i)} \cdot \boldsymbol{c})\boldsymbol{a}_{(i)} = \sum_{i=1}^{3} \alpha_{(i)}\boldsymbol{a}_{(i)}$$

with $\alpha_{(i)} \equiv \boldsymbol{b}_{(i)} \cdot \boldsymbol{c}$. The left dot product is defined by

$$\boldsymbol{c} \cdot \boldsymbol{A} = \boldsymbol{c} \cdot \left(\sum_{i=1}^{3} \boldsymbol{a}_{(i)} \otimes \boldsymbol{b}_{(i)}\right) = \sum_{i=1}^{3} (\boldsymbol{c} \cdot \boldsymbol{a}_{(i)})\boldsymbol{b}_{(i)} = \sum_{i=1}^{3} \beta_{(i)}\boldsymbol{b}_{(i)}$$

with $\beta_{(i)} \equiv \boldsymbol{c} \cdot \boldsymbol{a}_{(i)}$. The results of these operations are vectors. One can verify that

$$\boldsymbol{A} \cdot \boldsymbol{c} \neq \boldsymbol{c} \cdot \boldsymbol{A}, \quad \boldsymbol{A} \cdot \boldsymbol{c} = \boldsymbol{c} \cdot \boldsymbol{A}^T$$

A.4.7 Cross Products of a Second Rank Tensor and a Vector

The right cross product of a second rank tensor $\boldsymbol{A}$ and a vector $\boldsymbol{c}$ is defined by

$$\boldsymbol{A} \times \boldsymbol{c} = \left(\sum_{i=1}^{3} \boldsymbol{a}_{(i)} \otimes \boldsymbol{b}_{(i)}\right) \times \boldsymbol{c} = \sum_{i=1}^{3} \boldsymbol{a}_{(i)} \otimes (\boldsymbol{b}_{(i)} \times \boldsymbol{c}) = \sum_{i=1}^{3} \boldsymbol{a}_{(i)} \otimes \boldsymbol{d}_{(i)}$$

with $\boldsymbol{d}_{(i)} \equiv \boldsymbol{b}_{(i)} \times \boldsymbol{c}$. The left cross product is defined by

$$\boldsymbol{c} \times \boldsymbol{A} = \boldsymbol{c} \times \left(\sum_{i=1}^{3} \boldsymbol{a}_{(i)} \otimes \boldsymbol{b}_{(i)}\right) = \sum_{i=1}^{3} (\boldsymbol{c} \times \boldsymbol{a}_{(i)}) \otimes \boldsymbol{b}_{(i)} = \sum_{i=1}^{3} \boldsymbol{e}_{(i)} \otimes \boldsymbol{b}_{(i)}$$

with $\boldsymbol{e}_{(i)} \equiv \boldsymbol{b}_{(i)} \times \boldsymbol{c}$. The results of these operations are second rank tensors. It can be shown that

$$\boldsymbol{A} \times \boldsymbol{c} = -[\boldsymbol{c} \times \boldsymbol{A}^T]^T$$

A.4.8 Trace

The trace of a second rank tensor is defined by

$$\text{tr}\,\boldsymbol{A} = \text{tr}\left(\sum_{i=1}^{3} \boldsymbol{a}_{(i)} \otimes \boldsymbol{b}_{(i)}\right) = \sum_{i=1}^{3} \boldsymbol{a}_{(i)} \cdot \boldsymbol{b}_{(i)}$$

By taking the trace of a second rank tensor the dyadic product is replaced by the dot product. It can be shown that

$$\text{tr}\,\boldsymbol{A} = \text{tr}\,\boldsymbol{A}^T, \quad \text{tr}\,(\boldsymbol{A}\cdot\boldsymbol{B}) = \text{tr}\,(\boldsymbol{B}\cdot\boldsymbol{A}) = \text{tr}\,(\boldsymbol{A}^T\cdot\boldsymbol{B}^T) = \boldsymbol{A}\cdot\cdot\,\boldsymbol{B}$$

A.4.9 Symmetric Tensors

A second rank tensor is said to be symmetric if it satisfies the following equality

$$\boldsymbol{A} = \boldsymbol{A}^T$$

An alternative definition of the symmetric tensor can be given as follows. A second rank tensor is said to be symmetric if for any vector $\boldsymbol{c} \neq \boldsymbol{0}$ the following equality is valid

$$\boldsymbol{c}\cdot\boldsymbol{A} = \boldsymbol{A}\cdot\boldsymbol{c}$$

An important example of a symmetric tensor is the unit or identity tensor $\boldsymbol{I}$, which is defined by such a way that for any vector $\boldsymbol{c}$

$$\boldsymbol{c}\cdot\boldsymbol{I} = \boldsymbol{I}\cdot\boldsymbol{c} = \boldsymbol{c}$$

The representations of the identity tensor are

$$\boldsymbol{I} = \boldsymbol{e}_k \otimes \boldsymbol{e}^k = \boldsymbol{e}^k \otimes \boldsymbol{e}_k$$

for any basis $\boldsymbol{e}_k$ and $\boldsymbol{e}^k$, $\boldsymbol{e}_k\cdot\boldsymbol{e}^m = \delta_k^m$. For three orthonormal vectors $\boldsymbol{m}$, $\boldsymbol{n}$ and $\boldsymbol{p}$ the identity tensor has the form

$$\boldsymbol{I} = \boldsymbol{n}\otimes\boldsymbol{n} + \boldsymbol{m}\otimes\boldsymbol{m} + \boldsymbol{p}\otimes\boldsymbol{p}$$

A symmetric second rank tensor $\boldsymbol{P}$ satisfying the condition $\boldsymbol{P}\cdot\boldsymbol{P} = \boldsymbol{P}$ is called projector. Examples of projectors are

$$\boldsymbol{m}\otimes\boldsymbol{m}, \quad \boldsymbol{n}\otimes\boldsymbol{n} + \boldsymbol{p}\otimes\boldsymbol{p} = \boldsymbol{I} - \boldsymbol{m}\otimes\boldsymbol{m},$$

where $\boldsymbol{m}$, $\boldsymbol{n}$ and $\boldsymbol{p}$ are orthonormal vectors. The result of the dot product of the tensor $\boldsymbol{m}\otimes\boldsymbol{m}$ with any vector $\boldsymbol{a}$ is the projection of the vector $\boldsymbol{a}$ onto the line spanned on the vector $\boldsymbol{m}$, i.e. $\boldsymbol{m}\otimes\boldsymbol{m}\cdot\boldsymbol{a} = (\boldsymbol{a}\cdot\boldsymbol{m})\boldsymbol{m}$. The result of $(\boldsymbol{n}\otimes\boldsymbol{n} + \boldsymbol{p}\otimes\boldsymbol{p})\cdot\boldsymbol{a}$ is the projection of the vector $\boldsymbol{a}$ onto the plane spanned on the vectors $\boldsymbol{n}$ and $\boldsymbol{p}$.

A.4.10 Skew-Symmetric Tensors

A second rank tensor is said to be skew-symmetric if it satisfies the following equality

$$\boldsymbol{A} = -\boldsymbol{A}^T$$

or if for any vector $\boldsymbol{c} \neq \boldsymbol{0}$

$$\boldsymbol{c} \cdot \boldsymbol{A} = -\boldsymbol{A} \cdot \boldsymbol{c}$$

Any skew-symmetric tensor $\boldsymbol{A}$ can be represented by

$$\boldsymbol{A} = \boldsymbol{a} \times \boldsymbol{I} = \boldsymbol{I} \times \boldsymbol{a}$$

The vector $\boldsymbol{a}$ is called the associated vector. Any second rank tensor can be uniquely decomposed into the symmetric and skew-symmetric parts

$$\begin{aligned}
\boldsymbol{A} &= \frac{1}{2}\left(\boldsymbol{A} + \boldsymbol{A}^T\right) + \frac{1}{2}\left(\boldsymbol{A} - \boldsymbol{A}^T\right) = \boldsymbol{A}_1 + \boldsymbol{A}_2, \\
\boldsymbol{A}_1 &= \frac{1}{2}\left(\boldsymbol{A} + \boldsymbol{A}^T\right), \quad \boldsymbol{A}_1 = \boldsymbol{A}_1^T, \\
\boldsymbol{A}_2 &= \frac{1}{2}\left(\boldsymbol{A} - \boldsymbol{A}^T\right), \quad \boldsymbol{A}_2 = -\boldsymbol{A}_2^T
\end{aligned}$$

A.4.11 Vector Invariant

The vector invariant or "Gibbsian Cross" of a second rank tensor $\boldsymbol{A}$ is defined by

$$\boldsymbol{A}_\times = \left(\sum_{i=1}^{3} \boldsymbol{a}_{(i)} \otimes \boldsymbol{b}_{(i)}\right)_\times = \sum_{i=1}^{3} \boldsymbol{a}_{(i)} \times \boldsymbol{b}_{(i)}$$

The result of this operation is a vector. The vector invariant of a symmetric tensor is the zero vector. The following identities can be verified

$$\begin{aligned}
(\boldsymbol{a} \times \boldsymbol{I})_\times &= -2\boldsymbol{a}, \\
\boldsymbol{a} \times \boldsymbol{I} \times \boldsymbol{b} &= \boldsymbol{b} \otimes \boldsymbol{a} - (\boldsymbol{a} \cdot \boldsymbol{b})\boldsymbol{I}
\end{aligned}$$

A.4.12 Linear Transformations of Vectors

A vector valued function of a vector argument $\boldsymbol{f}(\boldsymbol{a})$ is called to be linear if $\boldsymbol{f}(\alpha_1\boldsymbol{a}_1 + \alpha_2\boldsymbol{a}_2) = \alpha_1\boldsymbol{f}(\boldsymbol{a}_1) + \alpha_2\boldsymbol{f}(\boldsymbol{a}_2)$ for any two vectors $\boldsymbol{a}_1$ and $\boldsymbol{a}_2$ and any two scalars α_1 and α_2. It can be shown that any linear vector valued function can be represented by $\boldsymbol{f}(\boldsymbol{a}) = \boldsymbol{A} \cdot \boldsymbol{a}$, where $\boldsymbol{A}$ is a second rank tensor. In many textbooks, e.g. [36, 300] a second rank tensor $\boldsymbol{A}$ is defined to be the linear transformation of the vector space into itself.

A.4.13 Determinant and Inverse of a Second Rank Tensor

Let $\boldsymbol{a}$, $\boldsymbol{b}$ and $\boldsymbol{c}$ be arbitrary linearly-independent vectors. The determinant of a second rank tensor $\boldsymbol{A}$ is defined by

$$\det \boldsymbol{A} = \frac{(\boldsymbol{A} \cdot \boldsymbol{a}) \cdot [(\boldsymbol{A} \cdot \boldsymbol{b}) \times (\boldsymbol{A} \cdot \boldsymbol{c})]}{\boldsymbol{a} \cdot (\boldsymbol{b} \times \boldsymbol{c})}$$

The following identities can be verified

$$\det(\boldsymbol{A}^T) = \det(\boldsymbol{A}),$$
$$\det(\boldsymbol{A} \cdot \boldsymbol{B}) = \det(\boldsymbol{A}) \det(\boldsymbol{B})$$

The inverse of a second rank tensor $\boldsymbol{A}^{-1}$ is introduced as the solution of the following equation

$$\boldsymbol{A}^{-1} \cdot \boldsymbol{A} = \boldsymbol{A} \cdot \boldsymbol{A}^{-1} = \boldsymbol{I}$$

$\boldsymbol{A}$ is invertible if and only if $\det \boldsymbol{A} \neq 0$. A tensor $\boldsymbol{A}$ with $\det \boldsymbol{A} = 0$ is called singular. Examples for singular tensors are projectors.

A.4.14 Principal Values and Directions of Symmetric Second Rank Tensors

Consider a dot product of a second rank tensor $\boldsymbol{A}$ and a unit vector $\boldsymbol{n}$. The resulting vector $\boldsymbol{a} = \boldsymbol{A} \cdot \boldsymbol{n}$ differs in general from $\boldsymbol{n}$ both by the length and the direction. However, one can find those unit vectors $\boldsymbol{n}$, for which $\boldsymbol{A} \cdot \boldsymbol{n}$ is collinear with $\boldsymbol{n}$, i.e. only the length of $\boldsymbol{n}$ is changed. Such vectors can be found from the equation

$$\boldsymbol{A} \cdot \boldsymbol{n} = \lambda \boldsymbol{n} \quad \text{or} \quad (\boldsymbol{A} - \lambda \boldsymbol{I}) \cdot \boldsymbol{n} = \boldsymbol{0} \tag{A.4.3}$$

The unit vector $\boldsymbol{n}$ is called the principal vector (principal direction) and the scalar λ the principal value of the tensor $\boldsymbol{A}$. The problem to find the principal values and principal directions of Eq. (A.4.3) is the eigen-value problem for $\boldsymbol{A}$. The principal values are the eigen-values, the principal directions are the eigen-directions.

Let $\boldsymbol{A}$ be a symmetric tensor. In this case the principal values are real numbers and there exist at least three mutually orthogonal principal vectors. The principal values can be found as roots of the characteristic polynomial

$$\det(\boldsymbol{A} - \lambda \boldsymbol{I}) = -\lambda^3 + J_1(\boldsymbol{A})\lambda^2 - J_2(\boldsymbol{A})\lambda + J_3(\boldsymbol{A}) = 0 \tag{A.4.4}$$

Here $J_i(\boldsymbol{A}) (i = 1, 2, 3)$ are the principal invariants of the tensor $\boldsymbol{A}$

$$\begin{aligned} J_1(\boldsymbol{A}) &= \operatorname{tr} \boldsymbol{A}, \\ J_2(\boldsymbol{A}) &= \frac{1}{2}[(\operatorname{tr} \boldsymbol{A})^2 - \operatorname{tr} \boldsymbol{A}^2], \\ J_3(\boldsymbol{A}) &= \det \boldsymbol{A} = \frac{1}{6}(\operatorname{tr} \boldsymbol{A})^3 - \frac{1}{2}\operatorname{tr} \boldsymbol{A} \operatorname{tr} \boldsymbol{A}^2 + \frac{1}{3}\operatorname{tr} \boldsymbol{A}^3 \end{aligned} \tag{A.4.5}$$

The principal values are specified by $\lambda_I, \lambda_{II}, \lambda_{III}$. The following three cases can be introduced

- three distinct values $\lambda_i, i = I, II, III$, i.e. $\lambda_I \neq \lambda_{II} \neq \lambda_{III}$, or
- one single value and one double solution, e.g. $\lambda_I = \lambda_{II} \neq \lambda_{III}$, or
- one triple solution $\lambda_I = \lambda_{II} = \lambda_{III}$

For a fixed solution $\lambda_i, i = I, II, III$ the eigen-directions can be found from

$$(\boldsymbol{A} - \lambda_i \boldsymbol{I}) \cdot \boldsymbol{n}_{(i)} = \boldsymbol{0} \qquad \text{(A.4.6)}$$

The eigen-direction is defied with respect to the arbitrary scalar multiplier.

For known principal values and principal directions the second rank tensor can be represented as follows (spectral representation)

$$\begin{aligned} \boldsymbol{A} &= \lambda_I \boldsymbol{n}_I \otimes \boldsymbol{n}_I + \lambda_{II} \boldsymbol{n}_{II} \otimes \boldsymbol{n}_{II} + \lambda_{III} \boldsymbol{n}_{III} \otimes \boldsymbol{n}_{III} && \text{for} \quad \lambda_I \neq \lambda_{II} \neq \lambda_{III}, \\ \boldsymbol{A} &= \lambda_I (\boldsymbol{I} - \boldsymbol{n}_{III} \otimes \boldsymbol{n}_{III}) + \lambda_{III} \boldsymbol{n}_{III} \otimes \boldsymbol{n}_{III} && \text{for} \quad \lambda_I = \lambda_{II} \neq \lambda_{III}, \\ \boldsymbol{A} &= \lambda \boldsymbol{I} && \text{for} \quad \lambda_I = \lambda_{II} = \lambda_{III} = \lambda \end{aligned}$$

A.4.15 Cayley-Hamilton Theorem

Any second rank tensor satisfies the following equation

$$\boldsymbol{A}^3 - J_1(\boldsymbol{A})\boldsymbol{A}^2 + J_2(\boldsymbol{A})\boldsymbol{A} - J_3(\boldsymbol{A})\boldsymbol{I} = \boldsymbol{0}, \qquad \text{(A.4.7)}$$

where $\boldsymbol{A}^2 = \boldsymbol{A} \cdot \boldsymbol{A}$, $\boldsymbol{A}^3 = \boldsymbol{A} \cdot \boldsymbol{A} \cdot \boldsymbol{A}$. The Cayley-Hamilton theorem can be applied to compute the powers of a tensor higher than two or negative powers of a non-singular tensor. For example, the fourth power of $\boldsymbol{A}$ can be computed by multiplying (A.4.7) by $\boldsymbol{A}$

$$\boldsymbol{A}^4 - J_1(\boldsymbol{A})\boldsymbol{A}^3 + J_2(\boldsymbol{A})\boldsymbol{A}^2 - J_3(\boldsymbol{A})\boldsymbol{A} = \boldsymbol{0}$$

After eliminating the third power we get

$$\boldsymbol{A}^4 = \left(J_1^2(\boldsymbol{A}) - J_2(\boldsymbol{A})\right)\boldsymbol{A}^2 + \left(J_3(\boldsymbol{A}) - J_1(\boldsymbol{A})J_2(\boldsymbol{A})\right)\boldsymbol{A} + J_1(\boldsymbol{A})J_3(\boldsymbol{A})\boldsymbol{I}$$

For a non-singular tensor $\boldsymbol{A}$ ($\det \boldsymbol{A} \neq 0$) the inverse can be computed by multiplying (A.4.7) by $\boldsymbol{A}^{-1}$. As a result one obtains

$$\boldsymbol{A}^{-1} = \left[\boldsymbol{A}^2 - J_1(\boldsymbol{A})\boldsymbol{A} + J_2(\boldsymbol{A})\boldsymbol{I}\right] \frac{1}{J_3(\boldsymbol{A})}$$

A.4.16 Coordinates of Second Rank Tensors

Let $\boldsymbol{e}_i$ be a basis and $\boldsymbol{e}^k$ the dual basis. Any two vectors $\boldsymbol{a}$ and $\boldsymbol{b}$ can be represented as follows

$$\boldsymbol{a} = a^i \boldsymbol{e}_i = a_j \boldsymbol{e}^j, \quad \boldsymbol{b} = b^l \boldsymbol{e}_l = b_m \boldsymbol{e}^m$$

A dyad $\boldsymbol{a} \otimes \boldsymbol{b}$ has the following representations

$$\boldsymbol{a} \otimes \boldsymbol{b} = a^i b^j \boldsymbol{e}_i \otimes \boldsymbol{e}_j = a^i b_j \boldsymbol{e}_i \otimes \boldsymbol{e}^j = a_i b_j \boldsymbol{e}^i \otimes \boldsymbol{e}^j = a_i b^j \boldsymbol{e}^i \otimes \boldsymbol{e}_j$$

For the representation of a second rank tensor $\boldsymbol{A}$ one of the following four bases can be used

$$\boldsymbol{e}_i \otimes \boldsymbol{e}_j, \quad \boldsymbol{e}^i \otimes \boldsymbol{e}^j, \quad \boldsymbol{e}^i \otimes \boldsymbol{e}_j, \quad \boldsymbol{e}_i \otimes \boldsymbol{e}^j$$

With these bases one can write

$$\boldsymbol{A} = A^{ij} \boldsymbol{e}_i \otimes \boldsymbol{e}_j = A_{ij} \boldsymbol{e}^i \otimes \boldsymbol{e}^j = A^{i*}_{*j} \boldsymbol{e}_i \otimes \boldsymbol{e}^j = A^{*j}_{i*} \boldsymbol{e}^i \otimes \boldsymbol{e}_j$$

For a selected basis the coordinates of a second rank tensor can be computed as follows

$$\begin{aligned} A^{ij} &= \boldsymbol{e}_i \cdot \boldsymbol{A} \cdot \boldsymbol{e}_j, \quad & A_{ij} &= \boldsymbol{e}^i \cdot \boldsymbol{A} \cdot \boldsymbol{e}^j, \\ A^{i*}_{*j} &= \boldsymbol{e}_i \cdot \boldsymbol{A} \cdot \boldsymbol{e}^j, \quad & A^{*j}_{i*} &= \boldsymbol{e}^i \cdot \boldsymbol{A} \cdot \boldsymbol{e}_j \end{aligned}$$

A.4.17 Orthogonal Tensors

A second rank tensor $\boldsymbol{Q}$ is said to be orthogonal if it satisfies the equation

$$\boldsymbol{Q}^T \cdot \boldsymbol{Q} = \boldsymbol{I}$$

If $\boldsymbol{Q}$ operates on a vector, its length remains unchanged, i.e. let $\boldsymbol{b} = \boldsymbol{Q} \cdot \boldsymbol{a}$, then

$$|\boldsymbol{b}|^2 = \boldsymbol{b} \cdot \boldsymbol{b} = \boldsymbol{a} \cdot \boldsymbol{Q}^T \cdot \boldsymbol{Q} \cdot \boldsymbol{a} = \boldsymbol{a} \cdot \boldsymbol{a} = |\boldsymbol{a}|^2$$

Furthermore, the orthogonal tensor does not change the scalar product of two arbitrary vectors. For two vectors $\boldsymbol{a}$ and $\boldsymbol{b}$ as well as $\boldsymbol{a}' = \boldsymbol{Q} \cdot \boldsymbol{a}$ and $\boldsymbol{b}' = \boldsymbol{Q} \cdot \boldsymbol{b}$ one can calculate

$$\boldsymbol{a}' \cdot \boldsymbol{b}' = \boldsymbol{a} \cdot \boldsymbol{Q}^T \cdot \boldsymbol{Q} \cdot \boldsymbol{b} = \boldsymbol{a} \cdot \boldsymbol{b}$$

From the definition of the orthogonal tensor follows

$$\begin{aligned} &\boldsymbol{Q}^T = \boldsymbol{Q}^{-1}, \quad \boldsymbol{Q}^T \cdot \boldsymbol{Q} = \boldsymbol{Q} \cdot \boldsymbol{Q}^T = \boldsymbol{I}, \\ &\det(\boldsymbol{Q} \cdot \boldsymbol{Q}^T) = (\det \boldsymbol{Q})^2 = \det \boldsymbol{I} = 1 \quad \Rightarrow \quad \det \boldsymbol{Q} = \pm 1 \end{aligned}$$

Orthogonal tensors with $\det \boldsymbol{Q} = 1$ are called proper orthogonal or rotation tensors. The rotation tensors are widely used in the rigid body dynamics, e.g. [342], and in the theories of rods, plates and shells, e.g. [28, 36].

Any orthogonal tensor is either the rotation tensor or the composition of the rotation and the tensor $-\boldsymbol{I}$. Let $\boldsymbol{P}$ be a rotation tensor, $\det \boldsymbol{P} = 1$, then an orthogonal tensor $\boldsymbol{Q}$ with $\det \boldsymbol{Q} = -1$ can be composed by

$$\boldsymbol{Q} = (-\boldsymbol{I}) \cdot \boldsymbol{P} = \boldsymbol{P} \cdot (-\boldsymbol{I}), \quad \det \boldsymbol{Q} = \det(-\boldsymbol{I}) \det \boldsymbol{P} = -1$$

For any two orthogonal tensors $\boldsymbol{Q}_1$ and $\boldsymbol{Q}_2$ the composition $\boldsymbol{Q}_3 = \boldsymbol{Q}_1 \cdot \boldsymbol{Q}_2$ is the orthogonal tensor, too. This property is used in the theory of symmetry and symmetry groups, e.g. [239, 340]. Two important examples for orthogonal tensors are the

- rotation tensor about a fixed axis

$$\boldsymbol{Q}(\varphi\boldsymbol{m}) = \boldsymbol{m}\otimes\boldsymbol{m} + \cos\varphi(\boldsymbol{I} - \boldsymbol{m}\otimes\boldsymbol{m}) + \sin\varphi\boldsymbol{m}\times\boldsymbol{I}, \quad \det\boldsymbol{Q} = 1, \tag{A.4.8}$$

  where the unit vector $\boldsymbol{m}$ represents the axis and φ is the angle of rotation,
- reflection tensor

$$\boldsymbol{Q} = \boldsymbol{I} - 2\boldsymbol{n}\otimes\boldsymbol{n}, \quad \det\boldsymbol{Q} = -1, \tag{A.4.9}$$

  where the unit vector $\boldsymbol{n}$ represents the normal to the mirror plane.

One can prove the following identities [343]

$$(\boldsymbol{Q}\cdot\boldsymbol{a})\times(\boldsymbol{Q}\cdot\boldsymbol{b}) = \det\boldsymbol{Q}\boldsymbol{Q}\cdot(\boldsymbol{a}\times\boldsymbol{b}), \tag{A.4.10}$$

$$\boldsymbol{Q}\cdot(\boldsymbol{a}\times\boldsymbol{Q}^T) = \boldsymbol{Q}\cdot(\boldsymbol{a}\times\boldsymbol{I})\cdot\boldsymbol{Q}^T = \det\boldsymbol{Q}\,[(\boldsymbol{Q}\cdot\boldsymbol{a})\times\boldsymbol{I}] \tag{A.4.11}$$

B Elements of Tensor Analysis

B.1 Coordinate Systems

The vector $\boldsymbol{r}$ characterizing the position of a point $\boldsymbol{P}$ can be represented by use of the Cartesian coordinates x_i as follows, Fig. B.1,

$$\boldsymbol{r}(x^1,x^2,x^3) = x^1\boldsymbol{e}_1 + x^2\boldsymbol{e}_2 + x^3\boldsymbol{e}_3 = x^i\boldsymbol{e}_i$$

Instead of coordinates x^i one can introduce any triple of curvilinear coordinates q^1,q^2,q^3 by means of one-to-one transformations

$$x^k = x^k(q^1,q^2,q^3) \quad \Leftrightarrow \quad q^k = q^k(x^1,x^2,x^3)$$

It is assumed that the above transformations are continuous and continuous differentiable as many times as necessary and for the Jacobians

$$\det\left(\frac{\partial x^k}{\partial q^i}\right) \neq 0, \quad \det\left(\frac{\partial q^i}{\partial x^k}\right) \neq 0$$

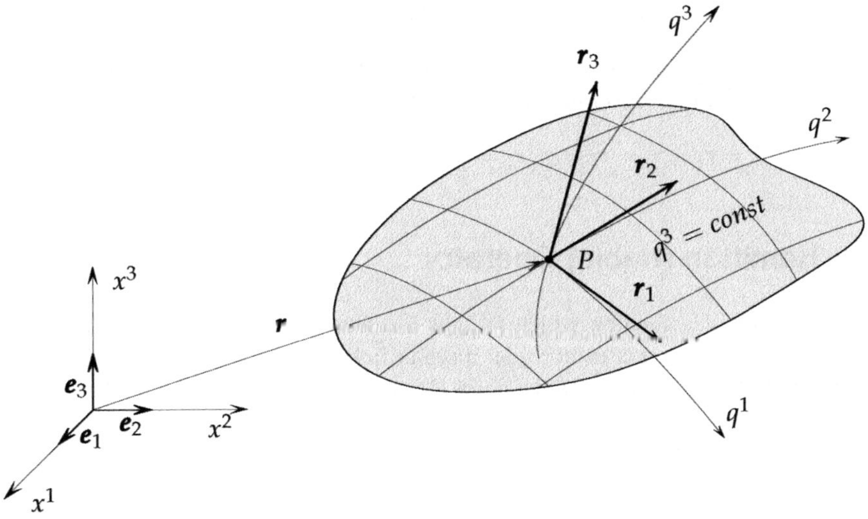

Fig. B.1 Cartesian and curvilinear coordinates

must be valid. With these assumptions the position vector can be considered as a function of curvilinear coordinates q^i, i.e. $\boldsymbol{r} = \boldsymbol{r}(q^1, q^2, q^3)$. Surfaces $q^1 = const$, $q^2 = const$, and $q^3 = const$, Fig. B.1, are called coordinate surfaces. For given fixed values $q^2 = q^2_*$ and $q^3 = q^3_*$ a curve can be obtained along which only q^1 varies. This curve is called the q^1-coordinate line, Fig. B.1. Analogously, one can obtain the q^2- and q^3-coordinate lines.

The partial derivatives of the position vector with respect the to selected coordinates

$$\boldsymbol{r}_1 = \frac{\partial \boldsymbol{r}}{\partial q^1}, \quad \boldsymbol{r}_2 = \frac{\partial \boldsymbol{r}}{\partial q^2}, \quad \boldsymbol{r}_3 = \frac{\partial \boldsymbol{r}}{\partial q^3}, \quad \boldsymbol{r}_1 \cdot (\boldsymbol{r}_2 \times \boldsymbol{r}_3) \neq 0$$

define the tangential vectors to the coordinate lines in a point P, Fig. B.1. The vectors $\boldsymbol{r}_i$ are used as the local basis in the point P. By use of (A.3.1) the dual basis $\boldsymbol{r}^k$ can be introduced. The vector $d\boldsymbol{r}$ connecting the point P with a point P' in the differential neighborhood of P is defined by

$$d\boldsymbol{r} = \frac{\partial \boldsymbol{r}}{\partial q^1} dq^1 + \frac{\partial \boldsymbol{r}}{\partial q^2} dq^2 + \frac{\partial \boldsymbol{r}}{\partial q^3} dq^3 = \boldsymbol{r}_k dq^k$$

The square of the arc length of the line element in the differential neighborhood of P is calculated by

$$ds^2 = d\boldsymbol{r} \cdot d\boldsymbol{r} = (\boldsymbol{r}_i dq^i) \cdot (\boldsymbol{r}_k dq^k) = g_{ik} dq^i dq^k,$$

where $g_{ik} \equiv \boldsymbol{r}_i \cdot \boldsymbol{r}_k$ are the so-called contravariant components of the metric tensor. With g_{ik} one can represent the basis vectors $\boldsymbol{r}_i$ by the dual basis vectors $\boldsymbol{r}^k$ as follows

$$\boldsymbol{r}_i = (\boldsymbol{r}_i \cdot \boldsymbol{r}_k)\boldsymbol{r}^k = g_{ik}\boldsymbol{r}^k$$

Similarly

$$\boldsymbol{r}^i = (\boldsymbol{r}^i \cdot \boldsymbol{r}^k)\boldsymbol{r}_k = g^{ik}\boldsymbol{r}_k, \quad g^{ik} \equiv \boldsymbol{r}^i \cdot \boldsymbol{r}^k,$$

where g^{ik} are termed covariant components of the metric tensor. For the selected bases $\boldsymbol{r}_i$ and $\boldsymbol{r}^k$ the second rank unit tensor has the following representations

$$\boldsymbol{I} = \boldsymbol{r}_i \otimes \boldsymbol{r}^i = \boldsymbol{r}_i \otimes g^{ik}\boldsymbol{r}_k = g^{ik}\boldsymbol{r}_i \otimes \boldsymbol{r}_k = g_{ik}\boldsymbol{r}^i \otimes \boldsymbol{r}^k = \boldsymbol{r}^i \otimes \boldsymbol{r}_i$$

B.2 Hamilton (Nabla) Operator

A scalar field is a function which assigns a scalar to each spatial point P for the domain of definition. Let us consider a scalar field $\varphi(\boldsymbol{r}) = \varphi(q^1, q^2, q^3)$. The total differential of φ by moving from a point P to a point P' in the differential neighborhood is

$$d\varphi = \frac{\partial \varphi}{\partial q^1} dq^1 + \frac{\partial \varphi}{\partial q^2} dq^2 + \frac{\partial \varphi}{\partial q^3} dq^3 = \frac{\partial \varphi}{\partial q^k} dq^k$$

Taking into account that $dq^k = d\boldsymbol{r} \cdot \boldsymbol{r}^k$

$$d\varphi = d\boldsymbol{r} \cdot \boldsymbol{r}^k \frac{\partial \varphi}{\partial q^k} = d\boldsymbol{r} \cdot \boldsymbol{\nabla} \varphi$$

The vector $\boldsymbol{\nabla}\varphi$ is called the gradient of the scalar field φ and the invariant operator $\boldsymbol{\nabla}$ (the Hamilton or nabla operator) is defined by

$$\boldsymbol{\nabla} = \boldsymbol{r}^k \frac{\partial}{\partial q^k}$$

For a vector field $\boldsymbol{a}(\boldsymbol{r})$ one may write

$$d\boldsymbol{a} = (d\boldsymbol{r} \cdot \boldsymbol{r}^k) \frac{\partial \boldsymbol{a}}{\partial q^k} = d\boldsymbol{r} \cdot \boldsymbol{r}^k \otimes \frac{\partial \boldsymbol{a}}{\partial q^k} = d\boldsymbol{r} \cdot \boldsymbol{\nabla} \otimes \boldsymbol{a} = (\boldsymbol{\nabla} \otimes \boldsymbol{a})^T \cdot d\boldsymbol{r},$$

$$\boldsymbol{\nabla} \otimes \boldsymbol{a} = \boldsymbol{r}^k \otimes \frac{\partial \boldsymbol{a}}{\partial q^k}$$

The gradient of a vector field is a second rank tensor. The operation $\boldsymbol{\nabla}$ can be applied to tensors of any rank. For vectors the following additional operations are defined

$$\operatorname{div} \boldsymbol{a} \equiv \boldsymbol{\nabla} \cdot \boldsymbol{a} = \boldsymbol{r}^k \cdot \frac{\partial \boldsymbol{a}}{\partial q^k},$$

$$\operatorname{rot} \boldsymbol{a} \equiv \boldsymbol{\nabla} \times \boldsymbol{a} = \boldsymbol{r}^k \times \frac{\partial \boldsymbol{a}}{\partial q^k}$$

The divergence and the rotation (curl) of tensor fields of any rank higher 1 can be calculated in a similar manner.

The following identities can be verified

$$\boldsymbol{\nabla} \otimes \boldsymbol{r} = \boldsymbol{r}^k \otimes \frac{\partial \boldsymbol{r}}{\partial q^k} = \boldsymbol{r}^k \otimes \boldsymbol{r}_k = \boldsymbol{I}, \quad \boldsymbol{\nabla} \cdot \boldsymbol{r} = 3$$

For a scalar α, a vector $\boldsymbol{a}$ and for a second rank tensor $\boldsymbol{A}$ the following identities are valid

$$\boldsymbol{\nabla}(\alpha \boldsymbol{a}) = \boldsymbol{r}^k \otimes \frac{\partial (\alpha \boldsymbol{a})}{\partial q^k} = \left(\boldsymbol{r}^k \frac{\partial \alpha}{\partial q^k} \right) \otimes \boldsymbol{a} + \alpha \boldsymbol{r}^k \otimes \frac{\partial \boldsymbol{a}}{\partial q^k} = (\boldsymbol{\nabla}\alpha) \otimes \boldsymbol{a} + \alpha \boldsymbol{\nabla} \otimes \boldsymbol{a}, \tag{B.2.1}$$

$$\begin{aligned} \boldsymbol{\nabla} \cdot (\boldsymbol{A} \cdot \boldsymbol{a}) &= \boldsymbol{r}^k \cdot \frac{\partial (\boldsymbol{A} \cdot \boldsymbol{a})}{\partial q^k} = \boldsymbol{r}^k \cdot \frac{\partial \boldsymbol{A}}{\partial q^k} \cdot \boldsymbol{a} + \boldsymbol{r}^k \cdot \boldsymbol{A} \cdot \frac{\partial \boldsymbol{a}}{\partial q^k} \\ &= (\boldsymbol{\nabla} \cdot \boldsymbol{A}) \cdot \boldsymbol{a} + \boldsymbol{A} \cdot\cdot \left(\frac{\partial \boldsymbol{a}}{\partial q^k} \otimes \boldsymbol{r}^k \right) \\ &= (\boldsymbol{\nabla} \cdot \boldsymbol{A}) \cdot \boldsymbol{a} + \boldsymbol{A} \cdot\cdot (\boldsymbol{\nabla} \otimes \boldsymbol{a})^T \end{aligned} \tag{B.2.2}$$

For a second rank tensor $\boldsymbol{A}$ and a position vector $\boldsymbol{r}$ one can prove the following identity

$$\begin{aligned}\nabla\cdot(\boldsymbol{A}\times\boldsymbol{r}) &= \boldsymbol{r}^k\cdot\frac{\partial(\boldsymbol{A}\times\boldsymbol{r})}{\partial q^k}=\boldsymbol{r}^k\cdot\frac{\partial\boldsymbol{A}}{\partial q^k}\times\boldsymbol{r}+\boldsymbol{r}^k\cdot\boldsymbol{A}\times\frac{\partial\boldsymbol{r}}{\partial q^k}\\ &= (\nabla\cdot\boldsymbol{A})\times\boldsymbol{r}+\boldsymbol{r}^k\cdot\boldsymbol{A}\times\boldsymbol{r}_k=(\nabla\cdot\boldsymbol{A})\times\boldsymbol{r}-\boldsymbol{A}_\times\end{aligned}\tag{B.2.3}$$

Here we used the definition of the vector invariant as follows

$$\boldsymbol{A}_\times=\left(\boldsymbol{r}_k\otimes\boldsymbol{r}^k\cdot\boldsymbol{A}\right)_\times=\boldsymbol{r}_k\times(\boldsymbol{r}^k\cdot\boldsymbol{A})=-\boldsymbol{r}^k\cdot\boldsymbol{A}\times\boldsymbol{r}_k$$

B.3 Integral Theorems

Let $\varphi(\boldsymbol{r})$, $\boldsymbol{a}(\boldsymbol{r})$ and $\boldsymbol{A}(\boldsymbol{r})$ be scalar, vector and second rank tensor fields. Let V be the volume of a bounded domain with a regular surface $A(V)$ and $\boldsymbol{n}$ be the outer unit normal to the surface at $\boldsymbol{r}$. The integral theorems can be summarized as follows

- Gradient Theorems

$$\begin{aligned}\int_V\nabla\varphi\,dV &= \int_{A(V)}\boldsymbol{n}\varphi\,dA,\\ \int_V\nabla\otimes\boldsymbol{a}\,dV &= \int_{A(V)}\boldsymbol{n}\otimes\boldsymbol{a}\,dA,\\ \int_V\nabla\otimes\boldsymbol{A}\,dV &= \int_{A(V)}\boldsymbol{n}\otimes\boldsymbol{A}\,dA\end{aligned}\tag{B.3.1}$$

- Divergence Theorems

$$\begin{aligned}\int_V\nabla\cdot\boldsymbol{a}\,dV &= \int_{A(V)}\boldsymbol{n}\cdot\boldsymbol{a}\,dA,\\ \int_V\nabla\cdot\boldsymbol{A}\,dV &= \int_{A(V)}\boldsymbol{n}\cdot\boldsymbol{A}\,dA\end{aligned}\tag{B.3.2}$$

- Curl Theorems

$$\begin{aligned}\int_V\nabla\times\boldsymbol{a}\,dV &= \int_{A(V)}\boldsymbol{n}\times\boldsymbol{a}\,dA,\\ \int_V\nabla\times\boldsymbol{A}\,dV &= \int_{A(V)}\boldsymbol{n}\times\boldsymbol{A}\,dA\end{aligned}\tag{B.3.3}$$

Based on the first equation in (B.3.2) and Eq. (B.2.2) the following formula can be derived

$$\int\limits_{A(V)} \boldsymbol{n} \cdot \boldsymbol{A} \cdot \boldsymbol{a}\, dA = \int\limits_{V} \boldsymbol{\nabla} \cdot (\boldsymbol{A} \cdot \boldsymbol{a}) dV = \int\limits_{V} \left[(\boldsymbol{\nabla} \cdot \boldsymbol{A}) \cdot \boldsymbol{a} + \boldsymbol{A} \cdot\cdot (\boldsymbol{\nabla} \otimes \boldsymbol{a})^T \right] dV \tag{B.3.4}$$

With the second equation in (B.3.2) and Eq. (B.2.3) the following relation can be obtained

$$\begin{aligned} \int\limits_{A(V)} \boldsymbol{r} \times (\boldsymbol{n} \cdot \boldsymbol{A}) dA &= -\int\limits_{A(V)} \boldsymbol{n} \cdot \boldsymbol{A} \times \boldsymbol{r} dA \\ &= -\int\limits_{V} \boldsymbol{\nabla} \cdot (\boldsymbol{A} \times \boldsymbol{r}) dV \\ &= \int\limits_{V} \left[\boldsymbol{r} \times (\boldsymbol{\nabla} \cdot \boldsymbol{A}) + \boldsymbol{A}_{\times} \right] dV \end{aligned} \tag{B.3.5}$$

B.4 Scalar-Valued Functions of Vectors and Second Rank Tensors

Let ψ be a scalar valued function of a vector $\boldsymbol{a}$ and a second rank tensor $\boldsymbol{A}$, i.e. $\psi = \psi(\boldsymbol{a}, \boldsymbol{A})$. Introducing a basis $\boldsymbol{e}_i$ the function ψ can be represented as follows

$$\psi(\boldsymbol{a}, \boldsymbol{A}) = \psi(a^i \boldsymbol{e}_i, A^{ij} \boldsymbol{e}_i \otimes \boldsymbol{e}_j) = \psi(a^i, A^{ij})$$

The partial derivatives of ψ with respect to $\boldsymbol{a}$ and $\boldsymbol{A}$ are defined according to the following rule

$$\begin{aligned} d\psi &= \frac{\partial \psi}{\partial a^i} da^i + \frac{\partial \psi}{\partial A^{ij}} dA^{ij} \\ &= d\boldsymbol{a} \cdot \boldsymbol{e}^i \frac{\partial \psi}{\partial a^i} + d\boldsymbol{A} \cdot\cdot \boldsymbol{e}^j \otimes \boldsymbol{e}^i \frac{\partial \psi}{\partial A^{ij}} \end{aligned} \tag{B.4.1}$$

In the coordinate-free form the above rule can be rewritten as follows

$$d\psi = d\boldsymbol{a} \cdot \frac{\partial \psi}{\partial \boldsymbol{a}} + d\boldsymbol{A} \cdot\cdot \left(\frac{\partial \psi}{\partial \boldsymbol{A}} \right)^T = d\boldsymbol{a} \cdot \psi_{,\boldsymbol{a}} + d\boldsymbol{A} \cdot\cdot (\psi_{,\boldsymbol{A}})^T \tag{B.4.2}$$

with

$$\psi_{,\boldsymbol{a}} = \frac{\partial \psi}{\partial \boldsymbol{a}} = \frac{\partial \psi}{\partial a^i} \boldsymbol{e}^i, \quad \psi_{,\boldsymbol{A}} \equiv \frac{\partial \psi}{\partial \boldsymbol{A}} = \frac{\partial \psi}{\partial A^{ij}} \boldsymbol{e}^i \otimes \boldsymbol{e}^j$$

It can be verified that $\psi_{,\boldsymbol{a}}$ and $\psi_{,\boldsymbol{A}}$ are independent from the choice of the basis. As an example let us calculate the partial derivatives of the function

$$\psi(\boldsymbol{a}, \boldsymbol{b}, \boldsymbol{A}) = \boldsymbol{a} \cdot \boldsymbol{A} \cdot \boldsymbol{b}$$

with respect to $\boldsymbol{a}$, $\boldsymbol{b}$ and $\boldsymbol{A}$. With

$$\begin{aligned} d\psi &= d\boldsymbol{a}\cdot\boldsymbol{A}\cdot\boldsymbol{b}+\boldsymbol{a}\cdot d\boldsymbol{A}\cdot\boldsymbol{b}+\boldsymbol{a}\cdot\boldsymbol{A}\cdot d\boldsymbol{b} \\ &= d\boldsymbol{a}\cdot\boldsymbol{A}\cdot\boldsymbol{b}+d\boldsymbol{A}\cdot\cdot\,(\boldsymbol{b}\otimes\boldsymbol{a})+d\boldsymbol{b}\cdot(\boldsymbol{a}\cdot\boldsymbol{A}) \\ &= d\boldsymbol{a}\cdot\psi_{,\boldsymbol{a}}+d\boldsymbol{A}\cdot\cdot\,(\psi_{,\boldsymbol{A}})^T+d\boldsymbol{b}\cdot\psi_{,\boldsymbol{b}} \end{aligned}$$

we obtain

$$\psi_{,\boldsymbol{a}}=\boldsymbol{A}\cdot\boldsymbol{b},\quad \psi_{,\boldsymbol{b}}=\boldsymbol{a}\cdot\boldsymbol{A},\quad \psi_{,\boldsymbol{A}}=\boldsymbol{a}\otimes\boldsymbol{b}$$

Let us calculate the derivatives of the functions $J_1(\boldsymbol{A}^k)=\operatorname{tr}\boldsymbol{A}^k$, $k=1,2,3$ with respect to $\boldsymbol{A}$. With

$$J_1(\boldsymbol{A})=\boldsymbol{A}\cdot\cdot\,\boldsymbol{I},\quad J_1(\boldsymbol{A}^2)=\boldsymbol{A}\cdot\cdot\,\boldsymbol{A},\quad J_1(\boldsymbol{A}^3)=\boldsymbol{A}\cdot\cdot\,(\boldsymbol{A}\cdot\boldsymbol{A})$$

we can write

$$dJ_1(\boldsymbol{A})=d\boldsymbol{A}\cdot\cdot\,\boldsymbol{I},\quad dJ_1(\boldsymbol{A}^2)=2d\boldsymbol{A}\cdot\cdot\,\boldsymbol{A},\quad dJ_1(\boldsymbol{A}^3)=3d\boldsymbol{A}\cdot\cdot\,(\boldsymbol{A}\cdot\boldsymbol{A})$$

Consequently

$$J_1(\boldsymbol{A})_{,\boldsymbol{A}}=\boldsymbol{I},\quad J_1(\boldsymbol{A}^2)_{,\boldsymbol{A}}=2\boldsymbol{A}^T,\quad J_1(\boldsymbol{A}^3)_{,\boldsymbol{A}}=3\boldsymbol{A}^{2^T} \tag{B.4.3}$$

With (B.4.3) the derivatives of principal invariants of a second rank tensor $\boldsymbol{A}$ can be calculated as follows

$$\begin{aligned} J_1(\boldsymbol{A})_{,\boldsymbol{A}} &= \boldsymbol{I}, \\ J_2(\boldsymbol{A})_{,\boldsymbol{A}} &= J_1(\boldsymbol{A})\boldsymbol{I}-\boldsymbol{A}^T, \\ J_3(\boldsymbol{A})_{,\boldsymbol{A}} &= \boldsymbol{A}^{2^T}-J_1(\boldsymbol{A})\boldsymbol{A}^T+J_2(\boldsymbol{A})\boldsymbol{I}=J_3(\boldsymbol{A})(\boldsymbol{A}^T)^{-1} \end{aligned} \tag{B.4.4}$$

To find the derivative of the function $\psi(\boldsymbol{A})=\psi(J_1(\boldsymbol{A}),J_2(\boldsymbol{A}),J_3(\boldsymbol{A}))$ with respect to $\boldsymbol{A}$ we may write

$$d\psi=d\boldsymbol{A}\cdot\cdot\left[\frac{\partial\psi}{\partial J_1}\left(J_1(\boldsymbol{A})_{,\boldsymbol{A}}\right)^T+\frac{\partial\psi}{\partial J_2}\left(J_2(\boldsymbol{A})_{,\boldsymbol{A}}\right)^T+\frac{\partial\psi}{\partial J_3}\left(J_3(\boldsymbol{A})_{,\boldsymbol{A}}\right)^T\right]$$

Taking into account (B.4.4) we obtain

$$\begin{aligned} \psi\Big(J_1(\boldsymbol{A}),J_2(\boldsymbol{A}),J_3(\boldsymbol{A})\Big)_{,\boldsymbol{A}} &= \left(\frac{\partial\psi}{\partial J_1}+J_1\frac{\partial\psi}{\partial J_2}+J_2\frac{\partial\psi}{\partial J_3}\right)\boldsymbol{I} \\ &- \left(\frac{\partial\psi}{\partial J_2}+J_1\frac{\partial\psi}{\partial J_3}\right)\boldsymbol{A}^T+\frac{\partial\psi}{\partial J_3}\boldsymbol{A}^{T2} \end{aligned} \tag{B.4.5}$$

C Orthogonal Transformations and Orthogonal Invariants

C.1 Definitions

An application of the theory of tensor functions is to find a basic set of scalar invariants for a given group of symmetry transformations, such that each invariant relative to the same group is expressible as a single-valued function of the basic set. The basic set of invariants is called functional basis. To obtain a compact representation for invariants, it is required that the functional basis is irreducible in the sense that removing any one invariant from the basis will imply that a complete representation for all the invariants is no longer possible.

Such a problem arises in the formulation of constitutive equations for a given group of material symmetries. For example, the strain energy density of an elastic non-polar material is a scalar valued function of the second rank symmetric strain tensor. In the theory of the Cosserat continuum two strain measures are introduced, where the first strain measure is a polar tensor while the second one is an axial tensor, e.g. [111]. The strain energy density of a thin elastic shell is a function of two second rank tensors and one vector, e.g. [28]. In all cases the problem is to find a minimum set of functionally independent invariants for the considered tensorial arguments.

For the theory of tensor functions we refer to [74]. Representations of tensor functions are reviewed in [287, 339]. An orthogonal transformation of a scalar α, a vector $\boldsymbol{a}$ and a second rank tensor $\boldsymbol{A}$ is defined by [28, 341]

$$\alpha' \equiv (\det \boldsymbol{Q})^{\zeta}\alpha, \quad \boldsymbol{a}' \equiv (\det \boldsymbol{Q})^{\zeta}\boldsymbol{Q}\cdot\boldsymbol{a}, \quad \boldsymbol{A}' \equiv (\det \boldsymbol{Q})^{\zeta}\boldsymbol{Q}\cdot\boldsymbol{A}\cdot\boldsymbol{Q}^T, \qquad \text{(C.1.1)}$$

where $\boldsymbol{Q}$ is an orthogonal tensor, $\zeta = 0$ for absolute (polar) scalars, vectors and tensors and $\zeta = 1$ for axial ones. An example of the axial scalar is the mixed product of three polar vectors, i.e. $\alpha = \boldsymbol{a}\cdot(\boldsymbol{b}\times\boldsymbol{c})$. A typical example of the axial vector is the cross product of two polar vectors, i.e. $\boldsymbol{c} = \boldsymbol{a}\times\boldsymbol{b}$. An example of the second rank axial tensor is the skew-symmetric tensor $\boldsymbol{W} = \boldsymbol{a}\times\boldsymbol{I}$, where $\boldsymbol{a}$ is a polar vector. Consider a group of orthogonal transformations S (e.g., the material symmetry transformations) characterized by a set of orthogonal tensors $\boldsymbol{Q}$. A scalar-valued function of a second rank tensor $f = f(\boldsymbol{A})$ is called to be an orthogonal invariant under the group S if

$$\forall \boldsymbol{Q} \in S: \quad f(\boldsymbol{A}') = (\det \boldsymbol{Q})^{\eta} f(\boldsymbol{A}), \qquad \text{(C.1.2)}$$

where $\eta = 0$ if values of f are absolute scalars and $\eta = 1$ if values of f are axial scalars.

Any second rank tensor $\boldsymbol{B}$ can be decomposed into a symmetric and a skew-symmetric part, i.e. $\boldsymbol{B} = \boldsymbol{A} + \boldsymbol{a} \times \boldsymbol{I}$, where $\boldsymbol{A}$ is a symmetric tensor and $\boldsymbol{a}$ is an associated vector. Therefore $f(\boldsymbol{B}) = f(\boldsymbol{A}, \boldsymbol{a})$. If $\boldsymbol{B}$ is a polar (axial) tensor, then $\boldsymbol{a}$ is an axial (polar) vector. For the set of symmetric second rank tensors and vectors the definition of an orthogonal invariant (C.1.2) can be generalized as follows

$$\begin{aligned}\forall \boldsymbol{Q} \in \mathcal{S}: \quad & f(\boldsymbol{A}_1', \boldsymbol{A}_2', \ldots, \boldsymbol{A}_n', \boldsymbol{a}_1', \boldsymbol{a}_2', \ldots, \boldsymbol{a}_k') \\ & = (\det \boldsymbol{Q})^\eta f(\boldsymbol{A}_1, \boldsymbol{A}_2, \ldots \boldsymbol{A}_n, \boldsymbol{a}_1, \boldsymbol{a}_2, \ldots, \boldsymbol{a}_k)\end{aligned} \qquad \text{(C.1.3)}$$

C.2 Invariants for the Full Orthogonal Group

In [344] orthogonal invariants for different sets of second rank tensors and vectors with respect to the full orthogonal group are presented. It is shown that orthogonal invariants are integrals of a generic partial differential equation (basic equations for invariants). Let us present two following examples

- Orthogonal invariants of a symmetric second rank tensor $\boldsymbol{A}$ are

$$I_k = \operatorname{tr} \boldsymbol{A}^k, \quad k = 1, 2, 3$$

  Instead of I_k it is possible to use the principal invariants J_k defined by (A.4.5).
- Orthogonal invariants of a symmetric second rank tensor $\boldsymbol{A}$ and a vector $\boldsymbol{a}$ are

$$\begin{aligned}& I_k = \operatorname{tr} \boldsymbol{A}^k, \quad k = 1, 2, 3, \quad I_4 = \boldsymbol{a} \cdot \boldsymbol{a}, \quad I_5 = \boldsymbol{a} \cdot \boldsymbol{A} \cdot \boldsymbol{a}, \\ & I_6 = \boldsymbol{a} \cdot \boldsymbol{A}^2 \cdot \boldsymbol{a}, \quad I_7 = \boldsymbol{a} \cdot \boldsymbol{A}^2 \cdot (\boldsymbol{a} \times \boldsymbol{A} \cdot \boldsymbol{a})\end{aligned} \qquad \text{(C.2.1)}$$

  In the above set of invariants only 6 are functionally independent. The relation between the invariants (so-called syzygy, [74]) can be formulated as follows

$$I_7^2 = \begin{vmatrix} I_4 & I_5 & I_6 \\ I_5 & I_6 & \boldsymbol{a} \cdot \boldsymbol{A}^3 \cdot \boldsymbol{a} \\ I_6 & \boldsymbol{a} \cdot \boldsymbol{A}^3 \cdot \boldsymbol{a} & \boldsymbol{a} \cdot \boldsymbol{A}^4 \cdot \boldsymbol{a} \end{vmatrix}, \qquad \text{(C.2.2)}$$

  where $\boldsymbol{a} \cdot \boldsymbol{A}^3 \cdot \boldsymbol{a}$ and $\boldsymbol{a} \cdot \boldsymbol{A}^4 \cdot \boldsymbol{a}$ can be expressed by I_l, $l = 1, \ldots 6$ applying the Cayley-Hamilton theorem (A.4.7).

The set of invariants for a symmetric second rank tensor $\boldsymbol{A}$ and a vector $\boldsymbol{a}$ can be applied for a non-symmetric second rank tensor $\boldsymbol{B}$ since it can be represented by $\boldsymbol{B} = \boldsymbol{A} + \boldsymbol{a} \times \boldsymbol{I}, \quad \boldsymbol{A} = \boldsymbol{A}^T$.

C.3 Invariants for the Transverse Isotropy Group

Transverse isotropy is an important type of the symmetry transformation due to a variety of applications. Transverse isotropy is usually assumed in constitutive modeling of fiber reinforced materials, e.g. [24], fiber suspensions, e.g. [25], directionally solidified alloys, e.g. [219], deep drawing sheets, e.g. [54, 61] and piezoelectric materials, e.g. [292]. The invariants and generating sets for tensor-valued functions with respect to different cases of transverse isotropy are discussed in [82, 337] (see also relevant references therein). In what follows we analyze the problem of a functional basis within the theory of linear first order partial differential equations rather than the algebra of polynomials. We develop the idea proposed in [344] for the invariants with respect to the full orthogonal group to the case of transverse isotropy. The invariants will be found as integrals of the generic partial differential equations. Although a functional basis formed by these invariants does not include any redundant element, functional relations between them may exist. It may be therefore useful to find out simple forms of such relations. We show that the proposed approach may supply results in a direct, natural manner.

C.3.1 Invariants for a Single Second Rank Symmetric Tensor

Consider the proper orthogonal tensor which represents a rotation about a fixed axis, i.e.

$$\boldsymbol{Q}(\varphi\boldsymbol{m}) = \boldsymbol{m}\otimes\boldsymbol{m} + \cos\varphi(\boldsymbol{I} - \boldsymbol{m}\otimes\boldsymbol{m}) + \sin\varphi\boldsymbol{m}\times\boldsymbol{I}, \quad \det\boldsymbol{Q}(\varphi\boldsymbol{m}) = 1, \tag{C.3.1}$$

where $\boldsymbol{m}$ is assumed to be a constant unit vector (axis of rotation) and φ denotes the angle of rotation about $\boldsymbol{m}$. The symmetry transformation defined by this tensor corresponds to the transverse isotropy, whereby five different cases are possible, e.g. [307, 340]. Let us find scalar-valued functions of a second rank symmetric tensor $\boldsymbol{A}$ satisfying the condition

$$f\big(\boldsymbol{A}'(\varphi)\big) = f\big(\boldsymbol{Q}(\varphi\boldsymbol{m})\cdot\boldsymbol{A}\cdot\boldsymbol{Q}^T(\varphi\boldsymbol{m})\big) = f(\boldsymbol{A}),\ \boldsymbol{A}'(\varphi) \equiv \boldsymbol{Q}(\varphi\boldsymbol{m})\cdot\boldsymbol{A}\cdot\boldsymbol{Q}^T(\varphi\boldsymbol{m}) \tag{C.3.2}$$

Equation (C.3.2) must be valid for any angle of rotation φ. In (C.3.2) only the left-hand side depends on φ. Therefore its derivative with respect to φ can be set to zero, i.e.

$$\frac{df}{d\varphi} = \frac{d\boldsymbol{A}'}{d\varphi}\cdot\cdot\left(\frac{\partial f}{\partial\boldsymbol{A}'}\right)^T = 0 \tag{C.3.3}$$

The derivative of $\boldsymbol{A}'$ with respect to φ can be calculated by the following rules

$$\begin{aligned} d\boldsymbol{A}'(\varphi) &= d\boldsymbol{Q}(\varphi\boldsymbol{m})\cdot\boldsymbol{A}\cdot\boldsymbol{Q}^T(\varphi\boldsymbol{m}) + \boldsymbol{Q}(\varphi\boldsymbol{m})\cdot\boldsymbol{A}\cdot d\boldsymbol{Q}^T(\varphi\boldsymbol{m}),\\ d\boldsymbol{Q}(\varphi\boldsymbol{m}) &= \boldsymbol{m}\times\boldsymbol{Q}(\varphi\boldsymbol{m})d\varphi \quad\Rightarrow\quad d\boldsymbol{Q}^T(\varphi\boldsymbol{m}) = -\boldsymbol{Q}^T(\varphi\boldsymbol{m})\times\boldsymbol{m}\,d\varphi \end{aligned} \tag{C.3.4}$$

By inserting the above equations into (C.3.3) we obtain

$$(\boldsymbol{m}\times\boldsymbol{A} - \boldsymbol{A}\times\boldsymbol{m})\cdot\cdot\left(\frac{\partial f}{\partial\boldsymbol{A}}\right)^T = 0 \tag{C.3.5}$$

Equation (C.3.5) is classified in [95] to be the linear homogeneous first order partial differential equation. The characteristic system of (C.3.5) is

$$\frac{d\boldsymbol{A}}{ds} = (\boldsymbol{m} \times \boldsymbol{A} - \boldsymbol{A} \times \boldsymbol{m}) \tag{C.3.6}$$

Any system of n linear ordinary differential equations has not more than $n-1$ functionally independent integrals [95]. By introducing a basis $\boldsymbol{e}_i$ the tensor $\boldsymbol{A}$ can be written down in the form $\boldsymbol{A} = A^{ij}\boldsymbol{e}_i \otimes \boldsymbol{e}_j$ and (C.3.6) is a system of six ordinary differential equations with respect to the coordinates A^{ij}. The five integrals of (C.3.6) may be written down as follows

$$g_i(\boldsymbol{A}) = c_i, \quad i = 1,2,\ldots,5,$$

where c_i are integration constants. Any function of the five integrals g_i is the solution of the partial differential equation (C.3.5). Therefore the five integrals g_i represent the invariants of the symmetric tensor $\boldsymbol{A}$ with respect to the symmetry transformation (C.3.1). The solutions of (C.3.6) are

$$\boldsymbol{A}^k(s) = \boldsymbol{Q}(s\boldsymbol{m}) \cdot \boldsymbol{A}_0^k \cdot \boldsymbol{Q}^T(s\boldsymbol{m}), \quad k = 1,2,3, \tag{C.3.7}$$

where $\boldsymbol{A}_0$ plays the role of the initial condition. In order to find the integrals, the variable s must be eliminated from (C.3.7). Taking into account the following identities

$$\begin{aligned} &\operatorname{tr}(\boldsymbol{Q} \cdot \boldsymbol{A}^k \cdot \boldsymbol{Q}^T) = \operatorname{tr}(\boldsymbol{Q}^T \cdot \boldsymbol{Q} \cdot \boldsymbol{A}^k) = \operatorname{tr} \boldsymbol{A}^k, \quad \boldsymbol{m} \cdot \boldsymbol{Q}(s\boldsymbol{m}) = \boldsymbol{m}, \\ &(\boldsymbol{Q} \cdot \boldsymbol{a}) \times (\boldsymbol{Q} \cdot \boldsymbol{b}) = (\det \boldsymbol{Q})\boldsymbol{Q} \cdot (\boldsymbol{a} \times \boldsymbol{b}) \end{aligned} \tag{C.3.8}$$

and using the notation $\boldsymbol{Q}_m \equiv \boldsymbol{Q}(s\boldsymbol{m})$ the integrals can be found as follows

$$\begin{aligned} \operatorname{tr}(\boldsymbol{A}^k) &= \operatorname{tr}(\boldsymbol{A}_0^k), \quad k = 1,2,3, \\ \boldsymbol{m} \cdot \boldsymbol{A}^l \cdot \boldsymbol{m} &= \boldsymbol{m} \cdot \boldsymbol{Q}_m \cdot \boldsymbol{A}_0^l \cdot \boldsymbol{Q}_m^T \cdot \boldsymbol{m} \\ &= \boldsymbol{m} \cdot \boldsymbol{A}_0^l \cdot \boldsymbol{m}, \quad l = 1,2, \\ \boldsymbol{m} \cdot \boldsymbol{A}^2 \cdot (\boldsymbol{m} \times \boldsymbol{A} \cdot \boldsymbol{m}) &= \boldsymbol{m} \cdot \boldsymbol{Q}_m \cdot \boldsymbol{A}_0^2 \cdot \boldsymbol{Q}_m^T \cdot (\boldsymbol{m} \times \boldsymbol{Q}_m \cdot \boldsymbol{A}_0 \cdot \boldsymbol{Q}_m^T \cdot \boldsymbol{m}) \\ &= \boldsymbol{m} \cdot \boldsymbol{A}_0^2 \cdot \boldsymbol{Q}_m^T \cdot \left[(\boldsymbol{Q}_m \cdot \boldsymbol{m}) \times (\boldsymbol{Q}_m \cdot \boldsymbol{A}_0 \cdot \boldsymbol{m})\right] \\ &= \boldsymbol{m} \cdot \boldsymbol{A}_0^2 \cdot (\boldsymbol{m} \times \boldsymbol{A}_0 \cdot \boldsymbol{m}) \end{aligned} \tag{C.3.9}$$

As a result we can formulate the six invariants of the tensor $\boldsymbol{A}$ with respect to the symmetry transformation (C.3.1) as follows

$$\begin{aligned} &I_k = \operatorname{tr}(\boldsymbol{A}^k), \quad k = 1,2,3, \quad I_4 = \boldsymbol{m} \cdot \boldsymbol{A} \cdot \boldsymbol{m}, \\ &I_5 = \boldsymbol{m} \cdot \boldsymbol{A}^2 \cdot \boldsymbol{m}, \quad I_6 = \boldsymbol{m} \cdot \boldsymbol{A}^2 \cdot (\boldsymbol{m} \times \boldsymbol{A} \cdot \boldsymbol{m}) \end{aligned} \tag{C.3.10}$$

The invariants with respect to various symmetry transformations are discussed in [82]. For the case of the transverse isotropy six invariants are derived in [82] by the use of another approach. In this sense our result coincides with the result given in [82]. However, from the derivations presented here it follows that only five invariants listed in (C.3.10) are functionally independent. Taking into account that I_6 is the mixed product of vectors $\boldsymbol{m}$, $\boldsymbol{A} \cdot \boldsymbol{m}$ and $\boldsymbol{A}^2 \cdot \boldsymbol{m}$ the relation between the invariants can be written down as follows

$$
I_6^2 = \det \begin{bmatrix} 1 & I_4 & I_5 \\ I_4 & I_5 & \boldsymbol{m} \cdot \boldsymbol{A}^3 \cdot \boldsymbol{m} \\ I_5 & \boldsymbol{m} \cdot \boldsymbol{A}^3 \cdot \boldsymbol{m} & \boldsymbol{m} \cdot \boldsymbol{A}^4 \cdot \boldsymbol{m} \end{bmatrix} \qquad \text{(C.3.11)}
$$

One can verify that $\boldsymbol{m} \cdot \boldsymbol{A}^3 \cdot \boldsymbol{m}$ and $\boldsymbol{m} \cdot \boldsymbol{A}^4 \cdot \boldsymbol{m}$ are transversely isotropic invariants, too. However, applying the the Cayley-Hamilton theorem (A.4.7) they can be uniquely expressed by $I_1, I_2, \ldots I_5$ in the following way [58]

$$
\begin{aligned}
\boldsymbol{m} \cdot \boldsymbol{A}^3 \cdot \boldsymbol{m} &= J_1 I_5 + J_2 I_4 + J_3, \\
\boldsymbol{m} \cdot \boldsymbol{A}^4 \cdot \boldsymbol{m} &= (J_1^2 + J_2) I_5 + (J_1 J_2 + J_3) I_4 + J_1 J_3,
\end{aligned}
$$

where J_1, J_2 and J_3 are the principal invariants of $\boldsymbol{A}$ defined by (A.4.5). Let us note that the invariant I_6 cannot be dropped. In order to verify this, it is enough to consider two different tensors

$$
\boldsymbol{A} \quad \text{and} \quad \boldsymbol{B} = \boldsymbol{Q}_n \cdot \boldsymbol{A} \cdot \boldsymbol{Q}_n^T,
$$

where

$$
\boldsymbol{Q}_n \equiv \boldsymbol{Q}(\pi \boldsymbol{n}) = 2\boldsymbol{n} \otimes \boldsymbol{n} - \boldsymbol{I}, \quad \boldsymbol{n} \cdot \boldsymbol{n} = 1, \quad \boldsymbol{n} \cdot \boldsymbol{m} = 0, \quad \det \boldsymbol{Q}_n = 1
$$

One can prove that the tensor $\boldsymbol{A}$ and the tensor $\boldsymbol{B}$ have the same invariants $I_1, I_2, \ldots, I_5$. Taking into account that $\boldsymbol{m} \cdot \boldsymbol{Q}_n = -\boldsymbol{m}$ and applying the last identity in (C.3.8) we may write

$$
\begin{aligned}
I_6(\boldsymbol{B}) &= \boldsymbol{m} \cdot \boldsymbol{B}^2 \cdot (\boldsymbol{m} \times \boldsymbol{B} \cdot \boldsymbol{m}) = \boldsymbol{m} \cdot \boldsymbol{A}^2 \cdot \boldsymbol{Q}_n^T \cdot (\boldsymbol{m} \times \boldsymbol{Q}_n \cdot \boldsymbol{A} \cdot \boldsymbol{m}) \\
&= -\boldsymbol{m} \cdot \boldsymbol{A}^2 \cdot (\boldsymbol{m} \times \boldsymbol{A} \cdot \boldsymbol{m}) = -I_6(\boldsymbol{A})
\end{aligned}
$$

We observe that the only difference between the two considered tensors is the sign of I_6. Therefore, the triples of vectors $\boldsymbol{m}$, $\boldsymbol{A} \cdot \boldsymbol{m}$, $\boldsymbol{A}^2 \cdot \boldsymbol{m}$ and $\boldsymbol{m}$, $\boldsymbol{B} \cdot \boldsymbol{m}$, $\boldsymbol{B}^2 \cdot \boldsymbol{m}$ have different orientations and cannot be combined by a rotation. It should be noted that the functional relation (C.3.11) would in no way imply that the invariant I_6 should be "dependent" and hence "redundant", namely should be removed from the basis (C.3.10). In fact, the relation (C.3.11) determines the magnitude but not the sign of I_6.

To describe yielding and failure of oriented solids a dyad $\boldsymbol{M} = \boldsymbol{v} \otimes \boldsymbol{v}$ has been used in [57, 78], where the vector $\boldsymbol{v}$ specifies a privileged direction. A plastic potential is assumed to be an isotropic function of the symmetric Cauchy stress tensor and the tensor generator $\boldsymbol{M}$. Applying the representation of isotropic functions the integrity basis including ten invariants was found. In the special case $\boldsymbol{v} = \boldsymbol{m}$ the number of invariants reduces to the five $I_1, I_2, \ldots I_5$ defined by (C.3.10). Further details of this approach and applications in continuum mechanics are given in [62, 74].

However, the problem statement to find an integrity basis of a symmetric tensor $\boldsymbol{A}$ and a dyad $\boldsymbol{M}$, i.e. to find scalar valued functions $f(\boldsymbol{A},\boldsymbol{M})$ satisfying the condition

$$\begin{aligned} & f(\boldsymbol{Q}\cdot\boldsymbol{A}\cdot\boldsymbol{Q}^T,\boldsymbol{Q}\cdot\boldsymbol{M}\cdot\boldsymbol{Q}^T) = (\det\boldsymbol{Q})^\eta f(\boldsymbol{A},\boldsymbol{M}), \\ & \forall\boldsymbol{Q}, \quad \boldsymbol{Q}\cdot\boldsymbol{Q}^T = \boldsymbol{I}, \quad \det\boldsymbol{Q} = \pm 1 \end{aligned} \tag{C.3.12}$$

essentially differs from the problem statement (C.3.2). In order to show this we take into account that the symmetry group of a dyad $\boldsymbol{M}$, i.e. the set of orthogonal solutions of the equation $\boldsymbol{Q}\cdot\boldsymbol{M}\cdot\boldsymbol{Q}^T = \boldsymbol{M}$ includes the following elements

$$\begin{aligned} \boldsymbol{Q}_{1,2} &= \pm\boldsymbol{I}, \\ \boldsymbol{Q}_3 &= \boldsymbol{Q}(\varphi\boldsymbol{m}), \quad \boldsymbol{m} = \frac{\boldsymbol{v}}{|\boldsymbol{v}|}, \\ \boldsymbol{Q}_4 &= \boldsymbol{Q}(\pi\boldsymbol{n}) = 2\boldsymbol{n}\otimes\boldsymbol{n} - \boldsymbol{I}, \quad \boldsymbol{n}\cdot\boldsymbol{n} = 1, \quad \boldsymbol{n}\cdot\boldsymbol{v} = 0, \end{aligned} \tag{C.3.13}$$

where $\boldsymbol{Q}(\varphi\boldsymbol{m})$ is defined by (C.3.1). The solutions of the problem (C.3.12) are at the same time the solutions of the following problem

$$f(\boldsymbol{Q}_i\cdot\boldsymbol{A}\cdot\boldsymbol{Q}_i^T,\boldsymbol{M}) = (\det\boldsymbol{Q}_i)^\eta f(\boldsymbol{A},\boldsymbol{M}), \quad i = 1,2,3,4,$$

i.e. the problem to find the invariants of $\boldsymbol{A}$ relative to the symmetry group (C.3.13). However, (C.3.13) includes much more symmetry elements if compared to the problem statement (C.3.2).

An alternative set of transversely isotropic invariants can be formulated by the use of the following decomposition

$$\boldsymbol{A} = \alpha\boldsymbol{m}\otimes\boldsymbol{m} + \beta(\boldsymbol{I} - \boldsymbol{m}\otimes\boldsymbol{m}) + \boldsymbol{A}_{pD} + \boldsymbol{t}\otimes\boldsymbol{m} + \boldsymbol{m}\otimes\boldsymbol{t}, \tag{C.3.14}$$

where α, β, $\boldsymbol{A}_{pD}$ and $\boldsymbol{t}$ are projections of $\boldsymbol{A}$. With the projectors $\boldsymbol{P}_1 = \boldsymbol{m}\otimes\boldsymbol{m}$ and $\boldsymbol{P}_2 = \boldsymbol{I} - \boldsymbol{m}\otimes\boldsymbol{m}$ we may write

$$\begin{aligned} \alpha &= \boldsymbol{m}\cdot\boldsymbol{A}\cdot\boldsymbol{m} = \operatorname{tr}(\boldsymbol{A}\cdot\boldsymbol{P}_1), \\ \beta &= \frac{1}{2}(\operatorname{tr}\boldsymbol{A} - \boldsymbol{m}\cdot\boldsymbol{A}\cdot\boldsymbol{m}) = \frac{1}{2}\operatorname{tr}(\boldsymbol{A}\cdot\boldsymbol{P}_2), \\ \boldsymbol{A}_{pD} &= \boldsymbol{P}_2\cdot\boldsymbol{A}\cdot\boldsymbol{P}_2 - \beta\boldsymbol{P}_2, \\ \boldsymbol{t} &= \boldsymbol{m}\cdot\boldsymbol{A}\cdot\boldsymbol{P}_2 \end{aligned} \tag{C.3.15}$$

The decomposition (C.3.14) is the analogue to the following representation of a vector $\boldsymbol{a}$

$$\boldsymbol{a} = \boldsymbol{I}\cdot\boldsymbol{a} = \boldsymbol{m}\otimes\boldsymbol{m}\cdot\boldsymbol{a} + (\boldsymbol{I} - \boldsymbol{m}\otimes\boldsymbol{m})\cdot\boldsymbol{a} = \psi\boldsymbol{m} + \boldsymbol{\tau}, \quad \psi = \boldsymbol{a}\cdot\boldsymbol{m}, \quad \boldsymbol{\tau} = \boldsymbol{P}_2\cdot\boldsymbol{a} \tag{C.3.16}$$

Decompositions of the type (C.3.14) are applied in [71, 82]. The projections introduced in (C.3.15) have the following properties

$$\operatorname{tr}(\boldsymbol{A}_{pD}) = 0, \quad \boldsymbol{A}_{pD}\cdot\boldsymbol{m} = \boldsymbol{m}\cdot\boldsymbol{A}_{pD} = \boldsymbol{0}, \quad \boldsymbol{t}\cdot\boldsymbol{m} = 0 \tag{C.3.17}$$

With (C.3.14) and (C.3.17) the tensor equation (C.3.6) can be transformed to the following system of equations

$$
\begin{cases}
\dfrac{d\alpha}{ds} = 0, \\
\dfrac{d\beta}{ds} = 0, \\
\dfrac{d\boldsymbol{A}_{pD}}{ds} = \boldsymbol{m} \times \boldsymbol{A}_{pD} - \boldsymbol{A}_{pD} \times \boldsymbol{m}, \\
\dfrac{d\boldsymbol{t}}{ds} = \boldsymbol{m} \times \boldsymbol{t}
\end{cases}
\tag{C.3.18}
$$

From the first two equations we observe that α and β are transversely isotropic invariants. The third equation can be transformed to one scalar and one vector equation as follows

$$
\frac{d\boldsymbol{A}_{pD}}{ds} \cdot\cdot \boldsymbol{A}_{pD} = 0 \quad \Rightarrow \frac{d(\boldsymbol{A}_{pD} \cdot\cdot \boldsymbol{A}_{pD})}{ds} = 0, \quad \frac{d\boldsymbol{b}}{ds} = \boldsymbol{m} \times \boldsymbol{b}
$$

with $\boldsymbol{b} \equiv \boldsymbol{A}_{pD} \cdot \boldsymbol{t}$. We observe that $\operatorname{tr}\,(\boldsymbol{A}_{pD}^2) = \boldsymbol{A}_{pD} \cdot\cdot \boldsymbol{A}_{pD}$ is a transversely isotropic invariant, too. Finally, we have to find the integrals of the following system

$$
\begin{cases}
\dfrac{d\boldsymbol{t}}{ds} = \boldsymbol{t} \times \boldsymbol{m}, \\
\dfrac{d\boldsymbol{b}}{ds} = \boldsymbol{b} \times \boldsymbol{m}
\end{cases}
\tag{C.3.19}
$$

The solutions of (C.3.19) are

$$
\boldsymbol{t}(s) = \boldsymbol{Q}(s\boldsymbol{m}) \cdot \boldsymbol{t}_0, \quad \boldsymbol{b}(s) = \boldsymbol{Q}(s\boldsymbol{m}) \cdot \boldsymbol{b}_0,
$$

where $\boldsymbol{t}_0$ and $\boldsymbol{b}_0$ play the role of initial conditions. The vectors $\boldsymbol{t}$ and $\boldsymbol{b}$ belong to the plane of isotropy, i.e. $\boldsymbol{t} \cdot \boldsymbol{m} = 0$ and $\boldsymbol{b} \cdot \boldsymbol{m} = 0$. Therefore, one can verify the following integrals

$$
\boldsymbol{t} \cdot \boldsymbol{t} = \boldsymbol{t}_0 \cdot \boldsymbol{t}_0, \quad \boldsymbol{b} \cdot \boldsymbol{b} = \boldsymbol{b}_0 \cdot \boldsymbol{b}_0, \quad \boldsymbol{t} \cdot \boldsymbol{b} = \boldsymbol{t}_0 \cdot \boldsymbol{b}_0, \quad (\boldsymbol{t} \times \boldsymbol{b}) \cdot \boldsymbol{m} = (\boldsymbol{t}_0 \times \boldsymbol{b}_0) \cdot \boldsymbol{m}
\tag{C.3.20}
$$

We found seven integrals, but only five of them are functionally independent. In order to formulate the relation between the integrals we compute

$$
\boldsymbol{b} \cdot \boldsymbol{b} = \boldsymbol{t} \cdot \boldsymbol{A}_{pD}^2 \cdot \boldsymbol{t}, \quad \boldsymbol{t} \cdot \boldsymbol{b} = \boldsymbol{t} \cdot \boldsymbol{A}_{pD} \cdot \boldsymbol{t}
$$

For any plane tensor $\boldsymbol{A}_p$ satisfying the equations $\boldsymbol{A}_p \cdot \boldsymbol{m} = \boldsymbol{m} \cdot \boldsymbol{A}_p = \boldsymbol{0}$ the Cayley-Hamilton theorem can be formulated as follows, see e.g. [74]

$$
\boldsymbol{A}_p^2 - (\operatorname{tr} \boldsymbol{A}_p)\boldsymbol{A}_p + \frac{1}{2}\left[(\operatorname{tr} \boldsymbol{A}_p)^2 - \operatorname{tr}\,(\boldsymbol{A}_p^2)\right](\boldsymbol{I} - \boldsymbol{m} \otimes \boldsymbol{m}) = \boldsymbol{0}
$$

Since $\operatorname{tr} \boldsymbol{A}_{pD} = 0$ we have

$$2\boldsymbol{A}_{pD}^2 = \text{tr}\,(\boldsymbol{A}_{pD}^2)(\boldsymbol{I} - \boldsymbol{m} \otimes \boldsymbol{m}), \quad \boldsymbol{t} \cdot \boldsymbol{A}_{pD}^2 \cdot \boldsymbol{t} = \frac{1}{2}\text{tr}\,(\boldsymbol{A}_{pD}^2)(\boldsymbol{t} \cdot \boldsymbol{t})$$

Because tr $(\boldsymbol{A}_{pD}^2)$ and $\boldsymbol{t} \cdot \boldsymbol{t}$ are already defined, the invariant $\boldsymbol{b} \cdot \boldsymbol{b}$ can be omitted. The vector $\boldsymbol{t} \times \boldsymbol{b}$ is spanned on the axis $\boldsymbol{m}$. Therefore

$$\boldsymbol{t} \times \boldsymbol{b} = \gamma \boldsymbol{m}, \quad \gamma = (\boldsymbol{t} \times \boldsymbol{b}) \cdot \boldsymbol{m},$$
$$\gamma^2 = (\boldsymbol{t} \times \boldsymbol{b}) \cdot (\boldsymbol{t} \times \boldsymbol{b}) = (\boldsymbol{t} \cdot \boldsymbol{t})(\boldsymbol{b} \cdot \boldsymbol{b}) - (\boldsymbol{t} \cdot \boldsymbol{b})^2$$

Now we can summarize six invariants and one relation between them as follows

$$\begin{aligned} &\bar{I}_1 = \alpha, \quad \bar{I}_2 = \beta, \quad \bar{I}_3 = \frac{1}{2}\text{tr}\,(\boldsymbol{A}_{pD}^2), \quad \bar{I}_4 = \boldsymbol{t} \cdot \boldsymbol{t} = \boldsymbol{t} \cdot \boldsymbol{A} \cdot \boldsymbol{m}, \\ &\bar{I}_5 = \boldsymbol{t} \cdot \boldsymbol{A}_{pD} \cdot \boldsymbol{t}, \quad \bar{I}_6 = (\boldsymbol{t} \times \boldsymbol{A}_{pD} \cdot \boldsymbol{t}) \cdot \boldsymbol{m}, \\ &\bar{I}_6^2 = \bar{I}_4^2 \bar{I}_3 - \bar{I}_5^2 \end{aligned} \qquad \text{(C.3.21)}$$

Let us assume that the symmetry transformation $\boldsymbol{Q}_n \equiv \boldsymbol{Q}(\pi \boldsymbol{n})$ belongs to the symmetry group of the transverse isotropy, as it was made in [62, 74]. In this case $f(\boldsymbol{A}') = f(\boldsymbol{Q}_n \cdot \boldsymbol{A} \cdot \boldsymbol{Q}_n^T) = f(\boldsymbol{A})$ must be valid. With $\boldsymbol{Q}_n \cdot \boldsymbol{m} = -\boldsymbol{m}$ we can write

$$\alpha' = \alpha, \quad \beta' = \beta, \quad \boldsymbol{A}'_{pD} = \boldsymbol{A}_{pD}, \quad \boldsymbol{t}' = -\boldsymbol{Q}_n \cdot \boldsymbol{t}$$

Therefore in (C.3.21) $\bar{I}'_k = \bar{I}_k$, $k = 1, 2, \ldots, 5$ and

$$\begin{aligned} \bar{I}'_6 &= (\boldsymbol{t}' \times \boldsymbol{A}'_{pD} \cdot \boldsymbol{t}') \cdot \boldsymbol{m} = ((\boldsymbol{Q}_n \cdot \boldsymbol{t}) \times \boldsymbol{Q}_n \cdot \boldsymbol{A}_{pD} \cdot \boldsymbol{t}) \cdot \boldsymbol{m} \\ &= (\boldsymbol{t} \times \boldsymbol{A}_{pD} \cdot \boldsymbol{t}) \cdot \boldsymbol{Q}_n \cdot \boldsymbol{m} = -(\boldsymbol{t} \times \boldsymbol{A}_{pD} \cdot \boldsymbol{t}) \cdot \boldsymbol{m} = -\bar{I}_6 \end{aligned}$$

Consequently

$$f(\boldsymbol{A}') = f(\bar{I}'_1, \bar{I}'_2, \ldots, \bar{I}'_5, \bar{I}'_6) = f(\bar{I}_1, \bar{I}_2, \ldots, \bar{I}_5, -\bar{I}_6)$$
$$\Rightarrow \quad f(\boldsymbol{A}) = f(\bar{I}_1, \bar{I}_2, \ldots, \bar{I}_5, \bar{I}_6^2)$$

and $\bar{I}_6^2$ can be omitted due to the last relation in (C.3.21).

C.3.2 Invariants for a Set of Vectors and Second Rank Tensors

By setting $\boldsymbol{Q} = \boldsymbol{Q}(\varphi \boldsymbol{m})$ in (C.1.3) and taking the derivative of (C.1.3) with respect to φ results in the following generic partial differential equation

$$\sum_{i=1}^{n} \left(\frac{\partial f}{\partial \boldsymbol{A}_i} \right)^T \cdot\cdot\, (\boldsymbol{m} \times \boldsymbol{A}_i - \boldsymbol{A}_i \times \boldsymbol{m}) + \sum_{j=1}^{k} \frac{\partial f}{\partial \boldsymbol{a}_j} \cdot (\boldsymbol{m} \times \boldsymbol{a}_j) = 0 \qquad \text{(C.3.22)}$$

The characteristic system of (C.3.22) is

$$\begin{cases} \dfrac{d\boldsymbol{A}_i}{ds} = (\boldsymbol{m} \times \boldsymbol{A}_i - \boldsymbol{A}_i \times \boldsymbol{m}), \quad i = 1,2,\ldots,n, \\ \dfrac{d\boldsymbol{a}_j}{ds} = \boldsymbol{m} \times \boldsymbol{a}_j, \quad j = 1,2,\ldots,k \end{cases} \tag{C.3.23}$$

The above system is a system of N ordinary differential equations, where $N = 6n + 3k$ is the total number of coordinates of $\boldsymbol{A}_i$ and $\boldsymbol{a}_j$ for a selected basis. The system (C.3.23) has not more then $N - 1$ functionally independent integrals. Therefore we can formulate:

Theorem C.3.1. *A set of n symmetric second rank tensors and k vectors with $N = 6n + 3k$ independent coordinates for a given basis has not more than $N - 1$ functionally independent invariants for $N > 1$ and one invariant for $N = 1$ with respect to the symmetry transformation $\boldsymbol{Q}(\varphi\boldsymbol{m})$.*

In essence, the proof of this theorem is given within the theory of linear first order partial differential equations [95].

As an example let us consider the set of a symmetric second rank tensor $\boldsymbol{A}$ and a vector $\boldsymbol{a}$. This set has eight independent invariants. For a visual perception it is useful to keep in mind that the considered set is equivalent to

$$\boldsymbol{A}, \quad \boldsymbol{a}, \quad \boldsymbol{A}\cdot\boldsymbol{a}, \quad \boldsymbol{A}^2\cdot\boldsymbol{a}$$

Therefore it is necessary to find the list of invariants, whose fixation determines this set as a rigid whole. The generic equation (C.3.22) takes the form

$$\left(\frac{\partial f}{\partial \boldsymbol{A}}\right)^T \cdot\cdot (\boldsymbol{m} \times \boldsymbol{A} - \boldsymbol{A} \times \boldsymbol{m}) + \frac{\partial f}{\partial \boldsymbol{a}} \cdot (\boldsymbol{m} \times \boldsymbol{a}) = 0 \tag{C.3.24}$$

The characteristic system of (C.3.24) is

$$\frac{d\boldsymbol{A}}{ds} = \boldsymbol{m} \times \boldsymbol{A} - \boldsymbol{A} \times \boldsymbol{m}, \qquad \frac{d\boldsymbol{a}}{ds} = \boldsymbol{m} \times \boldsymbol{a} \tag{C.3.25}$$

This system of ninth order has eight independent integrals. Six of them are invariants of $\boldsymbol{A}$ and $\boldsymbol{a}$ with respect to the full orthogonal group. They fix the considered set as a rigid whole. The orthogonal invariants are defined by Eqs (C.2.1) and (C.2.2).

Let us note that the invariant I_7 in (C.2.1) cannot be ignored. To verify this it is enough to consider two different sets

$$\boldsymbol{A}, \quad \boldsymbol{a} \quad \text{and} \quad \boldsymbol{B} = \boldsymbol{Q}_p \cdot \boldsymbol{A} \cdot \boldsymbol{Q}_p^T, \quad \boldsymbol{a},$$

where $\boldsymbol{Q}_p = \boldsymbol{I} - 2\boldsymbol{p} \otimes \boldsymbol{p}$, $\boldsymbol{p} \cdot \boldsymbol{p} = 1$, $\boldsymbol{p} \cdot \boldsymbol{a} = 0$. One can prove that the invariants $I_1, I_2, \ldots, I_6$ are the same for these two sets. The only difference is the invariant I_7, i.e. $\boldsymbol{a} \cdot \boldsymbol{B}^2 \cdot (\boldsymbol{a} \times \boldsymbol{B} \cdot \boldsymbol{a}) = -\boldsymbol{a} \cdot \boldsymbol{A}^2 \cdot (\boldsymbol{a} \times \boldsymbol{A} \cdot \boldsymbol{a})$ Therefore the triples of vectors $\boldsymbol{a}$, $\boldsymbol{A}\cdot\boldsymbol{a}$, $\boldsymbol{A}^2\cdot\boldsymbol{a}$ and $\boldsymbol{a}$, $\boldsymbol{B}\cdot\boldsymbol{a}$, $\boldsymbol{B}^2\cdot\boldsymbol{a}$ have different orientations and cannot be combined by a rotation. In order to fix the considered set with respect to the unit vector $\boldsymbol{m}$ it is enough to fix the next two invariants

$$I_8 = \boldsymbol{m} \cdot \boldsymbol{A} \cdot \boldsymbol{m}, \qquad I_9 = \boldsymbol{m} \cdot \boldsymbol{a} \tag{C.3.26}$$

The eight independent transversely isotropic invariants are (C.2.1), (C.2.2) and (C.3.26).

C.4 Invariants for the Orthotropic Symmetry Group

The orthogonal tensors

$$\boldsymbol{Q}_1 = 2\boldsymbol{n}_1 \otimes \boldsymbol{n}_1 - \boldsymbol{I}, \quad \boldsymbol{Q}_2 \equiv \boldsymbol{n}_2 \otimes \boldsymbol{n}_2 - \boldsymbol{I}, \quad \det \boldsymbol{Q}_1 = \det \boldsymbol{Q}_2 = 1$$

represent the rotations on the angle π about the axes $\boldsymbol{n}_1$ and $\boldsymbol{n}_2$. These tensors are the symmetry elements of the orthotropic (orthorhombic) symmetry group. Let us find the scalar-valued functions of a symmetric tensor $\boldsymbol{A}$ satisfying the following conditions

$$f(\boldsymbol{Q}_1 \cdot \boldsymbol{A} \cdot \boldsymbol{Q}_1^T) = f(\boldsymbol{Q}_2 \cdot \boldsymbol{A} \cdot \boldsymbol{Q}_2^T) = f(\boldsymbol{A}) \tag{C.4.1}$$

Replacing the tensor $\boldsymbol{A}$ by the tensor $\boldsymbol{Q}_2 \cdot \boldsymbol{A} \cdot \boldsymbol{Q}_2^T$ we find that

$$f(\boldsymbol{Q}_1 \cdot \boldsymbol{Q}_2 \cdot \boldsymbol{A} \cdot \boldsymbol{Q}_2^T \cdot \boldsymbol{Q}_1^T) = f(\boldsymbol{Q}_2 \cdot \boldsymbol{A} \cdot \boldsymbol{Q}_2^T) = f(\boldsymbol{A}) \tag{C.4.2}$$

Consequently the tensor $\boldsymbol{Q}_3 = \boldsymbol{Q}_1 \cdot \boldsymbol{Q}_2 = 2\boldsymbol{n}_3 \otimes \boldsymbol{n}_3 - \boldsymbol{I} = \boldsymbol{Q}(\pi \boldsymbol{n}_3)$ belongs to the symmetry group, where the unit vector $\boldsymbol{n}_3$ is orthogonal to $\boldsymbol{n}_1$ and $\boldsymbol{n}_2$. Consider three tensors $\boldsymbol{A}'_i$ formed from the tensor $\boldsymbol{A}$ by three symmetry transformations i.e., $\boldsymbol{A}'_i \equiv \boldsymbol{Q}_i \cdot \boldsymbol{A} \cdot \boldsymbol{Q}_i^T$. Taking into account that $\boldsymbol{Q}_i \cdot \boldsymbol{n}_i = \boldsymbol{n}_i$ (no summation over i) and $\boldsymbol{Q}_i \cdot \boldsymbol{n}_j = -\boldsymbol{n}_j$, $i \neq j$ we can write

$$\begin{aligned}
\operatorname{tr}(\boldsymbol{A}'^k_i) &= \operatorname{tr}(\boldsymbol{A}^k), \quad k = 1,2,3, \quad i = 1,2,3 \\
\boldsymbol{n}_i \cdot \boldsymbol{A}'_i \cdot \boldsymbol{n}_i &= \boldsymbol{n}_i \cdot \boldsymbol{Q}_i \cdot \boldsymbol{A} \cdot \boldsymbol{Q}_i^T \cdot \boldsymbol{n}_i \\
&= \boldsymbol{n}_i \cdot \boldsymbol{A} \cdot \boldsymbol{n}_i, \quad i = 1,2,3 \\
\boldsymbol{n}_i \cdot \boldsymbol{A}'^2_i \cdot \boldsymbol{n}_i &= \boldsymbol{n}_i \cdot \boldsymbol{Q}_i \cdot \boldsymbol{A}^2 \cdot \boldsymbol{Q}_i^T \cdot \boldsymbol{n}_i \\
&= \boldsymbol{n}_i \cdot \boldsymbol{A}^2 \cdot \boldsymbol{n}_i, \quad i = 1,2,3
\end{aligned} \tag{C.4.3}$$

The above set of includes 9 scalars. The number can be reduced to 7 due to the obvious relations

$$\operatorname{tr}(\boldsymbol{A}^k) = \boldsymbol{n}_1 \cdot \boldsymbol{A}^k \cdot \boldsymbol{n}_1 + \boldsymbol{n}_2 \cdot \boldsymbol{A}^k \cdot \boldsymbol{n}_2 + \boldsymbol{n}_3 \cdot \boldsymbol{A}^k \cdot \boldsymbol{n}_3, \quad k = 1,2$$

Therefore the orthotropic scalar-valued function of the symmetric second rank tensor can be represented as a function of the following seven arguments

$$\begin{aligned}
&I_1 = \boldsymbol{n}_1 \cdot \boldsymbol{A} \cdot \boldsymbol{n}_1, \quad I_2 = \boldsymbol{n}_2 \cdot \boldsymbol{A} \cdot \boldsymbol{n}_2, \quad I_3 = \boldsymbol{n}_3 \cdot \boldsymbol{A} \cdot \boldsymbol{n}_3, \\
&I_4 = \boldsymbol{n}_1 \cdot \boldsymbol{A}^2 \cdot \boldsymbol{n}_1, \quad I_5 = \boldsymbol{n}_2 \cdot \boldsymbol{A}^2 \cdot \boldsymbol{n}_2, \quad I_6 = \boldsymbol{n}_3 \cdot \boldsymbol{A}^2 \cdot \boldsymbol{n}_3, \quad I_7 = \operatorname{tr} \boldsymbol{A}^3
\end{aligned} \tag{C.4.4}$$

Instead of I_4, I_5, I_6 and I_7 in (C.4.4) one may use the following list of arguments [205]

$$\begin{aligned}
&J_1 = (\boldsymbol{n}_1 \cdot \boldsymbol{A} \cdot \boldsymbol{n}_2)^2, \quad J_2 = (\boldsymbol{n}_2 \cdot \boldsymbol{A} \cdot \boldsymbol{n}_3)^2, \quad J_3 = (\boldsymbol{n}_1 \cdot \boldsymbol{A} \cdot \boldsymbol{n}_3)^2, \\
&J_4 = (\boldsymbol{n}_1 \cdot \boldsymbol{A} \cdot \boldsymbol{n}_2)(\boldsymbol{n}_1 \cdot \boldsymbol{A} \cdot \boldsymbol{n}_3)(\boldsymbol{n}_2 \cdot \boldsymbol{A} \cdot \boldsymbol{n}_3)
\end{aligned} \tag{C.4.5}$$

The invariants J_1, J_2, J_3, J_4 can be uniquely expressed through $I_1, \ldots, I_7$ by use of the following relations

$$
\begin{aligned}
&I_4 = I_1^2 + J_1 + J_3, \quad I_5 = I_2^2 + J_1 + J_2, \quad I_6 = I_3^2 + J_2 + J_3, \\
&I_7 = 2I_1(I_4 - I_1^2) + 2I_2(I_5 - I_2^2) + 2I_3(I_6 - I_3^2) + J_4
\end{aligned}
\tag{C.4.6}
$$

Let us note that if $\boldsymbol{A}$ is the polar tensor, then the lists of invariants (C.4.4) and (C.4.5) are also applicable to the class of the orthotropic symmetry characterized by the following eight symmetry elements

$$
\boldsymbol{Q} = \pm \boldsymbol{n}_1 \otimes \boldsymbol{n}_1 \pm \boldsymbol{n}_2 \otimes \boldsymbol{n}_2 \pm \boldsymbol{n}_3 \otimes \boldsymbol{n}_3 \tag{C.4.7}
$$

In Sect. C.3 we derived the generic partial differential equation for the case of the transverse isotropy. Applying this approach one may find the list of functionally independent invariants among all possible invariants. Let us formulate the generic partial differential equation for the case of orthotropic symmetry. To this end let us find the scalar valued arguments of the tensor $\boldsymbol{A}$ from the following condition

$$
f(\boldsymbol{A}', \boldsymbol{n}_1' \otimes \boldsymbol{n}_1', \boldsymbol{n}_2' \otimes \boldsymbol{n}_2', \boldsymbol{n}_3' \otimes \boldsymbol{n}_3') = f(\boldsymbol{A}, \boldsymbol{n}_1 \otimes \boldsymbol{n}_1, \boldsymbol{n}_2 \otimes \boldsymbol{n}_2, \boldsymbol{n}_3 \otimes \boldsymbol{n}_3), \tag{C.4.8}
$$

where $\boldsymbol{A}' = \boldsymbol{Q} \cdot \boldsymbol{A} \cdot \boldsymbol{Q}^T$, $\boldsymbol{n}_i' = \boldsymbol{Q} \cdot \boldsymbol{n}_i$, $\forall \boldsymbol{Q}$, $\det \boldsymbol{Q} = 1$. The symmetry group of a single dyad is given by Eqs. (C.3.13). It can be shown that the symmetry group of three dyads $\boldsymbol{n}_i \otimes \boldsymbol{n}_i$ includes eight elements (C.4.7). Among all rotation tensors $\boldsymbol{Q}$ the three rotations $\boldsymbol{Q}_1$, $\boldsymbol{Q}_2$ and $\boldsymbol{Q}_3$ belong to the symmetry group of $\boldsymbol{n}_i \otimes \boldsymbol{n}_i$. Therefore Eq. (C.4.8) is equivalent to the following three equations

$$
\begin{aligned}
f(\boldsymbol{Q}_1 \cdot \boldsymbol{A} \cdot \boldsymbol{Q}_1^T, \boldsymbol{n}_1 \otimes \boldsymbol{n}_1, \boldsymbol{n}_2 \otimes \boldsymbol{n}_2, \boldsymbol{n}_3 \otimes \boldsymbol{n}_3) &= f(\boldsymbol{A}, \boldsymbol{n}_1 \otimes \boldsymbol{n}_1, \boldsymbol{n}_2 \otimes \boldsymbol{n}_2, \boldsymbol{n}_3 \otimes \boldsymbol{n}_3), \\
f(\boldsymbol{Q}_2 \cdot \boldsymbol{A} \cdot \boldsymbol{Q}_2^T, \boldsymbol{n}_1 \otimes \boldsymbol{n}_1, \boldsymbol{n}_2 \otimes \boldsymbol{n}_2, \boldsymbol{n}_3 \otimes \boldsymbol{n}_3) &= f(\boldsymbol{A}, \boldsymbol{n}_1 \otimes \boldsymbol{n}_1, \boldsymbol{n}_2 \otimes \boldsymbol{n}_2, \boldsymbol{n}_3 \otimes \boldsymbol{n}_3), \\
f(\boldsymbol{Q}_3 \cdot \boldsymbol{A} \cdot \boldsymbol{Q}_3^T, \boldsymbol{n}_1 \otimes \boldsymbol{n}_1, \boldsymbol{n}_2 \otimes \boldsymbol{n}_2, \boldsymbol{n}_3 \otimes \boldsymbol{n}_3) &= f(\boldsymbol{A}, \boldsymbol{n}_1 \otimes \boldsymbol{n}_1, \boldsymbol{n}_2 \otimes \boldsymbol{n}_2, \boldsymbol{n}_3 \otimes \boldsymbol{n}_3),
\end{aligned}
$$

Consequently, the scalar-valued arguments of $\boldsymbol{A}$ found from (C.4.8) satisfy three Eqs. (C.4.1) and (C.4.2). To derive the generic partial differential equation for invariants we follow the approach presented in [344, 345]. Let $\boldsymbol{Q}(\tau)$ be a continuous set of rotations depending on the real parameter τ. In this case

$$
\begin{aligned}
&\frac{d}{d\tau}\boldsymbol{Q}(\tau) = \boldsymbol{\omega}(\tau) \times \boldsymbol{Q}(\tau) \quad \Rightarrow \quad \frac{d}{d\tau}\boldsymbol{Q}^T(\tau) = -\boldsymbol{Q}^T(\tau) \times \boldsymbol{\omega}(\tau), \\
&\boldsymbol{Q}(0) = \boldsymbol{I}, \quad \boldsymbol{\omega}(0) = \boldsymbol{\omega}_0,
\end{aligned}
$$

where the axial vector $\boldsymbol{\omega}$ has the sense of the angular velocity of rotation. Taking the derivative of Eq. (C.4.8) with respect to τ we obtain the following partial differential equation

$$
\begin{aligned}
&(\boldsymbol{\omega} \times \boldsymbol{A}' - \boldsymbol{A}' \times \boldsymbol{\omega}) \cdot\cdot \left(\frac{\partial f}{\partial \boldsymbol{A}'}\right)^T \\
&+ \sum_{i=1}^{3} (\boldsymbol{\omega} \times \boldsymbol{n}_i' \otimes \boldsymbol{n}_i' - \boldsymbol{n}_i' \otimes \boldsymbol{n}_i' \times \boldsymbol{\omega}) \cdot\cdot \left(\frac{\partial f}{\partial \boldsymbol{n}_i' \otimes \boldsymbol{n}_i'}\right)^T = 0,
\end{aligned}
\tag{C.4.9}
$$

where $\boldsymbol{A}'(\tau) = \boldsymbol{Q}(\tau) \cdot \boldsymbol{A} \cdot \boldsymbol{Q}^T(\tau)$, $\boldsymbol{n}_i'(\tau) = \boldsymbol{Q}(\tau) \cdot \boldsymbol{n}_i$. For $\tau = 0$ Eq. (C.4.9) takes the form

$$
\begin{aligned}
&(\boldsymbol{\omega}_0 \times \boldsymbol{A} - \boldsymbol{A} \times \boldsymbol{\omega}_0) \cdot\cdot \left(\frac{\partial f}{\partial \boldsymbol{A}}\right)^T \\
&+ \sum_{i=1}^{3} (\boldsymbol{\omega}_0 \times \boldsymbol{n}_i \otimes \boldsymbol{n}_i - \boldsymbol{n}_i \otimes \boldsymbol{n}_i \times \boldsymbol{\omega}_0) \cdot\cdot \left(\frac{\partial f}{\partial \boldsymbol{n}_i \otimes \boldsymbol{n}_i}\right)^T = 0,
\end{aligned}
\qquad (C.4.10)
$$

Taking into account the following identities

$$
(\boldsymbol{a} \times \boldsymbol{A}) \cdot\cdot \boldsymbol{B} = \boldsymbol{a} \cdot (\boldsymbol{A} \cdot \boldsymbol{B})_\times, \quad \boldsymbol{B} \cdot\cdot (\boldsymbol{A} \times \boldsymbol{a}) = \boldsymbol{a} \cdot (\boldsymbol{B} \cdot \boldsymbol{A})_\times,
$$

Eq. (C.4.10) can be transformed to

$$
\begin{aligned}
\boldsymbol{\omega}_0 \cdot &\left[\boldsymbol{A} \cdot \left(\frac{\partial f}{\partial \boldsymbol{A}}\right)^T - \left(\frac{\partial f}{\partial \boldsymbol{A}}\right)^T \cdot \boldsymbol{A} \right. \\
&\left. + \sum_{i=1}^{3} \boldsymbol{n}_i \otimes \boldsymbol{n}_i \cdot \left(\frac{\partial f}{\partial \boldsymbol{n}_i \otimes \boldsymbol{n}_i}\right)^T - \sum_{i=1}^{3} \left(\frac{\partial f}{\partial \boldsymbol{n}_i \otimes \boldsymbol{n}_i}\right)^T \cdot \boldsymbol{n}_i \otimes \boldsymbol{n}_i \right]_\times = 0
\end{aligned}
$$

Because $\boldsymbol{\omega}_0$ is the arbitrary vector we obtain

$$
\begin{aligned}
&\left[\boldsymbol{A} \cdot \left(\frac{\partial f}{\partial \boldsymbol{A}}\right)^T - \left(\frac{\partial f}{\partial \boldsymbol{A}}\right)^T \cdot \boldsymbol{A} \right. \\
&\left. + \sum_{i=1}^{3} \boldsymbol{n}_i \otimes \boldsymbol{n}_i \cdot \left(\frac{\partial f}{\partial \boldsymbol{n}_i \otimes \boldsymbol{n}_i}\right)^T - \sum_{i=1}^{3} \left(\frac{\partial f}{\partial \boldsymbol{n}_i \otimes \boldsymbol{n}_i}\right)^T \cdot \boldsymbol{n}_i \otimes \boldsymbol{n}_i \right]_\times = \boldsymbol{0}
\end{aligned}
\qquad (C.4.11)
$$

The vector partial differential equation (C.4.11) corresponds to three scalar differential equations. The total number of scalar arguments of the function f is 9 including 6 components of the symmetric tensor $\boldsymbol{A}$ and three parameters (e.g. three Euler angles) characterizing three dyads $\boldsymbol{n}_i \otimes \boldsymbol{n}_i$. Each of the scalar partial differential equations in (C.4.11) reduces the number of independent arguments by one. Therefore, the total number of independent arguments is 6. It can be shown that all seven arguments presented by Eqs. (C.4.4) or Eqs. (C.4.5) satisfies (C.4.11). Because only six of them are independent, one functional relation must exist. In the case of the list (C.4.5) the functional relation is obvious. Indeed, we can write

$$
J_4^2 = J_1 J_2 J_3 \qquad (C.4.12)
$$

To derive the functional relation for the list (C.4.4) one nay apply Eqs. (C.4.6) to express $J_1, \ldots, J_4$ through $I_1, \ldots I_7$. The result should be inserted into Eq. (C.4.12).

References

1. ABAQUS, Benchmarks Manual (2006): ABAQUS, Inc.
2. ABAQUS, User Subroutines Reference Manual (2006): ABAQUS, Inc.
3. Abe, F. (2001): Creep rates and strengthening mechanisms in tungsten-strengthened 9Cr steels. Materials Science and Engineering *A319 – A321*, 770 – 773
4. Altenbach, H. (1999): Classical and nonclassical creep models. In: Altenbach, H., Skrzypek, J. (eds.) Creep and Damage in Materials and Structures. Springer, Wien, New York, pp. 45 – 95. CISM Lecture Notes No. 399
5. Altenbach, H. (2000): On the determination of transverse shear stiffnesses of orthotropic plates. J. App. Math. Phys. (ZAMP) *51*, 629 – 649
6. Altenbach, H. (2001): A generalized limit criterion with application to strength, yielding, and damage of isotropic materials. In: Lemaitre, J. (ed.) Handbook of Materials Behaviour Models. Academic Press, San Diego, pp. 175 – 186
7. Altenbach, H., Altenbach, J., Kissing, W. (2004): Mechanics of composite structural elements. Springer, Berlin
8. Altenbach, H., Altenbach, J., Naumenko, K. (1997): On the prediction of creep damage by bending of thin-walled structures. Mechanics Time Dependent Mat. *1*, 181 – 193
9. Altenbach, H., Altenbach, J., Naumenko, K. (1998): Ebene Flächentragwerke. Springer, Berlin
10. Altenbach, H., Altenbach, J., Rikards, R. (1996): Einführung in die Mechanik der Laminat- und Sandwichtragwerke. Deutscher Verlag für Grundstoffindustrie, Stuttgart
11. Altenbach, H., Altenbach, J., Schieße, P. (1990): Konzepte der Schädigungsmechanik und ihre Anwendung bei der werkstoffmechanischen Bauteilanalyse. Technische Mechanik *11*, 2, 81 – 93
12. Altenbach, H., Altenbach, J., Zolochevsky, A. (1995): Erweiterte Deformationsmodelle und Versagenskriterien der Werkstoffmechanik. Deutscher Verlag für Grundstoffindustrie, Stuttgart
13. Altenbach, H., Blumenauer, H. (1989): Grundlagen und Anwendungen der Schädigungsmechanik. Neue Hütte *34*, 6, 214 – 219
14. Altenbach, H., Breslavsky, D., Morachkovsky, O., Naumenko, K. (2000): Cyclic creep damage in thin-walled structures. J. Strain Anal. *35*, 1, 1 – 11
15. Altenbach, H., Huang, C., Naumenko, K. (2001): Modelling of creep damage under the reversed stress states considering damage activation and deactivation. Technische Mechanik *21*, 4, 273 – 282
16. Altenbach, H., Huang, C., Naumenko, K. (2002): Creep damage predictions in thin-walled structures by use of isotropic and anisotropic damage models. J. Strain Anal. *37*, 3, 265 – 275
17. Altenbach, H., Kolarow, G., Morachkovsky, O., Naumenko, K. (2000): On the accuracy of creep-damage predictions in thinwalled structures using the finite element method. Comp. Mech. *25*, 87 – 98

18. Altenbach, H., Kolarow, G., Naumenko, K. (1999): Solution of creep-damage problems for beams and rectangular plates using the Ritz and finite element method. Technische Mechanik *19*, 249 – 258
19. Altenbach, H., Kushnevsky, V., Naumenko, K. (2001): On the use of solid- and shell-type finite elements in creep-damage predictions of thinwalled structures. Arch. Appl. Mech. *71*, 164 – 181
20. Altenbach, H., Morachkovsky, O., Naumenko, K., Sichov, A. (1996): Zum Kriechen dünner Rotationsschalen unter Einbeziehung geometrischer Nichtlinearität sowie der Asymmetrie der Werkstoffeigenschaften. Forschung im Ingenieurwesen *62*, 6, 47 – 57
21. Altenbach, H., Morachkovsky, O., Naumenko, K., Sychov, A. (1997): Geometrically nonlinear bending of thin-walled shells and plates under creep-damage conditions. Arch. Appl. Mech. *67*, 339 – 352
22. Altenbach, H., Naumenko, K. (1997): Creep bending of thin-walled shells and plates by consideration of finite deflections. Comp. Mech. *19*, 490 – 495
23. Altenbach, H., Naumenko, K. (2002): Shear correction factors in creep-damage analysis of beams, plates and shells. JSME Int. J. Series A *45*, 77 – 83
24. Altenbach, H., Naumenko, K., L'vov, G.I., Pylypenko, S. (2003): Numerical estimation of the elastic properties of thin-walled structures manufactured from short-fiber reinforced thermoplastics. Mechanics of Composite Materials *39*, 221 – 234
25. Altenbach, H., Naumenko, K., Zhilin, P. (2003): A micro-polar theory for binary media with application to phase-transitional flow of fiber suspensions. Continuum Mechanics and Thermodynamics *15*, 539 – 570
26. Altenbach, H., Naumenko, K., Zhilin, P.A. (2005): A direct approach to the formulation of constitutive equations for rods and shells. In: Pietraszkiewicz, W., Szymczak, C. (eds.) Shell Structures: Theory and Applications. Taylor & Francis, Leiden, pp. 87 – 90
27. Altenbach, H., Schieße, P., Zolochevsky, A. (1991): Zum Kriechen isotroper Werkstoffe mit komplizierten Eigenschaften. Rheol. Acta *30*, 388 – 399
28. Altenbach, H., Zhilin, P.A. (1988): Osnovnye uravneniya neklassicheskoi teorii uprugikh obolochek (Basic equations of a non-classical theory of elastic shells, in Russ.). Advances in Mechanics *11*, 107 – 148
29. Altenbach, H., Zhilin, P.A. (2004): The theory of simple elastic shells. In: Kienzler, R., Altenbach, H., Ott, I. (eds.) Theories of Plates and Shells. Critical Review and New Applications. Springer, Berlin, pp. 1 – 12
30. Altenbach, H., Zolochevsky, A. (1996): A generalized failure criterion for three-dimensional behaviour of isotropic materials. Engng Fracture Mechanics *54*, 1, 75 – 90
31. Altenbach, H., Zolochevsky, A.A. (1994): Eine energetische Variante der Theorie des Kriechens und der Langzeitfestigkeit für isotrope Werkstoffe mit komplizierten Eigenschaften. ZAMM *74*, 3, 189 – 199
32. Altenbach, J., Altenbach, H. (1994): Einführung in die Kontinuumsmechanik. Teubner Studienbücher Mechanik. Teubner, Stuttgart
33. Altenbach, J., Altenbach, H., Naumenko, K. (1997): Lebensdauerabschätzung dünnwandiger Flächentragwerke auf der Grundlage phänomenologischer Materialmodelle für Kriechen und Schädigung. Technische Mechanik *17*, 4, 353 – 364
34. Altenbach, J., Altenbach, H., Naumenko, K. (2004): Egde effects in moderately thick plates under creep damage conditions. Technische Mechanik *24*, 3 – 4, 254 – 263
35. ANSYS, Inc. Theory Manual (2001): Swanson Analysis Systems, Inc.
36. Antman, S. (1995): Nonlinear Problems of Elasticity. Springer, Berlin
37. Ashby, M.F., Gandhi, C., Taplin, D.M.R. (1979): Fracture-mechanism maps and their construction for f.c.c. metals and alloys. Acta Metall. *27*, 699 – 729
38. Aurich, D., Kloos, K.H., Lange, G., Macherauch, E. (1999): Eigenspannungen und Verzug durch Wärmeeinwirkung, DFG Forschungsbericht. Wiley-VCH, Weinheim
39. Backhaus, G. (1983): Deformationsgesetze. Akademie-Verlag, Berlin

40. Bassani, J.L., Hawk, D.E. (1990): Influence of damage on crack-tip fields under small-scale-creep conditions. Int. J. Fracture *42*, 157 – 172
41. Bathe, K.J. (1996): Finite Element Rocedures. Prentice-Hall, Englewood Cliffs, New Jersey
42. Becker, A.A., Hyde, T.H., Sun, W., Andersson, P. (2002): Benchmarks for finite element analysis of creep continuum damage mechanics. Comp. Mat. Sci. *25*, 34 – 41
43. Becker, A.A., Hyde, T.H., Xia, L. (1994): Numerical analysis of creep in components. J. Strain Anal. *29*, 3, 185 – 192
44. Bellmann, R. (1970): Introduction to Matrix Analysis. McGraw Hill, New York
45. Bend Tech Inc. (2005). Induction pipe bending, standard bend tolerances. WWW page, http://www.bendtec.com/bendtol.html
46. Bernhardi, O., Mücke, R. (2000): A lifetime prediction procedure for anisotropic materials. Commmunications in Numerical Methods in Engineering *16*, 519 – 527
47. Bernhardt, E.O., Hanemann, H. (1938): Über den Kriechvorgang bei dynamischer Belastung und den Begriff der dynamischen Kriechfestigkeit. Z. für Metallkunde *30*, 12, 401 – 409
48. Bertram, A. (1989): Axiomatische Einführung in die Kontinuumsmechanik. B.I. Wissenschaftsverlag, Mannheim
49. Bertram, A. (2003): Finite thermoplasticity based on isomorphismus. Int. J. of Plasticity *19*, 2027 – 2050
50. Bertram, A. (2005): Elasticity and Plasticity of Large Deformations. Springer, Berlin
51. Bertram, A., Olschewski, J. (1996): Anisotropic modelling of the single crystal superalloy SRR99. Comp. Mat. Sci. *5*, 12 – 16
52. Bertram, A., Olschewski, J. (2001): A phenomenological anisotropic creep model for cubic single crystals. In: Lemaitre, J. (ed.) Handbook of Materials Behaviour Models. Academic Press, San Diego, pp. 303 – 307
53. Besseling, J.F., van der Giessen, E. (1994): Mathematical Modelling of Inelastic Deformation. Chapman & Hall, London
54. Betten, J. (1976): Plastic anisotropy and Bauschinger-effect - general formulation and comparison with experimental yield curves. Acta Mechanica *25*, 1 – 2, 79 – 94
55. Betten, J. (1982): Zur Aufstellung einer Integritätsbasis für Tensoren zweiter und vierter Stufe. ZAMM *62*, 5, T274 – T275
56. Betten, J. (1984): Materialgleichungen zur Beschreibung des sekundären und tertiären Kriechverhaltens anisotroper Stoffe. ZAMM *64*, 211 – 220
57. Betten, J. (1985): On the representation of the plastic potential of anisotropic solids. In: Boehler, J. (ed.) Plastic Behavior of Anisotropic Solids. CNRS, Paris, pp. 213 – 228
58. Betten, J. (1987): Tensorrechnung für Ingenieure. Springer, Berlin
59. Betten, J. (1993): Kontinuumsmechanik. Springer, Berlin
60. Betten, J. (1998): Anwendungen von Tensorfunktionen in der Kontinuumsmechanik anisotroper Materialien. ZAMM *78*, 8, 507 – 521
61. Betten, J. (2001): Kontinuumsmechanik. Springer, Berlin
62. Betten, J. (2005): Creep Mechanics. Springer, Berlin
63. Betten, J., Borrmann, M. (1987): Stationäres Kriechverhalten innendruckbelasteter dünnwandiger Kreiszylinderschalen unter Berücksichtigung des orthotropen Werkstoffverhaltens und des CSD–Effektes. Forschung im Ingenieurwesen *53*, 3, 75 – 82
64. Betten, J., Borrmann, M., Butters, T. (1989): Materialgleichungen zur Beschreibung des primären Kriechverhaltens innendruckbeanspruchter Zylinderschalen aus isotropem Werkstoff. Ingenieur-Archiv *60*, 3, 99 – 109
65. Betten, J., Butters, T. (1990): Rotationssymmetrisches Kriechbeulen dünnwandiger Kreiszylinderschalen im primären Kriechbereich. Forschung im Ingenieurwesen *56*, 3, 84 – 89

66. Betten, J., El-Magd, E., Meydanli, S.C., Palmen, P. (1995): Untersuchnung des anisotropen Kriechverhaltens vorgeschädigter Werkstoffe am austenitischen Stahl X8CrNiMoNb 1616. Arch. Appl. Mech. *65*, 121 – 132
67. Bialkiewicz, J., Kuna, H. (1996): Shear effect in rupture mechanics of middle-thick plates plates. Engng Fracture Mechanics *54*, 3, 361 – 370
68. Bielski, J., Skrzypek, J. (1989): Failure modes of elastic-plastic curved tubes under external pressure with in-plane bending. Int. J. Mech. Sci. *31*, 435 – 458
69. Billington, E.W. (1985): The Poynting-Swift effect in relation to initial and post-yield deformation. Int. J. Solids and Structures *21*, 4, 355 – 372
70. Billington, E.W. (1986): Introduction to the Mechanics and Physics of Solids. Hilger, Bristol
71. Bischoff-Beiermann, B., Bruhns, O. (1994): A physically motivated set of invariants and tensor generators in the case of transverse isotropy. Int. J. Eng. Sci. *32*, 1531 – 1552
72. Bodnar, A., Chrzanowski, M. (1991): A non-unilateral damage in creeping plates. In: Życzkowski, M. (ed.) Creep in Structures. Springer, Berlin, Heidelberg, pp. 287 – 293
73. Bodnar, A., Chrzanowski, M. (2001): Cracking of creeping structures described by means of CDM. In: Murakami, S., Ohno, N. (eds.) IUTAM Symposium on Creep in Structures. Kluwer, Dordrecht, pp. 189 – 196
74. Boehler, J.P. (ed.) (1987): Application of Tensor Functions in Solid Mechanics. Springer, Wien
75. Boehler, J.P. (1987): On a rational formulation of isotropic and anisotropic hardening. In: Boehler, J.P. (ed.) Applications of tensor functions in solid mechanics. Springer, Wien, pp. 99 – 122. CISM Lecture Notes No. 292
76. Boehler, J.P. (1987): Representations for isotropic and anisotropic non-polynomial tensor functions. In: Boehler, J.P. (ed.) Applications of tensor functions in solid mechanics. Springer, Wien, pp. 31 – 53. CISM Lecture Notes No. 292
77. Boehler, J.P. (1987): Yielding and failure of transversely isotropic solids. In: Boehler, J.P. (ed.) Applications of tensor functions in solid mechanics. Springer, Wien, pp. 67 – 97. CISM Lecture Notes No. 292
78. Boehler, J.P., Sawczuk, A. (1977): On yielding of oriented solids. Acta Mechanica *27*, 185 – 206
79. Böhlke, T. (2000): Crystallographic Texture Evolution and Elastic Anisotropy. Simulation, Modelling and Applications, PhD-Thesis. Shaker Verlag, Aachen
80. Boyle, J.T., Spence, J. (1983): Stress Analysis for Creep. Butterworth, London
81. Brebbia, C.A., Telles, J.C.T., Wrobel, L.C. (1983): Boundary Element Techniques. Springer, Berlin
82. Bruhns, O., Xiao, H., Meyers, A. (1999): On representation of yield functions for crystals, quasicrystals and transversely isotropic solids. Eur. J. Mech. A/Solids *18*, 47 – 67
83. Burlakov, A.V., Lvov, G.I., Morachkovsky, O.K. (1977): Polzuchest' tonkikh obolochek (Creep of thin shells, in Russ.). Kharkov State Univ. Publ., Kharkov
84. Burlakov, A.V., Lvov, G.I., Morachkovsky, O.K. (1981): Dlitel'naya prochnost' obolochek (Long-term strength of shells, in Russ.). Vyshcha shkola, Kharkov
85. Byrne, T.P., Mackenzie, A.C. (1966): Secondary creep of a cylindrical thin shell subject to axisymmetric loading. J. Mech. Eng. Sci. *8*, 2, 215 – 225
86. Cane, B.J. (1981): Creep fracture of dispersion strengthened low alloy ferritic steels. Acta Metall. *29*, 1581 – 1591
87. Cazacu, O., Barlat, F. (2003): Application of the theory of representation to describe yielding of anisotropic aluminium alloys. Int. J. Eng. Sci. *41*, 1367 – 1385
88. Chaboche, J.L. (1988): Continuum damage mechanics: part I - general concepts. Trans. ASME. J. Appl. Mech. *55*, 59 – 64
89. Chaboche, J.L. (1988): Continuum damage mechanics: part II - damage growth, crack initiation, and crack growth. Trans. ASME. J. Appl. Mech. *55*, 65 – 71

90. Chaboche, J.L. (1989): Constitutive equations for cyclic plasticity and cyclic viscoplasticity. Int. J. Plasticity *5*, 247 – 302
91. Chaboche, J.L. (1999): Thermodynamically founded CDM models for creep and other conditions. In: Altenbach, H., Skrzypek, J. (eds.) Creep and Damage in Materials and Structures. Springer, Wien, New York, pp. 209 – 283. CISM Lecture Notes No. 399
92. Chawla, K.K. (1987): Composite Materials. Springer, New York
93. Choudhary, B.K., Paniraj, C., Rao, K., Mannan, S.L. (2001): Creep deformation behavior and kinetic aspects of 9Cr-1Mo ferritic steel. Iron and Steel Institut of Japan (ISIJ) International *41*, 73 – 80
94. Cordebois, J., Sidoroff, F. (1983): Damage induced elastic anisotropy. In: Boehler, J.P. (ed.) Mechanical Behaviours of Anisotropic Solids. Martinus Nijhoff Publishers, Boston, pp. 761 – 774
95. Courant, R., Hilbert, D. (1989): Methods of Mathematical Physics, Vol 2. Partial Differential Equations. Wiley Interscience Publication, New York
96. Curnier, A. (1994): Computational Methods in Solid Mechanics. Kluwer, Dordrecht
97. Cvelodub, I.Y. (1991): Postulat ustoichivosti i ego prilozheniya v teorii polzuchesti metallicheskikh materialov (On the stability postulate and its apllication in the creep theory of metallic materials, in Russ.). Institute Gidrodinamiki, Novosibirsk
98. da C. Andrade, E.N. (1910): On the viscous flow of metals, and allied phenomena. Proc. R. Soc. Lond. *ALXXXIV*, 1 – 12
99. de Boer, R. (1982): Vektor und Tensorrechnung für Ingenieure. Springer, Berlin
100. Diebels, S. (2000): Mikropolare Zweiphasenmodelle: Formulierung auf der Basis der Theorie Poröser Medien. Habilitationsschrift, Universität Stuttgart, Bericht Nr. II-4
101. Dunne, F.P.E., Hayhurst, D.R. (1992): Continuum damage based constitutive equations for copper under high temperature creep and cyclic plasticity. Proc. R. Soc. Lond. *A437*, 545 – 566
102. Dyson, B.F. (1992): Material data requirements, creep damage machanisms, and predictive models. In: Larson, L.H. (ed.) High Temperature Structural Design. Mechanical Engineering Publ., London, pp. 335 – 354
103. Dyson, B.F. (1996): Mechanical testing of high-temperature materials: modelling data-scatter. In: Cahn, R.W., Evans, A.G., McLean, M. (eds.) High-temperature structural materials. Chapman & Hall, London, pp. 160 – 192
104. Dyson, B.F., McLean, M. (1998): Microstructural evolution and its effects on the creep performance of high temperature alloys. In: Strang, A., Cawley, J., Greenwood, G.W. (eds.) Microstructural Stability of Creep Resistant Alloys for High Temperature Plant Applications. Cambridge University Press, Cambridge, pp. 371 – 393
105. Dyson, B.F., McLean, M. (2001): Micromechanism-quantification for creep constitutive equations. In: Murakami, S., Ohno, N. (eds.) IUTAM Symposium on Creep in Structures. Kluwer, Dordrecht, pp. 3 – 16
106. Eggeler, G., Rametke, A., Coleman, M., Chew, B., Peter, G., Burblies, A., Hald, J., Jefferey, C., Rantala, J., deWitte, M., Mohrmann, R. (1994): Analysis of creep in a welded P91 pressure vessel. Int. J. Pres. Ves. & Piping *60*, 237 – 257
107. Ehlers, W., Ramm, E., Diebels, S., D'Addeta, G. (2003): From particle ensembles to Cosserat continua: homogenisation of contact forces towards stresses and couple stresses. Int. J. Solids and Structures *40*, 6681 – 6702
108. El-Magd, E., Betten, J., Palmen, P. (1996): Auswirkung der Schädigungsanisotropie auf die Lebensdauer von Stählen bei Zeitstandbeanspruchung. Mat.-wiss. u. Werkstofftechn. *27*, 239 – 245
109. El-Magd, E., Nicolini, G. (1999): Creep Behaviour and microstructure of dispersion-strengthened PM-aluminium materials at elevated temperatures. In: Mugharbi, H., Gottstein, G., Mecking, H., Riedel, H., Tobolski, J. (eds.) Microstructure and Mechanical Properties of Metallic High-Temperature Materials: Research Report/DFG. Wiley-VCH, Weinheim, pp. 445 – 464

110. Engeln-Müllges, G., Reutter, F. (1991): Formelsammlung zur numerischen Mathematik mit QuickBASIC-Programmen. B.I. Wissenschaftsverlag, Mannheim
111. Eringen, A.C. (1999): Microcontinuum Field Theories, volume I: Foundations and Solids. Springer, New York
112. Eringen, A.C. (2001): Microcontinuum Field Theories, volume II: Fluent Media. Springer, New York
113. Estrin, Y. (1996): Dislocation-density-related constitutive modelling. In: Krausz, A.S., Krausz, K. (eds.) Unified Constitutive Laws of Plastic Deformation. Academic Press, San Diego, pp. 69 – 104
114. Faddejew, D.K., Faddejewa, W.N. (1964): Numerische Methoden der linearen Algebra. Deutscher Verlag der Wissenschaften, Berlin
115. Faria, S.H. (2001): Mixtures with continuous diversity: general theory and application to polymer solutions. Continuum Mechanics and Thermodynamics *13*, 91 – 120
116. Faruque, M.O., Zaman, M., Hossain, M.I. (1996): Creep constitutive modelling of an aluminium alloy under multiaxial and cyclic loading. Int. J. of Plasticity *12*, 6, 761 – 780
117. Fessler, H., Hyde, T.H. (1994): The use of model materials to simulate creep behavior. J. Strain Anal. *29*, 3, 193 – 200
118. FINOW Rohrleitungssystem- und Apparatebau Serviceleistungs GmbH (2005). Induktivrohrbiegungen. http://www.finow.de/de/induktivrohrbiegungen.html
119. François, D., Pineau, A., Zaoui, A. (1998): Mechanical Behaviour of Materials, Vol. II. Kluwer Academic Publishers, Dordrecht
120. Frost, H.J., Ashby, M.F. (1982): Deformation-Mechanism Maps. Pergamon, Oxford
121. Ganczarski, A., Skrzypek, J. (2000): Damage effect on thermo-mechanical fields in a mid-thick plate. J. Theor. Appl. Mech. *38*, 2, 271 – 284
122. Ganczarski, A., Skrzypek, J. (2001): Application of the modified Murakami's anisotropic creep-damage model to 3D rotationally-symmetric problem. Technische Mechanik *21*, 4, 251 – 260
123. Ganczarski, A., Skrzypek, J. (2004): Anisotropic thermo-creep-damage in 3D thick plate vs. Reissner's approach. In: Kienzler, R., Altenbach, H., Ott, I. (eds.) Theories of Plates and Shells. Critical Review and new Applications. Springer, Berlin, pp. 39 – 44
124. Gaudig, W., Kussmaul, K., Maile, K., Tramer, M., Granacher, J., Kloos, K.H. (1999): A microstructural model to predict the multiaxial creep and damage in 12 Cr Steel Grade at 550°C. In: Mugharbi, H., Gottstein, G., Mecking, H., Riedel, H., Tobolski, J. (eds.) Microstructure and Mechanical Properties of Metallic High-Temperature Materials: Research Report/DFG. Wiley-VCH, Weinheim, pp. 192 – 205
125. Gibson, R.F. (1994): Principles of composite materials. McGraw-Hill, New York
126. Giesekus, H. (1994): Phänomenologische Rheologie. Springer, Berlin
127. Gooch, D.J. (2003): Remnant creep life prediction in ferritic materials. In: Saxena, A. (ed.) Comprehensive Structural Integrity, Vol. 5, Creep and High-Temperature Failure. Elsevier, Amsterdam, pp. 309 – 359
128. Gorev, B.V., Rubanov, V.V., Sosnin, O.V. (1979): O polzuchesti materialov s raznymi svoistvami pri rastyazhenii i szhatii (On the creep of materials with different properties in tension and compression, in Russ.). Problemy prochnosti, 7, 62 – 67
129. Hahn, H.G. (1985): Elastizitätstheorie. B.G. Teubner, Stuttgart
130. Hairer, E., Norset, S.P., Wanner, G. (1987): Solving ordinary differential equations, volume I: Nonstiff Problems. Springer, Berlin
131. Hald, J. (1998): Service performance of a 12CrMoV steam pipe steel. In: Strang, A., Cawley, J., Greenwood, G.W. (eds.) Microstructural Stability of Creep Resistant Alloys for High Temperature Plant Applications. Cambridge University Press, Cambridge, pp. 173 – 184
132. Hall, F.R., Hayhurst, D.R. (1994): Continuum damage mechanics modelling of high temperature deformation and failure in a pipe weldment. Proc. R. Soc. Lond. A *433*, 383 – 403

133. Hartmann, F. (1987): Methode der Randelemente. Springer, Berlin
134. Haupt, P. (2002): Continuum Mechanics and Theory of Materials. Springer, Berlin
135. Hayhurst, D.R. (1972): Creep rupture under multiaxial states of stress. J. Mech. Phys. Solids *20*, 381 – 390
136. Hayhurst, D.R. (1994): The use of continuum damage mechanics in creep analysis for design. J. Strain Anal. *25*, 3, 233 – 241
137. Hayhurst, D.R. (1999): Materials data bases and mechanisms-based constitutive equations for use in design. In: Altenbach, H., Skrzypek, J. (eds.) Creep and Damage in Materials and Structures. Springer, Wien, New York, pp. 285 – 348. CISM Lecture Notes No. 399
138. Hayhurst, D.R. (2001): Computational continuum damage mechanics: its use in the prediction of creep fracture in structures - past, present and future. In: Murakami, S., Ohno, N. (eds.) IUTAM Symposium on Creep in Structures. Kluwer, Dordrecht, pp. 175 – 188
139. Hayhurst, D.R., Leckie, F.A. (1990): High temperature creep continuum damage in metals. In: Boehler, J.P. (ed.) Yielding, Damage and Failure of Anisotropic Solids. Mechanical Engineering Publ., London, pp. 445 – 464
140. Hayhurst, D.R., Wong, M.T., Vakili-Tahami, F. (2002): The use of CDM analysis techniques in high temperature creep failure of welded structures. JSME Int. J. Series A *45*, 90 – 97
141. Hill, R. (1950): The Mathematical Theory of Plasticity. Materials Research and Engineering. Oxford University Press, London
142. Hult, J.A. (1966): Creep in Engineering Structures. Blaisdell Publishing Company, Waltham
143. Hutchinson, J.R. (2001): Shear coefficients for Timoshenko beam theory. Trans. ASME J. Appl. Mech. *68*, 87 – 92
144. Hyde, T.H., Sun, W., Agyakwa, P.A., Shipeay, P.H., Williams, J.A. (2003): Anisotropic creep and fracture behaviour of a 9CrMoNbV weld metal at 650°C. In: Skrzypek, J.J., Ganczarski, A.W. (eds.) Anisotropic Behaviour of Damaged Materials. Springer, Berlin, pp. 295–316
145. Hyde, T.H., Sun, W., Becker, A.A. (2000): Failure prediction for multi-material creep test specimens using steady-state creep rupture stress. Int. J. Mech. Sci. *42*, 401 – 423
146. Hyde, T.H., Sun, W., Becker, A.A., Williams, J.A. (1997): Creep continuum damage constitutive equations for the base, weld and heat-affected zone materials of a service-aged 1/2Cr1/2Mo1/4V:2 1/4Cr1Mo multipass weld at 640°C. J. Strain Anal. *32*, 4, 273 – 285
147. Hyde, T.H., Sun, W., Williams, J.A. (1999): Creep behaviour of parent, weld and HAZ materials of new, service aged and repaired 1/2Cr1/2Mo1/4V: 21/4Cr1Mo pipe welds at 640°C. Materials at High Temperatures *16*, 3, 117 – 129
148. Hyde, T.H., Sun, W., Williams, J.A. (2003): Creep analysis of pressurized circumferential pipe weldments - a review. J. Strain Anal. *38*, 1, 1 – 29
149. Hyde, T.H., Xia, L., Becker, A.A. (1996): Prediction of creep failure in aeroengine materials under multi-axial stress states. Int. J. Mech. Sci. *38*, 4, 385 – 403
150. Hyde, T.H., Yaghi, A., Becker, A.A., Earl, P.G. (2002): Finite element creep continuum damage mechanics analysis of pressurised pipe bends with ovality. JSME Int. J. Series A *45*, 1, 84 – 89
151. Inoue, T. (1988): Inelastic constitutive models under plasticity-creep interaction condition – Theories and evaluations. JSME Int. J. Series I *31*, 4, 653 – 663
152. Inoue, T., Ohno, N., Suzuki, A., Igari, T. (1989): Evaluation of inelastic constitutive models under plasticity-creep interaction for 21/4Cr-1Mo steel at 600°C. Nucl. Engng Design *114*, 259 – 309
153. Kachanov, L.M. (1958): O vremeni razrusheniya v usloviyakh polzuchesti (On the time to rupture under creep conditions, in Russ.). Izv. AN SSSR. Otd. Tekh. Nauk , 8, 26 – 31

154. Kachanov, L.M. (1969): Osnovy teorii plastichnosti (Basics of the theory of plasticity, in Russ.). Nauka, Moskva
155. Kachanov, L.M. (1986): Introduction to Continuum Damage Mechanics. Martinus Nijhoff, Dordrecht
156. Kaliszky, S. (1984): Plastizitätslehre Theorie und Technische Anwendung. VDI-Verlag, Düsseldorf
157. Kaneko, T. (1975): On Timoshenko's correction for shear in vibrating beams. J. Phys. D *8*, 1927 – 1936
158. Kassner, M.E., Hayes, T.A. (2003): Creep cavitation in metals. Int. J. of Plasticity *19*, 1715 – 1748
159. Kassner, M.E., Pérez-Prado, M.T. (2004): Fundamentals of Creep in Melals and Alloys. Elsevier, Amsterdam
160. Kawai, M. (1989): Creep and plasticity of austenitic stainless steel under multiaxial non-proportional loadings at elevated temperatures. In: Hui, D., Kozik, T.J. (eds.) Visco-plastic behavior of new materials. ASME, PVP-Vol. 184, New York, pp. 85 – 93
161. Kawai, M. (1996): Coupled creep-damage modelling for isotropic hardening and kinematic hardening materials. In: Design and life assessment at high temperature. Mechanical Engineering Publ., London, pp. 91 – 100
162. Kawai, M. (1997): Coupled inelasticity and damage model for metal matrix composites. Int. J. Damage Mechanics *6*, 453 – 478
163. Kawai, M. (2002): Constitutive modelling of creep and damage behaviors of the non-Mises type for a class of polycrystalline metals. Int. J. Damage Mechanics *11*, 223 – 245
164. Khan, A.S., Huang, S. (1995): Continuum Theory of Plasticity. Wiley, New York
165. Kloc, L., Skienička, V., Ventruba, J. (2001): Comparison of low creep properties of ferritic and austenitic creep resistant steels. Material Science and Engineering *A319-A321*, 774 – 778
166. Kolupaev, V. (2006): Dreidimensionales Kriechverhalten von Bauteilen aus unverstärkten Thermoplasten. Dissertation, Math.-Naturw.-Techn. Fakultät (ingenieurw. Bereich), Martin-Luther-Universität Halle-Wittenberg
167. Konkin, V.N., Morachkovskij, O.K. (1987): Polzuchest' i dlitel'naya prochnost' legkikh splavov, proyavlyayushchikh anizotropnye svoistva (Creep and long-term strength of light alloys with anisotropic properties, in Russ.). Problemy prochnosti, 5, 38 – 42
168. Kostenko, Y., Lvov, G., Gorash, E., Altenbach, H., Naumenko, K. (2006): Power plant component design using creep-damage analysis. In: Proc. of IMECE2006, number IMECE2006-13710. ASME, Chicago, pp. 1 – 10
169. Koundy, V., Forgeron, T., Naour, F.L. (1997): Modeling of multiaxial creep behavior for incoloy 800 tubes under internal pressure. Trans. ASME. J. Pressure Vessel & Technology *119*, 313 – 318
170. Kowalewski, Z. (1987): The surface of constant rate of energy dissipation under creep and its experimental determination. Arch. Mech. *39*, 445 – 459
171. Kowalewski, Z. (1991): Creep behaviour of copper under plane stress state. Int. J. of Plasticity *7*, 387 – 400
172. Kowalewski, Z. (1992): The role of grain size on creep of copper under uniaxial tension. Archive of Metallurgy *37*, 65 – 76
173. Kowalewski, Z.L. (1995): Experimental evaluation of the influence of the stress state type on the creep characteristics of copper at 523 K. Arch. Mech. *47*, 1, 13 – 26
174. Kowalewski, Z.L. (1996): Creep rupture of copper under complex stress state at elevated temperature. In: Design and life assessment at high temperature. Mechanical Engineering Publ., London, pp. 113 – 122
175. Kowalewski, Z.L. (2001): Assessment of the multiaxial creep data based on the isochronous creep surface concept. In: Murakami, S., Ohno, N. (eds.) IUTAM Symposium on Creep in Structures. Kluwer, Dordrecht, pp. 401 – 418

176. Kowalewski, Z.L., Hayhurst, D.R., Dyson, B.F. (1994): Mechanisms-based creep constitutive equations for an aluminium alloy. J. Strain Anal. *29*, 4, 309 – 316
177. Krajcinovic, D. (1996): Damage Mechanics. Applied Mathematics and Mechanics Vol. 41. North-Holland, Amsterdam
178. Kraus, H. (1980): Creep Analysis. John Wiley & Sons, New York
179. Krausz, A.S., Krausz, K. (1996): Unified Constitutive Laws of Plastic Deformation. Academic Press, San Diego
180. Krawietz, A. (1986): Materialtheorie. Mathematische Beschreibung des phänomenologischen thermomechanischen Verhalten. Springer, Berlin
181. Krempl, E. (1999): Creep-plasticity interaction. In: Altenbach, H., Skrzypek, J. (eds.) Creep and Damage in Materials and Structures. Springer, Wien, New York, pp. 285 – 348. CISM Lecture Notes No. 399
182. Krieg, R. (1999): Reactor Pressure Vessel Under Severe Accident Loading. Final Report of EU-Project Contract FI4S-CT95-0002. Technical report, Forschungszentrum Karlsruhe, Karlsruhe
183. Lagally, M. (1962): Vorlesungen über Vektorrechnung. Geest & Portig, Leipzig
184. Lai, W.M., Rubin, D., Krempl, E. (1993): Introduction to Continuum Mechanics. Pergamon Press, Oxford
185. Lazan, B.J. (1949): Dynamic creep and rupture properties of temperature-resistant materials under tensile fatigue stress. Proc. ASTM *49*, 757 – 787
186. Le May, I., da Silveria, T.L., Cheung-Mak, S.K.P. (1994): Uncertainties in the evaluations of high temperature damage in power stations and petrochemical plant. Int. J. of Pressure Vessels & Piping *59*, 335 – 343
187. Leal, L.G., Hinch, E.J. (1973): Theoretical studies of a suspension of rigid particles affected by Brownian couples. Rheol. Acta *12*, 127–132
188. Leckie, F.A., Hayhurst, D.R. (1977): Constitutive equations for creep rupture. Acta Metall. *25*, 1059 – 1070
189. Lemaitre, J. (1996): A Course on Damage Mechanics. Springer, Berlin
190. Lemaitre, J. (ed.) (2001): Handbook of Materials Behavior Models. Academic Press, San Diego
191. Lemaitre, J., Chaboche, J.L. (1990): Mechanics of Solid Materials. Cambridge University Press, Cambridge
192. Levinson, M. (1981): A new rectangular beam theory. J. Sound Vibr. *74*, 81 – 87
193. Lewin, G., Lehmann, B. (1977): Ergebnisse über das Spannungs-Verformungsverhalten von PVC, dargestellt an einem zylindrischen Bauelement. Wiss. Z. TH Magdeburg *21*, 415 – 422
194. Libai, A., Simmonds, J.G. (1998): The Nonlinear Theory of Elastic Shells. Cambridge University Press, Cambridge
195. Lin, J., Dunne, F.P.E., Hayhurst, D.R. (1999): Aspects of testpiece design responsible for errors in cyclic plasticity experiments. Int. J. of Damage Mechanics *8*, 109 – 137
196. Lin, T.H. (1962): Bending of a plate with nonlinear strain hardening creep. In: Hoff, N.J. (ed.) Creep in Structures. Springer, Berlin, pp. 215 – 228
197. Lippmann, H. (1993): Angewandte Tensorrechnung. Springer, Berlin
198. Liu, Y., Murakami, S. (1998): Damage localization of conventional creep damage models and proposition of a new model for creep damage analysis. JSME Int. J. Series A *41*, 57 – 65
199. Liu, Y., Murakami, S., Kageyama, Y. (1994): Mesh-dependence and stress singularity in finite element analysis of creep crack growth by continuum damage mechanics approach. Eur. J. Mech. A/Solids *35*, 3, 147 – 158
200. Lo, K.H., Christensen, R.M., Wu, E.M. (1977): A high–order theory of plate deformation. Part I: Homogeneous Plates. Trans. ASME. J. Appl. Mech. *44*, 4, 663 – 668
201. Lucas, G.E., Pelloux, R.M.N. (1981): Texture and stress state dependent creep in Zircaloy-2. Metallurgical Transactions A *12A*, 1321 – 1331

202. Lüdeke, M., Schmidt, R. (1992): Current design practice for high temperature components in fossil power plants and chemical plants in Germany. In: Larson, L.H. (ed.) High Temperature Structural Design. Mechanical Engineering Publ., London, pp. 3 – 20
203. Lundin, C.D., Liu, P., Prager, M. (2001): Creep behavior of weld heat affected zone regions for modified 9Cr-1Mo Steel. In: Proceedings of CREEP7. Japan Society of Mechanical Engineers, Tsukuba, pp. 379 – 382
204. Lurie, A. (2005): Theory of Elasticity. Foundation of Engineering Mechanics. Springer, Berlin
205. Lurie, A.I. (1990): Nonlinear Theory of Elasticity. North-Holland, Dordrecht
206. Mahnken, R. (1998): Anisotropic creep modeling based on elastic projection operators with applications to CMSX-4 superalloy. Int. J. Mech. Sci. *14*, 235 – 280
207. Malinin, N.N. (1975): Prikladnaya teoriya plastichnosti i polzuchesti (Applied theory of plasticity and creep, in Russ.). Mashinostroenie, Moskva
208. Malinin, N.N. (1981): Raschet na polzuchest' konstrukcionnykh elementov (Creep calculations of structural elements, in Russ.). Mashinostroenie, Moskva
209. Malinin, N.N., Khadjinsky, G.M. (1969): K postroeniyu teorii polzuchesti s anizotropnym uprochneniem (On the formulation of a creep theory with anisotropic hardening, in Russ.). Izv. AN SSSR. Mechanika tverdogo tela *3*, 148 – 152
210. Malinin, N.N., Khadjinsky, G.M. (1972): Theory of creep with anisotropic hardening. Int. J. Mech. Sci. *14*, 235 – 246
211. Matsui, M., Tabuchi, M., Watanabe, T., Kubo, K., Kinugawa, J., Abe, F. (2001): Degradation of creep strength in welded joint of 9%Cr steel. Iron and Steel Institut of Japan (ISIJ) International *41*, S126 – S130
212. Maugin, G.A. (1992): The Thermomechanics of Plasticity and Fracture. Cambridge University Press, Cambridge
213. Meenen, J., Altenbach, H. (2001): A consistent deduction of von Kármán-type plate theories from threedimensional non-linear continuum mechanics. Acta Mechanica *147*, 1 – 17
214. Miller, A.K. (ed.) (1987): Unified Constitutive Equations for Creep and Plasticity. Elsevier, London, New York
215. Miyazaki, N. (1987): Creep buckling analyses of circular cylindrical shells under axial compression-bifurcation buckling analysis by the finite element method. Trans. ASME. J. Pressure Vessel & Technol. *109*, 179 – 183
216. Miyazaki, N. (1988): Creep buckling analyses of circular cylindrical shell under both axial compression and internal or external pressure. Computers & Struct. *28*, 437 – 441
217. Mohrmann, R., Sester, M. (1999): A multiaxial model for the evolution of the microstructure and for the mechanical behaviour of a 12% chromium steel under creep conditions. In: Mugharbi, H., Gottstein, G., Mecking, H., Riedel, H., Tobolski, J. (eds.) Microstructure and Mechanical Properties of Metallic High-Temperature Materials: Research Report/DFG. Wiley-VCH, Weinheim, pp. 206 – 221
218. Morishita, T., Hirao, M. (1997): Creep damage modelling based on ultrasonic velocities in copper. Int. J. Solids and Structures *34*, 1169 – 1182
219. Mücke, R., Bernhardi, O. (2003): A constitutive model for anisotropic materials based on Neuber's rule. Comput. Methods Appl. Mech. Engrg. *192*, 4237 – 4255
220. Mühlhaus, H.B. (ed.) (1995): Continuum Models for Materials with Microstructure. John Wiley & Sons, Chichester
221. Murakami, S. (1983): Notion of continuum damage mechanics and its application to anisotropic creep damage theory. ASME J. Engng. Mat. Tech. *105*, 99 – 105
222. Murakami, S., Liu, Y. (1995): Mesh-dependence in local approach to creep fracture. Int. J. of Damage Mech. *4*, 230 – 250
223. Murakami, S., Ohno, N. (1981): A continuum theory of creep and creep damage. In: Ponter, A.R.S., Hayhurst, D.R. (eds.) Creep in Structures. Springer, Berlin, pp. 422 – 444

224. Murakami, S., Sanomura, Y. (1985): Creep and creep damage of copper under multi-axial states of stress. In: Sawczuk, A., Bianchi, B. (eds.) Plasticity Today – Modelling, Methods and Applications. Elsevier, London, New York, pp. 535 – 551
225. Murakami, S., Sanomura, Y., Hattori, M. (1986): Modelling of the coupled effect of plastic damage and creep damage in Nimonic 80A. Int. J. Solids and Structures *22*, 4, 373 – 386
226. Murakami, S., Suzuki, K. (1971): On the creep analysis of pressurized circular cylindrical shells. Int. J. Non-Linear Mechanics *6*, 377 – 392
227. Murakami, S., Suzuki, K. (1973): Application of the extended Newton method to the creep analysis of shells of revolution. Ing.-Arch. *42*, 194 – 207
228. Nabarro, F.R.N., de Villiers, H.L. (1995): The Physics of Creep. Creep and Creep-resistant Alloys. Taylor & Francis, London
229. Naghdi, P.M. (1990): A critical review of the state of finite plasticity. J. App. Math. Phys. (ZAMP) *41*, 316 – 394
230. Naumenko, K. (1996): Modellierung und Berechnung der Langzeitfestigkeit dünnwandiger Flächentragwerke unter Berücksichtigung von Werkstoffkriechen und Schädigung. Dissertation, Fakultät für Maschinenbau, Otto-von-Guericke-Universität Magdeburg
231. Naumenko, K. (2000): On the use of the first order shear deformation models of beams, plates and shells in creep lifetime estimations. Technische Mechanik *20*, 3, 215 – 226
232. Naumenko, K. (2006): Modeling of High-Temperature Creep for Structural Analysis Applications. Habilitationschrift, Mathematisch-Naturwissenschaftlich-Technische Fakultät, Martin-Luther-Universität Halle-Wittenberg
233. Naumenko, K., Altenbach, H. (2005): A phenomenological model for anisotropic creep in a multi-pass weld metal. Arch. Appl. Mech. *75*, 808 – 819
234. Naumenko, K., Altenbach, J., Altenbach, H., Naumenko, V.K. (2001): Closed and approximate analytical solutions for rectangular Mindlin plates. Acta Mechanica *147*, 153 – 172
235. Nikitenko, A.F. (1984): Eksperimental'noe obosnovanie gipotezy suschestvovaniya poverkhnosti polzuchesti v usloviyakh slozhnogo nagruzheniya, (Experimental justification of the hypothsis on existence of the creep surface under complex loading conditions, in Russ.). Problemy prochnosti, 8, 3 – 8
236. Niu, L., Kobayashi, M., Takaku, H. (2002): Creep rupture properties of an austenitic steel with high ductility under multi-axial stresses. Iron and Steel Institut of Japan (ISIJ) International *42*, 1156 – 1181
237. Noll, W. (1972): A new mathematical theory of simple materials. Arch. Rat. Mech. Anal. *48*, 1 – 50
238. Nowacki, W. (1986): Theory of Asymmetric Elasticity. Pergamon Press, Oxford
239. Nye, J.F. (1992): Physical Properties of Crystals. Oxford Science Publications, Oxford
240. Odqvist, F.K.G. (1962): Applicability of the elastic analogue to creep problems of plates, membranes and beams. In: Hoff, N.J. (ed.) Creep in Structures. Springer, Berlin, pp. 137 – 160
241. Odqvist, F.K.G. (1974): Mathematical Theory of Creep and Creep Rupture. Oxford University Press, Oxford
242. Odqvist, F.K.G. (1981): Historical survay of the development of creep mechanics from its beginnings in the last century to 1970. In: Ponter, A.R.S., Hayhurst, D.R. (eds.) Creep in Structures. Springer, Berlin, pp. 1 – 12
243. Odqvist, F.K.G., Hult, J. (1962): Kriechfestigkeit metallischer Werkstoffe. Springer, Berlin u.a.
244. Ohno, N. (1990): Recent topics in constitutive modelling of cyclic plasticity and viscoplasticity. Appl. Mech. Rev. *43*, 283 – 295

245. Ohno, N., Kawabata, M., Naganuma, J. (1990): Aging effects on monotonic, stress-paused, and alternating creep of type 304 stainless steel. Int. J. of Plasticity *6*, 315 – 327
246. Ohno, N., Takeuchi, T. (1994): Anisotropy in multiaxial creep of nickel-based single-crystal superalloy CMSX-2 (experiments and identification of active slip systems). JSME Int. J. Series A *37*, 129 – 137
247. Onat, E.T., Leckie, F.A. (1988): Representation of mechanical behavior in the presence of changing internal structure. Trans. ASME. J. Appl. Mech. *55*, 1 – 10
248. Onck, P.R., Nguyen, B.N., der Giessen, E.V. (2000): Microstructural modelling of creep fracture in polycrystalline materials. In: Murakami, S., Ohno, N. (eds.) Creep in Structures. Kluwer Academic Publishers, Dordrecht, pp. 51 – 64
249. Orlová, A., Buršík, J., Kuchařová, K., Sklenička, V. (1998): Microstructural development during high temperature creep of 9% Cr steel. Material Science and Engineering *A245*, 39 – 48
250. Othman, A.M., Dyson, B.F., Hayhurst, D.R., Lin, J. (1994): Continuum damage mechanics modelling of circumferentially notched tension bars undergoing tertiary creep with physically-based constitutive equations. Acta Metall. Mater. *42*, 3, 597 – 611
251. Othman, A.M., Hayhurst, D.R., Dyson, B.F. (1993): Skeletal point stresses in circumferentially notched tension bars undergoing tertiary creep modelled with physically-based constitutive equations. Proc. R. Soc. Lond. *A441*, 343 – 358
252. Oytana, C., Delobelle, P., Mermet, A. (1982): Constitutive equations study in biaxial stress experimants. Trans. ASME. J. Engng Mat. & Techn. *104*, 3, 1 – 11
253. Palmov, V. (1998): Vibrations in Elasto-Plastic Bodies. Springer, Berlin
254. Peirce, D., Shih, C.F., Needleman, A. (1984): A tangent modulus method for rate dependent solids. Comp. & Structures *18*, 875 – 887
255. Penkalla, H.J., Schubert, F., Nickel, H. (1988): Torsional creep of alloy 617 tubes at elevated temperature. In: Reichman, S., Duhl, D.N., Maurer, G., Antolovich, S., Lund, C. (eds.) Superalloys 1988. The Metallurgical Society, pp. 643 – 652
256. Penny, R.K. (1964): Axisymmetrical bending of the general shell of revolution during creep. J. Mech. Eng. Sci. *6*, 44 – 45
257. Penny, R.K., Mariott, D.L. (1995): Design for Creep. Chapman & Hall, London
258. Perrin, I.J., Hayhurst, D.R. (1994): Creep constitutive equations for a 0.5Cr-0.5Mo-0.25V ferritic steel in the temperature range 600-675°C. J. Strain Anal. *31*, 4, 299 – 314
259. Perrin, I.J., Hayhurst, D.R. (1999): Continuum damage mechanics analyses of type IV creep failure in ferritic steel crossweld specimens. Int. J. Pressure Vessels Piping *76*, 599 – 617
260. Pilkey, W.D., Wunderlich, W. (1994): Mechanics of Structures: Variational and Computational Methods. CRC Press, Boca Raton
261. Pintschovius, L., Gering, E., Munz, D., Fett, T., Soubeyroux, J.L. (1989): Determination of non-symmetric secondary creep behaviour of ceramics by residual stress measurements using neutron diffractometry. J. Mater. Sci. Lett. *8*, 7, 811 – 813
262. Podgorny, A.N., Bortovoj, V.V., Gontarovsky, P.P., Kolomak, V.D., Lvov, G.I., Matyukhin, Y.J., Morachkovsky, O.K. (1984): Polzuchest' elementov mashinostroitel'nykh konstrykcij (Creep of mashinery structural members, in Russ.). Naukova dumka, Kiev
263. Powell, P.C. (1994): Engineering with Fiber-reinforced Laminates. Chapman & Hall, London
264. Prager, W. (1956): A new method of analyzing stresses and strains in work-hardening plastic solids. J. Appl. Mech. *23*, 4, 493 – 496
265. Qi, W. (1998): Modellierung der Kriechschädigung einkristalliner Superlegierungen im Hochtemperaturbereich, Fortschrittberichte VDI, Reihe 18, Nr. 230. VDI Verlag, Düsseldorf

266. Qi, W., Bertram, A. (1997): Anisotropic creep damage modelling of single crystal superalloys. Technische Mechanik *17*, 4, 313 – 322
267. Qi, W., Bertram, A. (1998): Damage modeling of the single crystal superalloy SRR99 under monotonous creep. Comp. Mater. Sci. *13*, 132 – 141
268. Qi, W., Bertram, A. (1999): Anisotropic continuum damage modeling for single crystals at high temperatures. Int. J. of Plasticity *15*, 1197 – 1215
269. Rabotnov, Y.N. (1959): O mechanizme dlitel'nogo razrusheniya (A mechanism of the long term fracture, in Russ.). Voprosy prochnosti materialov i konstruktsii, AN SSSR, 5 – 7
270. Rabotnov, Y.N. (1963): O razrushenii vsledstvie polzuchesti (On fracture as a consequence of creep, in Russ.). Prikladnaya mekhanika i tekhnicheskaya fizika *2*, 113 – 123
271. Rabotnov, Y.N. (1967): Vliyanie kontsentratsii napryazhenij na dlitel'nuyu prochnost' (Influence of stress concentration on the long-term strength, in Russ.). Prikladnaya mekhanika i tekhnicheskaya fizika *3*, 36 – 41
272. Rabotnov, Y.N. (1969): Creep Problems in Structural Members. North-Holland, Amsterdam
273. Rabotnov, Y.N. (1979): Mekhanika deformiruemogo tverdogo tela (Mechanics of the deformable solid, in Russ.). Nauka, Moskva
274. Reddy, J.N. (1984): A simple higher-order theory for laminated composite plate. Trans. ASME. J. Appl. Mech. *51*, 745 – 752
275. Reddy, J.N. (1997): Mechanics of Laminated Composite Plates: Theory and Analysis. CRC Press, Boca Raton
276. Reiner, M. (1969): Deformation and Flow. An Elementary Introduction to Rheology. 3rd edition. H.K. Lewis & Co., London
277. Reissner, E. (1950): A variational theorem in elasticity. J. Math. Phys. *29*, 90 – 95
278. Riedel, H. (1987): Fracture at High Temperatures. Materials Research and Engineering. Springer, Berlin
279. Robinson, D.N. (1984): Constitutive relationships for anisotropic high-temperature alloys. Nuclear Engng Design *83*, 389 – 396
280. Robinson, D.N., Binienda, W.K., Ruggles, M.B. (2003): Creep of polymer matrix composites. I: Norton/Bailey Creep Law for transverse isotropy. Trans. ASCE. J. Engng Mech. *129*, 3, 310 – 317
281. Robinson, D.N., Binienda, W.K., Ruggles, M.B. (2003): Creep of polymer matrix composites. II: Monkman-Grant failure relationship for transverse isotropy. Trans. ASCE. J. Engng Mech. *129*, 3, 318 – 323
282. Roche, R.L., Townley, C.H.A., Regis, V., Hübel, H. (1992): Structural analysis and available knowledge. In: Larson, L.H. (ed.) High Temperature Structural Design. Mechanical Engineering Publ., London, pp. 161 – 180
283. Rodin, G.J. (2000): Continuum damage mechanics and creep analysis. Trans. ASME. J. Appl. Mech. *67*, 193 – 196
284. Rodin, G.J., Parks, D.M. (1986): Constitutive models of a power-law matrix containing aligned penny-shaped cracks. Mechanics of Materials *5*, 221 – 228
285. Rodin, G.J., Parks, D.M. (1988): A self-consistent analysis of a creeping matrix with aligned cracks. J. Mech. Phys. Solids *36*, 237 – 249
286. Rogers, T.G. (1990): Yield criteria, flow rules and hardening in anisotropic plasticity. In: Boehler, J.P. (ed.) Yielding, Damage and Failure of Anisotropic Solids. Mechanical Engineering Publ., London, pp. 53 – 79
287. Rychlewski, J., Zhang, J. (1991): On representation of tensor functions: a rerview. Advances in Mechanics *14*, 75 – 94
288. Saanouni, K., Chaboche, J.L., Lense, P.M. (1989): On the creep crack-growth prediction by a non-local damage formulation. Eur. J. Mech. A/Solids *8*, 6, 437 – 459

289. Sakane, M., Hosokawa, T. (2001): Biaxial and triaxial creep testing of type 304 stainless steel at 923 K. In: Murakami, S., Ohno, N. (eds.) IUTAM Symposium on Creep in Structures. Kluwer, Dordrecht, pp. 411 – 418
290. Sakane, M., Tokura, H. (2002): Experimental study of biaxial creep damage for type 304 stainless steel. Int. J. of Damage Mechanics *11*, 247 – 262
291. Schiesse, P. (1994): Ein Beitrag zur Berechnung des Deformationsverhaltens anisotrop gescheädigter Kontinua unter Berücksichtigung der thermoplastischer Kopplung. Dissertation, Ruhr-Universität Bochum, Mitteilungen aus dem Institut für Mechanik 89
292. Schröder, J., Gross, D. (2004): Invariant formulation of the electromechanical enthalpy function of transversely isotropic piezoelectric materials. Arch. Appl. Mech *73*, 533 – 552
293. Schröder, J., Neff, P. (2003): Invariant formulation of hyperelastic transverse isotropy based on polyconvex free energy functions. Int. J. Solids and Structures *40*, 401 – 445
294. Schwetlick, H., Kretzschmar, H. (1991): Numerische Verfahren für Naturwissenschaftler und Ingenieure. Fachbuchverlag, Leipzig
295. Sdobyrev, V.P. (1959): Kriterij dlitel'noj prochnosti dlya nekotorykh zharoprochnykh splavov pri slozhnom napryazhennom sostoyanii, (Criterion of long term strength of some high-temperature alloys under multiaxial stress state, in Russ.). Izv. AN SSSR. OTN. Mekh. i Mashinostroenie, 6, 93 – 99
296. Segle, P., Tu, S.T., Storesund, J., Samuelson, L.A. (1996): Some issues in life assessment of longitudinal seam welds based on creep tests with cross-weld specimens. Int. J. of Pressure Vessels & Piping *66*, 199 – 222
297. Shibli, I.A. (2002): Performance of P91 thick section welds under steady and cyclic loading conditions: power plant and research experience. OMMI *1*, 3. Available from http://www.ommi.co.uk
298. Skrzypek, J., Ganczarski, A. (1998): Modelling of Material Damage and Failure of Structures. Foundation of Engineering Mechanics. Springer, Berlin
299. Skrzypek, J.J. (1993): Plasticity and Creep. CRC Press, Boca Raton
300. Smith, D.R. (1993): An Introduction to Continuum Mechanics. Kluwer, Dordrecht
301. Soločevskij, A., Konkin, V., Moračkowskij, O., Koczyk, S. (1985): Eine Theorie zur nichtlinearen Verformung anisotroper Körper und ihre Anwendung auf die Berechnung des Kriechverhaltens dünner Schalen. Technische Mechanik *6*, 4, 27 – 36
302. Sosnin, O.V. (1971): K voprosu o suschestvovanii potentsiala polzuchseti, (On the question on existense of creep potential, in Russ.). Izv. AN SSSR Mekhanika tverdogo tela, 5, 85 – 89
303. Sosnin, O.V. (1974): Energeticheskii variant teorii polzuchesti i dlitel'noi prochnosti. Polzuchest' i razrushenie neuprochnyayushikhsya materialov (Energetic variant of the creep and long–term strength theories. Creep and fracture of nonhardening materials, in Russ.). Problemy prochnosti, 5, 45 – 49
304. Sosnin, O.V., Gorev, B.V., Nikitenko, A.F. (1986): Energeticheskii variant teorii polzuchesti (Energetic variant of the creep theory, in Russ.). Institut Gidrodinamiki, Novosibirsk
305. Spence, J. (1973): Analysis of smooth curved pipes under in-plane bending. Journal of Mechanicsl Engineering Science *15*, 252 – 265
306. Spencer, A.J.M. (1984): Linear elastic constitutive equations for fibre-reinforced material. In: Spencer, A.J.M. (ed.) Continuum Theory of the Mechanics of Fibre-Reinforced Composites. Springer-Verlag, Wien, pp. 1 – 32. CISM Lecture Notes No. 282
307. Spencer, A.J.M. (1987): Isotropic polynomial invariants and tensor functions. In: Boehler, J. (ed.) Applications of tensor functions in solid mechanics. Springer, Wien, pp. 141 – 169. CISM Lecture Notes No. 292

308. Stamm, H., von Estroff, U. (1993): Determination of damage parameters for microckrack formation under creep conditions. In: Ainsworth, R.A., Skelton, R.P. (eds.) Behaviour of Defects at High Temperatures. Mechanical Engineering Publ., London, pp. 123 – 151
309. Stouffer, D.C., Dame, L.T. (1996): Inelastic Deformation of Metals. John Wiley & Sons, New York
310. Taira, S. (1962): Lifetime of structures subjected to varying load and temperature. In: Hoff, N.J. (ed.) Creep in Structures. Springer, Berlin, pp. 96 – 119
311. Taira, S., Koterazawa, R. (1962): Dynamic creep and fatigue of an 18-8-Mo-Nb Steel. Bull. JSME *5*, 17, 15 – 20
312. Taira, S., Ohtani, R. (1986): Teorija vysokotemperaturnoj prochnosti materialov (Theory of high-temperature strength of materials, in Russ.). Metallurgija, Moscow
313. Takezono, S., Fujoka, S. (1981): The creep of moderately thick shells of revolution under axisymmetrical load. In: Ponter, A.R.S., Hayhurst, D.R. (eds.) Creep in Structures. Springer-Verlag, Berlin, pp. 128 – 143
314. Takezono, S., Migita, K., Hirakawa, A. (1988): Elastic/Visco-plastic deformation of multi-layered shells of revolution. JSME, Ser. 1 *31*, 3, 536 – 544
315. Tamuzh, V.P., Kuksenko, V.S. (1978): Mikromeckanika razrusheniya polimernykh materialov (Micromechanics of fracture of polymeric materials, in Russ.). Zinatne, Riga
316. Timoshenko, S.P. (1953): History of Strength of Materials. Dover Publ., New York
317. Ting, T.C.T. (1996): Anisotropic Elasticity. Theory and Applications. Oxford University Press, Oxford
318. Trivaudey, F., Delobelle, P. (1993): Experimental study and modelization of creep damage under multi-axial loadings at high temperature. In: Wilshire, B., Evans, R.W. (eds.) Creep and Fracture of Engineering Materials and Structures. Institute of Materials, London, pp. 137–147
319. Trostel, R. (1993): Mathematische Grundlagen der Technischen Mechanik, volume I: Vektor- und Tensoralgebra. Vieweg, Braunschweig
320. Truesdell, C. (1964): Second-order effects in the mechanics of materials. In: Reiner, M., Abir, D. (eds.) Second-Order Effects in Elasticity, Plasticity and Fluid Dynamics. Pergamon Press, Oxford, pp. 228 – 251
321. Truesdell, C., Noll, W. (1992): The Non-Linear Field Theories of Mechanics. 2nd edition. Springer, Berlin
322. Trunin, I.I. (1965): Kriterii prochnosti v usloviyakh polzuchesti pri slozhnom napryazhennom sostoyanii (Failure criteria under creep conditions in multiaxial stress state, in Russ.). Prikl. Mekhanika *1*, 7, 77 – 83
323. Tvergaard, V. (1990): Material failure by void growth to coalescence. Advances in Applied Mechanics *27*, 83 – 51
324. Vakulenko, A.A., Kachanov, M.L. (1971): Kontinual'naya teoria sredy s treshchinami (Continuum theory of a medium with cracks, in Russ.). Izv. AN SSSR. Mekhanika tverdogo tela, 4, 159 – 166
325. van der Giessen, E., Tvergaard, V. (1995): Development of final creep failure in polycrystalline aggregates. Acta Metall. Mater. *42*, 959 – 973
326. van der Giessen, E., van der Burg, M.W.D., Needleman, A., Tvergaard, V. (1995): Void growth due to creep and grain boundary diffusion at high triaxialities. J. Mech. Phys. Solids *43*, 123 – 165
327. von Kármán, T. (1911): Festigkeitsprobleme im Maschinenbau. In: Encyklop. d. math. Wissensch. IV/2. Teubner, Leipzig, pp. 311 – 385
328. von Mises, R. (1928): Mechanik der plastischen Formänderung von Kristallen. ZAMM *8*, 3, 161 – 185
329. Washizu, K. (1982): Variational Methods in Elasticity and Plasticity. Pergamon Press, Oxford

330. Weber, J., Klenk, A., Rieke, M. (2005): A new method of strength calculation and lifetime prediction of pipe bends operating in the creep range. Int. J. Pressure Vessels & Piping *82*, 77 – 84
331. Wilson, E.B. (1901): Vector Analysis, Founded upon the Lectures of G. W. Gibbs. Yale University Press, New Haven
332. Winstone, M.R. (1998): Microstructure and alloy developments in nickel-based superalloys. In: Strang, A., Cawley, J., Greenwood, G.W. (eds.) Microstructural Stability of Creep Resistant Alloys for High Temperature Plant Applications. Cambridge University Press, Cambridge, pp. 27 – 47
333. Wohlfahrt, H., Brinkmann, D. (2001): Consideration of inhomogeneties in application of deformation models, describing the inelastic behaviour of welded joints. In: Steck, E., Ritter, R., Peil, U., Ziegenbein, A. (eds.) Plasticity of Metals: Experiments, Models, Computation. Final Report of the Coll. Res. Centre 319. Wiley-VCH, Weinheim, pp. 361 – 382
334. Wriggers, P. (2001): Nichtlineare Finite-Element-Methoden. Springer, Berlin
335. Wu, R., Sandström, R., Seitisleam, F. (2004): Influence of extra coarse grains on the creep properties of 9 percent CrMoV (P91) steel weldment. Trans. ASME Journal of Engineering Materials and Technology *26*, 87 – 94
336. Wu, Z. (1993): Kriechen und Kriechschädigung von Eis unter mehrachsiger Belastung, Fortschrittberichte VDI, Reihe 18, Nr. 135. VDI Verlag, Düsseldorf
337. Xiao, H., Bruhns, O., Meyers, A. (2000): Minimal generating sets for symmetric 2nd-order tensor-valued transversely isotropic functions of vectors and 2nd-order tensors. ZAMM *80*, 565 – 569
338. Yang, H.T.Y., Saigal, S., Masud, A., Kapania, R.K. (2000): A survey of recent shell finite elements. Int. J. Numer. Meth. Engng. *47*, 101 – 127
339. Zheng, Q.S. (1994): Theory of representations of tensor functions. Appl. Mech. Rev. *47*, 11, 545 – 587
340. Zheng, Q.S., Boehler, J.P. (1994): The description, classification and reality of material and physical symmetries. Acta Mechanica *102*, 73 – 89
341. Zhilin, P.A. (1982): Osnovnye uravneniya neklassicheskoi teorii obolochek (Basic equations of a non-classical theory of shells, in Russ.). Dinamika i prochnost' mashin, Trudy LPI *386*, 29 – 46
342. Zhilin, P.A. (1996): A new approach to the analysis of free rotations of rigid bodies. ZAMM *76*, 4, 187 – 204
343. Zhilin, P.A. (2001): Vektory i tensory vtorogo ranga v trekhmernom prostranstve (Vectors and second rank tensors in three-dimensional space, in Russ.). Nestor, St. Petersburg
344. Zhilin, P.A. (2003): Modifizirovannaya teoriya simmetrii tenzorov i tenzornykh invariantov (A modified theory of symmetry for tensors and tensor invariants, in Russ.). Izvestiya vuzov. Estestvennye Nauki, 176 – 195
345. Zhilin, P.A. (2006). Symmetries and orthogonal invariance in oriented space
346. Zhilin, P.A., Ivanova, E.A. (1995): Modifitsirovannyi funktsional energii v teorii plastin tipa Reissnera (A modified energy functional in the Reissner type plate theory, in Russ.). Izv. RAS. Mekhanika tverdogo tela *2*, 120 – 128
347. Ziegler, H. (1963): Some extremum principles in irreversible thermodynamics with application to continuum mechanics. In: Sneddon, I.N., Hill, R. (eds.) Progress in Solid Mechanics. North-Holland, Amsterdam, pp. 93 – 192
348. Zienkiewicz, O.C., Taylor, R.L. (1991): The Finite Element Method. McGraw-Hill, London
349. Zolochevskij, A.A. (1988): Kriechen von Konstruktionselementen aus Materialien mit von der Belastung abhängigen Charakteristiken. Technische Mechanik *9*, 177 – 184
350. Zurmühl, R., Falk, S. (1992): Matrizen und ihre Anwendungen. Springer, Berlin
351. Życzkowski, M. (2000): Creep damage evolution equations expressed in terms of dissipated power. Int. J. Mech. Sci. *42*, 755 – 769

Index

A

B

C

D

E

F

G

H

I

K

L

T

U

V

W

Foundations of Engineering Mechanics

Series Editors: Vladimir I. Babitsky, Loughborough University, UK
Jens Wittenburg, Karlsruhe University, Germany

Further volumes of this series can be found on our homepage: springer.com

(Continued from page ii)

Alfutov, N.A.
Stability of Elastic Structures, 2000
ISBN 3-540-65700-2

Astashev, V.K., Babitsky, V.I., Kolovsky, M.Z., Birkett, N.
Dynamics and Control of Machines, 2000
ISBN 3-540-63722-2

Kolovsky, M.Z., Evgrafov, A.N., Semenov Y.A, Slousch, A.V.
Advanced Theory of Mechanisms and Machines, 2000
ISBN 3-540-67168-4